T0181901

A Project-Based Introduction to Computational Statics

Andreas Öchsner

A Project-Based Introduction to Computational Statics

Second Edition

 Springer

Andreas Öchsner
Faculty of Mechanical Engineering
Esslingen University of Applied Sciences
Esslingen am Neckar, Baden-Württemberg
Germany

ISBN 978-3-030-58773-4   ISBN 978-3-030-58771-0   (eBook)
https://doi.org/10.1007/978-3-030-58771-0

This Springer imprint is published by the registered company Springer Nature Switzerland AG
The registered company address is: Gewerbestrasse 11, 6330 Cham, Switzerland

Persönlichkeiten werden nicht durch schöne Reden geformt, sondern durch Arbeit und eigene Leistung.

Albert Einstein (1879–1955)

Preface to the Second Edition

The pedagogical methodology proposed in the first edition has been well accepted by students and the scientific community. The second edition has been extended by more than 40 pages. Many additional examples on the refined analysis and different modifications of extensometers have been included. Last but not least, the entire content and the graphical illustrations have been thoroughly revised and updated.

Esslingen, Germany
July 2020

Andreas Öchsner

Preface to the First Edition

This book presents a novel concept for introducing the finite element method, applied in the context of solid mechanics. It presents a major conceptual shift, i.e., taking away lengthy theoretical derivations from the face-to-face interaction with students, focusing on the summary of key equations and concepts, and to practice these on well-chosen example problems. The theoretical derivations are provided as additional reading and students must study and review the derivations in a self-study approach. The theoretical foundation is provided to solve a comprehensive design project in the context of tensile testing. A classical clip-on extensometer serves as the demonstrator on which to apply the provided concepts. The major goal is to derive the calibration curve based on different approaches, i.e., analytical mechanics and based on the finite element method, and to consider further design questions such as technical drawing, manufacturing, and cost assessment. Working with two concepts, i.e., analytical and computational mechanics, strengthens the vertical integration of knowledge and allows the student to compare and understand the different concepts, as well as highlighting the essential need for benchmarking any numerical result. It is beyond question that such an approach can serve only as a first introduction to this powerful and complex method and that further in-depth study is required for a reliable and confident application of the finite element method.

Southport, Australia
September 2017

Andreas Öchsner

Preface to the First Edition

Acknowledgements

It is important to highlight the contribution of the students which helped to develop the concept and content of this book. Their questions, comments, and struggles during difficult lectures, assignments, and final exams helped me to develop the idea for this new teaching approach. Furthermore, I would like to express my sincere appreciation to Springer, especially to Dr. Christoph Baumann, for giving me the opportunity to release this book.

Contents

Symbols and Abbreviations

Latin Symbols (capital letters)

A	Area
A_s	Shear area
E	YOUNG's modulus
F	Force
F_a	Auxiliary force
F^R	Reaction force
G	Shear modulus
I	Second moment of area
I_p	Polar second moment of area
K	Global stiffness matrix
K^e	Elemental stiffness matrix
L	Length
M	Moment
M_a	Auxiliary moment
M^R	Reaction moment
N	Normal force (internal), interpolation function
Q	Shear force (internal)
R_p	Initial (tensile) yield stress
T	Transformation matrix
V	Volume
X	Global Cartesian coordinate
Y	Global Cartesian coordinate
Z	Global Cartesian coordinate

Latin Symbols (small letters)

a Geometric dimension
b Geometric dimension
c Constant of integration
f Global column matrix of loads
f^e Elemental column matrix of loads
g Standard gravity
h Geometric dimension
k_s Shear correction factor
m Distributed moment (external), mass
n Node number
p Distributed load in x-direction (tension)
q Distributed load in z-direction (bending)
u Displacement
u_p Global column matrix of nodal unknowns
u_p^e Elemental column matrix of nodal unknowns
x Local Cartesian coordinate
y Local Cartesian coordinate
z Local Cartesian coordinate

Greek Symbols (capital letters)

Γ Boundary
Π Total strain energy
Ω Domain

Greek Symbols (small letters)

α Rotation angle, scale factor
γ Shear strain (engineering definition)
ε Normal strain
κ Curvature
ν POISSON's ratio

ξ Natural coordinate
π Volumetric strain energy
ϱ Mass density
σ Normal stress
τ Shear stress
ϕ Rotation (TIMOSHENKO beam)
φ Rotation (BERNOULLI beam)

Mathematical Symbols

$\times$ Multiplication sign (used where essential)
$\langle\dots\rangle$ MACAULAY's bracket
$\mathcal{L}\{\dots\}$ Differential operator
d   Derivative symbol
∂ Partial derivative symbol (rounded d)

Indices, Superscripted

$\dots^e$ Element
$\dots^R$ Reaction

Indices, Subscripted

$\dots_p$ Point
$\dots_{lim}$ Limit
$\dots_{max}$ Maximum
$\dots_{sp}$ Specimen

Abbreviations

1D   One-dimensional
2D   Two-dimensional
3D   Three-dimensional
BC   Boundary condition

CT	Compact tension
DOF	Degree(s) of freedom
FE	Finite element
FEM	Finite element method
PBL	Project-based learning
PDE	Partial differential equation

Chapter 1
Introduction and Problem Formulation

Abstract This chapter briefly reviews different teaching approaches for computational statics. The major focus is on the presentation of the design project which serves to introduce the basic application of the finite element method. The project is taken from the context of tensile testing of engineering materials and relates to the design of a clip-on extensometer. The mechanical model of this sensor can be simplified to a ⊔-shaped frame structure which allows the application of classical analytical mechanics as well as a computational approach. Thus, the design problem also serves to review and strengthen classical applied mechanics and its comparison with modern numerical approaches.

The incorporation of projects in the classical curriculum structure of higher education dates back to the late 1960s. Nowadays, some institutions even have moved to project-based curricula in engineering [1]. Traditional teaching approaches would introduce the finite element method based on lectures, which focus on the underlying theory [12], and tutorials, which deepen the topic, mainly based on hand calculations of simple problems [3]. Some universities may additionally offer computer laboratories where either a classical programming language (e.g., FORTRAN or C++) [14] or a multi-paradigm numerical computing environment (e.g., Maxima, Maple, or MATLAB) [11] is used to write small finite element routines. Alternatively, a commercial package might be used [9, 10]. Among the commercial packages, some allow the implementation of own computer code (e.g., new element formulations) based on user-subroutines [2]. A more recent approach, i.e., project-based learning (PBL), tries to incorporate these different teaching elements and to focus around a certain design project. Such approaches are believed to facilitate the learning process and are closer to engineering practice. In many cases, commercial finite element packages are used and quite complex structures are investigated or even optimized [5, 15, 16]. The analysis of complex structures is normally linked to the application of a larger number of elements and the evaluation of results is, many times, based on contour plots of field quantities ('colored pictures'). The merit of this approach lies in showing the strength of the method and to illustrate how the theory is transferred to a commercial program. Despite the fact that there are numerous commercial finite element packages available, the general steps to perform a finite element analysis do

A. Öchsner, *A Project-Based Introduction to Computational Statics*,
https://doi.org/10.1007/978-3-030-58771-0_1

not change and only the graphical interface needs to be mastered when moving from one package to another one.

However, solely focusing on commercial packages and complex structures involves many dangers for the finite element beginner. Without the knowledge of the underlying theory, a generation of an appropriate and accurate computational model (e.g., element type, mesh size, and refinement) might be difficult. Furthermore, to rely on displayed colors or values might be misleading if, for example, the difference between nodal and integration point results is not known. Thus, a serious introduction to the finite element method must incorporate the corresponding theory and must enable the user to judge the quality of the obtained results.

The project-based introduction to the finite element method in this book focuses on a simple but real engineering structure which allows to connect analytical mechanics to the computational approach based on the finite element method. Doing so, the basics of applied mechanics are reviewed, strengthened, and linked to a numerical approach. Furthermore, the project allows to address further questions from many other subjects such as material selection, manufacturing, costs and lightweight potential. Nevertheless, it must be highlighted that this approach is only a first introduction to the finite element method and a reliable application requires further studies into the underlying theory, as well as comprehensive practice.

1.1 Project Outline

The proposed design project is related to the classical tensile test (see, for example, [8]), i.e., the most important and common test to characterize the mechanical behavior of engineering materials. The quantities to be measured are normally the applied force and a distance in loading direction[1]. The measurements are then converted to the acting stress and strain, normally expressed as so-called engineering quantities. The force measurement is normally based on a load cell and does not imply major problems if the capacity is chosen according to the expected force range. The measurement of the strain is more demanding since the data recording should happen in the gage section on the specimen and not, for example, be based on the movement of the machine crosshead. The application of sensors, such as strain gages or extensometers, directly on the specimen's surface or noncontacting optical approaches such as video or laser extensometers are the general options for this task [4, 13]. Figure 1.1 shows a typical clip-on extensometer with knife edges used to attach the sensor on the specimens via two clip-springs. These knife edges ensure that the ends of the sensor legs and the corresponding part of the specimen (gage length) perform the same movement, i.e., displacement but more or less 'free' rotation. Other configurations are possible where the feet would be fixed with small screws on the specimen (quite common in fracture mechanics). The set-up of this

---

[1]Some evaluation procedures require also a distance measurement perpendicular to the loading direction in order to evaluate POISSON's ratio.

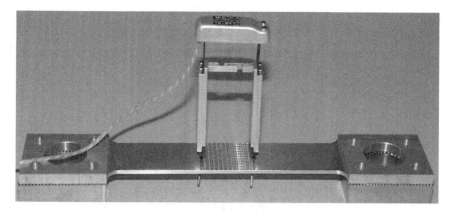

Fig. 1.1 Flat tensile specimen with extensometer (SANDNER Messtechnik GmbH, Germany) [6]

type of sensor looks like a ⊔-shaped frame structure[2] with an additional horizontal mechanical protective mechanism to avoid overexpansion (elongation failure) of the sensor.

The extensometer shown in Fig. 1.1 can be simplified to a mechanical model as shown in Fig. 1.2. From this representation it can be seen that all three members of the frame undergo at least a bending deformation as soon as the specimen is elongated. Based on the configuration with knife edges, a free rotation of ends of the extensometers legs is assumed. Furthermore, the measuring principle is indicated, i.e., the strain in the horizontal member is recorded via a strain gage ($\varepsilon_{\text{strain gage}}$) and must be related to the strain ($\varepsilon_{\text{specimen}}$) in the actual specimen (so-called calibration). Thus, the engineering task is to relate the recorded signal in the extensometer to the real strain in the specimen based on a factor or some kind of equation.

The design concept of the extensometer shown in Fig. 1.1 is clearly visible in the three photos of Fig. 1.3. After removing the casing, it is possible to identify the strain gages on the top beam and the corresponding wiring.

An alternative configuration based on the same measuring principle is shown in Fig. 1.4. The beam with strain gages on both sides (i.e., one in the tension and the other one in the compressive strain regime) does not span over the entire gage length of this extensometer model.

A similar design to Fig. 1.1 is shown in Fig. 1.5 where the knife edges and the mechanical protective mechanism can be clearly identified.

Based on the explanations regarding project-based introductions of the finite element method presented at the beginning of this chapter and the peculiarities of the sensor design as outlined in Figs. 1.1, 1.2, 1.3, 1.4 and 1.5, a structure of the design process as outlined in Fig. 1.6 has been created. The project has been split in two design phases. The initial design phase is restricted to classical analytical mechanics in order to derive a general expression of the calibration curve. It is worth noting

---

[2]Civil engineers would call it a portal frame.

(a)

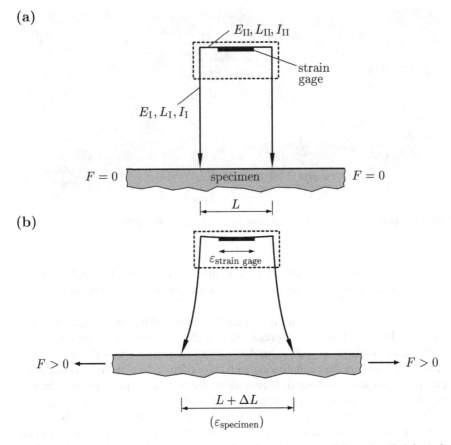

(b)

Fig. 1.2 Schematic sketch of the extensometer and tensile specimen: **a** undeformed and **b** deformed state

that no specific numbers should be assigned to the derivation and a general, i.e., as a function of the design variables (geometrical and material properties of the sensor), expression should be derived. This will allow to later easily check different design proposals in order to have a reasonable ratio between both strain values. In addition, reviewing analytical mechanics and applying it to a practical design problem allows to strengthen the vertical knowledge integration and students should be able to understand the different benefits and drawbacks to each method. In addition, it was decided to avoid lengthly derivations in this first introduction to the finite element method and to provide the derivations as additional reading for weekly self-study. The textbooks mentioned in the literature section may serve for this purpose. Thus, the following chapters collect only a summary of basic concepts and equations with a focus on their application to relatively simple problems.

Fig. 1.3 Technical details of an extensometer as shown in Fig. 1.1 (SANDNER Messtechnik GmbH, Germany): **a** global view, **b** global view with different orientation, and **c** details of the strain gages with wiring and solder posts

(a)

(b)

(c)

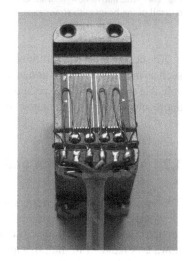

(a)

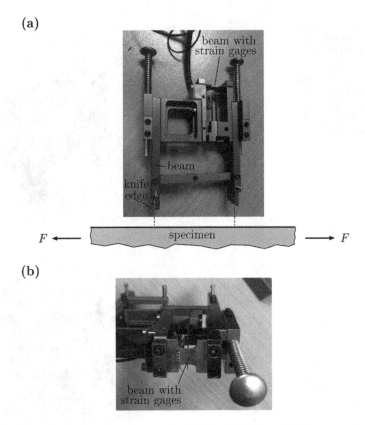

(b)

Fig. 1.4 Alternative design of an extensometer (Epsilon Technology, USA): **a** overall view and **b** detail of the beam with strain gages

The design process as outlined in Fig. 1.6 contains components from other courses such as material selection or manufacturing. This is not covered in this book but is a valuable addition to the expected design reports since it relates to a complete design approach as known from engineering practice.

1.2 Assessment Items and Marking Criteria

The proposed design project can be handled as a group or individual assignment. Obviously, there are different benefits and drawbacks to each method. Working in groups or teams is definitely closer to the industrial context where larger projects are nowadays handled in multidisciplinary teams. Thus, it is essential for an engineer to be trained to work in such a context. This ability is connected to many different soft-skills, ranging from communication and presentation skills to simplifying com-

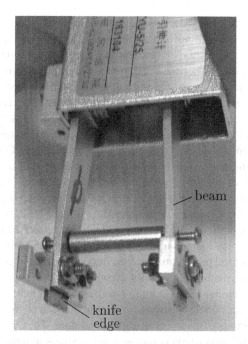

Fig. 1.5 Alternative design of an extensometer (NCS, China)

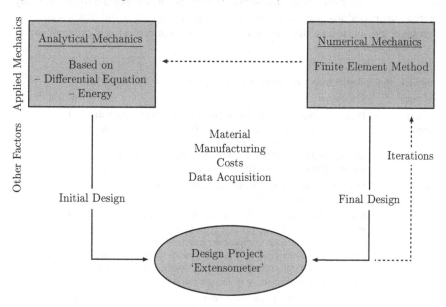

Fig. 1.6 Flowchart of the design process

plex circumstances for team members from other areas of expertise. If students are working in groups, it is important to include a self-assessment component to evaluate the contribution of each team member. It is sometimes not too uncommon that all team members give each other the maximum score in order to increase their final mark. Thus, the allocated weight for any self-assessment must be carefully chosen. Another typical occurrence is that very good students do not really like classical group work and the corresponding evaluation as a team. These students sometimes feel that they are carrying someone else's load in addition to their own work package. Another point to consider is the grouping of students in teams. Giving the students the freedom to form their own team of x students is in general quite popular while a random selection of students by the course convenor is less popular. However, the latter avoids the problem that some students are not able to join a team, or at least claim so. Once the project is completed or at least has progressed to a certain stage, the findings should be summarized and communicated [7]. In the case of group projects, this may happen in the form of group or individual reports and/or group presentations. Oral presentations are an important skill but group presentations involving all the team members with only a few minutes of presentation time (e.g., 2 to 3 min) for each member are difficult to evaluate if individual marks are required. The aforementioned comments also hold for the submission of group reports, i.e., the top students are normally less in favor of the procedure. Thus, reports should have at least an individual submission component for students to have the possibility to distinguish from the other team members. In any case, it is important to not only rely on non-supervised assessment items, i.e., design reports. To avoid that a student or a group could 'outsource' the preparation of the design project, a significant super-

Table 1.1 Initial design report: elements and marking criteria (20% of total mark)

Element	Comment	Weight (%)
Formal aspects	Punctual submission; maximum number of pages (10; including everything, e.g., appendix etc.); PDF format; each page with name and enrollment number; only electronic submission; suitable font size and line spacing; A4 format; naming of the file as `course code_family name_first name.pdf`	10
Initial sketches	Only freehand sketches of the sensor design; indicate all dimensions with variables	15
Analytical calculations	Use analytical mechanics (i.e., differential equation–based approach and energy approach) to characterize the deformation behavior of the sensor; compare the results of both approaches; derive a general expression, i.e., based on variables, to relate the specimen's deformation/strain to the measured strain in the sensor; justify the selection of the structural members	50
Literature	Sufficient and appropriate references	10
Peer-assessment	Assess the engagement of your group members; provide a small paragraph for each member and give a final mark	15

Table 1.2 Final design report: elements and marking criteria (30% of total mark)

Element	Comment	Weight (%)
Formal aspects	Punctual submission; maximum number of pages (20; including everything, e.g., appendix etc.); PDF format; each page with name and enrollment number; only electronic submission; suitable font size and line spacing; A4 format; naming of the file as course code_family name_first name.pdf	5
Technical drawings	Provide complete manufacturing drawings and specify the manufacturing process; if your sensor is composed of different parts, provide assembly instructions; explain in detail how the sensor measures and records the deformation (e.g., use of strain gages); explain the fixation on the specimen	10
FE calculations	Use finite-element 'hand calculations' to derive a general equation for the deformation behavior of your sensor; compare and validate your results with the analytical calculations and the commercial finite element package, which was introduced in the course; justify the element types and mesh density used; explain in detail all chosen dimensions of the sensor; estimate the weight of the sensor	50
Calibration	Provide clear instructions on the calibration process of your sensor	7.5
Cost estimate	Predict the entire costs for a prototype, including material, manufacturing, and electronic components	7.5
Literature	Sufficient and appropriate references	5
Peer-assessment	Assess the engagement of your group members; provide for each member a small paragraph and give a final mark	15

vised assessment item (quiz or exam) must be included. It is also recommended that the students must pass this assessment item (hurdle) to pass the entire course.

Based on the above reflections, the following assessment scheme is proposed:

- Initial design report (20% of total mark),
- final design report (30% of total mark), and
- final exam (50% of total mark).

The split into an initial and final design report allows to provide a qualified feedback to the students at an intermediate stage of the semester. Furthermore, it allows to split the calculation approach in the application of analytical mechanics (initial design) and the finite element method (final design).

The selected marking criteria for the initial design report are summarized in Table 1.1. It might be questioned why a course on computational methods should include a major reference to analytical mechanics, which is the topic of a few other courses. On the one hand, it definitely strengthens the vertical integration of knowledge and shows a practical application of the classical engineering mechanics. On the other hand, the analytical solution can serve to validate the results of the finite element approach. It should be noted that the validation of computational results is the most important and challenging task of an engineer. This becomes quite demanding for complex engineering structures where no analytical solutions are available.

The marking criteria for the final design report are summarized in Table 1.2. It can be seen that also areas of other courses are well represented (e.g., technical drawing, manufacturing, cost assessment) in order to generate a complete design approach.

References

1. Heitmann G (1996) Project-oriented study and project-organized curricula: a brief review of intentions and solutions. Eur J Eng Edu 21:121–131
2. Javanbakht Z, Öchsner A (2017) Advanced finite element simulation with MSC Marc: application of user subroutines. Springer, Cham
3. Javanbakht Z, Öchsner A (2018) Computational statics revision course. Springer, Cham
4. Loughlin C (1993) Sensors for industrial inspection. Springer, Dordrecht
5. Miner SM, Link RE (2000) A project based introduction to the finite element method. Paper presented at 2000 ASEE annual conference, St. Louis, Missouri. https://peer.asee.org/8643. Cited 10 June 2017
6. Öchsner A (2003) Experimentelle und numerische Untersuchung des elastoplastischen Verhaltens zellularer Modellwerkstoffe. VDI-Verlag, Düsseldorf
7. Öchsner A (2013) Introduction to scientific publishing: backgrounds, concepts, strategies. Springer, Berlin
8. Öchsner A (2016) Continuum damage and fracture mechanics. Springer, Singapore
9. Öchsner A, Öchsner M (2016) The finite element analysis program MSC Marc/Mentat. Springer, Singapore
10. Öchsner A, Öchsner M (2018) A first introduction to the finite element analysis program MSC Marc/Mentat. Springer, Cham
11. Öchsner A, Makvandi R (2019) Finite elements using Maxima: theory and routines for rods and beams. Springer, Cham

12. Öchsner A (2020) Computational statics and dynamics: an introduction based on the finite element method. Springer, Singapore
13. Sharpe WN (2008) Springer handbook of experimental solid mechanics. Springer, New York
14. Trapp M, Öchsner A (2018) Computational plasticity for finite elements: a Fortran-based introduction. Springer, Cham
15. Ural A (2013) A Hands-on finite element modeling experience in a multidisciplinary project-based freshman course. Comput Appl Eng Educ 21:294–299
16. Zhuge Y, Mills JE (2009) Teaching finite element modelling at the undergraduate level: a PBL approach. In: 20th annual conference of the Australian association of engineering education, 6–9 December 2009, Adelaide, Australia. https://eprints.usq.edu.au/6772/. Cited 10 June 2017

Chapter 2
Review of Analytical Mechanics

Abstract This chapter treats simple structural members based on two different analytical approaches. On the one hand based on fundamental equations of continuum mechanics, i.e., the kinematics, the equilibrium and the constitutive equation, the describing partial differential equations are provided, including their general solution based on constants of integration. As an alternative approach, the total strain energy of a system is introduced and applied in Castigliano's theorems. The covered structural members are rods (tensile deformation) as well as thin and thick beams (bending deformation). The provided concepts are finally applied to the extensometer design problem.

2.1 Overview: One-Dimensional Structural Members

2.2 Partial Differential Equation-Based Approaches

2.2.1 Rods

A rod is defined as a prismatic body whose axial dimension is much larger than its transverse dimensions [3, 4, 11, 16, 21, 27, 29]. This structural member is only loaded in the direction of the main body axes, see Fig. 2.1. As a result of this loading, the deformation occurs only along its main axis.

Derivations are restricted many times to the following simplifications:

- only applying to straight rods,
- displacements are (infinitesimally) small,
- strains are (infinitesimally) small, and
- the material is linear-elastic.

The three basic equations of continuum mechanics, i.e the kinematics relationship, the constitutive law and the equilibrium equation, as well as their combination to the describing partial differential equation (PDE) are summarized in Table 2.1.

Fig. 2.1 Schematic
representation of a
continuum rod

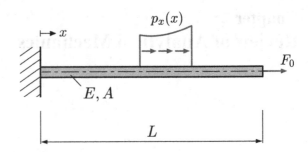

Table 2.1 Different formulations of the basic equations for a rod (x-axis along the principal rod axis), with $\mathcal{L}_1(\ldots) = \frac{d(\ldots)}{dx}$

Specific formulation	General formulation
Kinematics	
$\varepsilon_x(x) = \dfrac{du_x(x)}{dx}$	$\varepsilon_x(x) = \mathcal{L}_1(u_x(x))$
Constitution	
$\sigma_x(x) = E\varepsilon_x(x)$	$\sigma_x(x) = C\varepsilon_x(x)$
Equilibrium	
$\dfrac{d\sigma_x(x)}{dx} + \dfrac{p_x(x)}{A} = 0$	$\mathcal{L}_1^T(\sigma_x(x)) + b = 0$
PDE	
$\dfrac{d}{dx}\left(E(x)A(x)\dfrac{du_x}{dx}\right) + p_x(x) = 0$	$\mathcal{L}_1^T(EA\mathcal{L}_1(u_x(x))) + p_x = 0$

Under the assumption of constant material ($E = $ const.) and geometric ($A = $ const.) properties, the differential equation in Table 2.1 can be easily integrated twice for constant distributed load ($p_x = p_0 = $ const.) to obtain the general solution of the problem [20]:

$$u_x(x) = \frac{1}{EA}\left(-\frac{1}{2}p_0 x^2 + c_1 x + c_2\right), \qquad (2.1)$$

where the two constants of integration c_i ($i = 1, 2$) must be determined based on the boundary conditions (see Table 2.2). The following equation for the internal normal force N_x was obtained based on one-time integration of the PDE and might be useful to determine some of the constants of integration:

$$N_x(x) = EA\frac{du_x(x)}{dx} = -p_0 x + c_1. \qquad (2.2)$$

Table 2.2 Different boundary conditions and corresponding reactions for a continuum rod (deformation occurs along the x-axis)

Case	Boundary Condition	Reaction
$\longmapsto x$	$u_x(x=0) = 0$	$\longmapsto x$ $F^{\mathrm{R}} \rightarrow$
$\longrightarrow u_0$ L	$u_x(x=L) = u_0$	$\longrightarrow F^{\mathrm{R}}$ L
$\longrightarrow F_0$ L	$EA\frac{\mathrm{d}u_x(L)}{\mathrm{d}x} = N_x(L) = F_0$	$\longrightarrow u_x$ L

Fig. 2.2 Internal reactions for a continuum rod

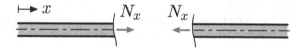

The internal reactions in a rod become visible if one cuts—at an arbitrary location x—the member in two parts. As a result, two opposite oriented normal forces N_x can be indicated. Summing up the internal reactions from both parts must result in zero. Their positive direction is connected with the direction of the outward surface normal vector and the orientation of the positive x-axis, see Fig. 2.2.

Once the internal normal force N_x is known, the normal stress σ_x can be calculated:

$$\sigma_x(x) = \frac{N_x(x)}{A}. \tag{2.3}$$

Application of HOOKE's law (see Table 2.1) allows us to calculate the normal strain ε_x. Typical distributions of stress and strain in a rod element are shown in Fig. 2.3. It can be seen that both distributions are constant over the cross section.

To be able to realize a closed-form presentation with discontinuities (e.g. load, material, or geometry), the so-called MACAULAY bracket[1] can be used for closed-form representations. This mathematical notation has the following meaning:

$$\langle x - a \rangle^n = \begin{cases} 0 & \text{for } x < a \\ (x-a)^n & \text{for } x \geq a. \end{cases} \tag{2.4}$$

---

[1]In the German literature, this approach is named after August Otto FÖPPL (1854–1942).

Fig. 2.3 Axially loaded rod:
a strain and **b** stress
distribution

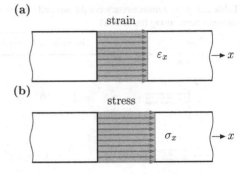

In particular with the case $n = 0$

$$\langle x - a \rangle^0 = \begin{cases} 0 \text{ for } x < a \\ 1 \text{ for } x \geq a \end{cases} \tag{2.5}$$

the closed-form presentation of jumps can be realized. The first three discontinuous functions are shown in Fig. 2.4. Furthermore, derivations and integrals are defined by regarding the triangular bracket symbol as classical round brackets:

$$\frac{\mathrm{d}}{\mathrm{d}x} \langle x - a \rangle^n = n \langle x - a \rangle^{n-1}, \tag{2.6}$$

$$\int \langle x - a \rangle^n \mathrm{d}x = \frac{1}{n+1} \langle x - a \rangle^{n+1} + c. \tag{2.7}$$

Table 2.3 shows a few examples of discontinuous loads and their corresponding representations due to the discontinuous function given in Eq. (2.4).

In regards to the first case in Table 2.3, it should be noted that a positive singe force ($F_0 > 0$) results in a negative jump in the normal force distribution (N_x).

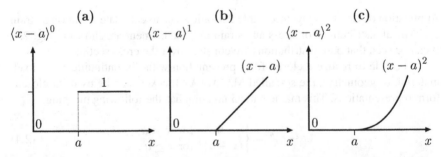

Fig. 2.4 Graphical representation of the first three discontinuous functions according to Eq. (2.4): **a** jump ($n = 0$); **b** kink ($n = 1$); **c** smooth transition ($n = 2$). Adapted from [4]

Table 2.3 Discontinuous loads expressed due to discontinuous functions (deformation occurs along the x-axis). Adapted from [4]

Case	Load (Discontinuity Function)
	$N_x(x) = -F_0 \langle x - a \rangle^0 + c$
	$p_x(x) = p_0 \left(\langle x - a_1 \rangle^0 - \langle x - a_2 \rangle^0 \right)$
	$p_x(x) = \dfrac{p_0}{a_2 - a_1}(x - a_1)\left(\langle x - a_1 \rangle^0 - \langle x - a_2 \rangle^0 \right)$

Fig. 2.5 Rod under different loading conditions: **a** displacement and **b** force

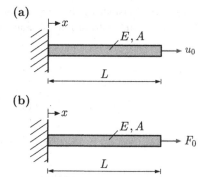

If no single closed-form representation is required, all the previous equations (see Table 2.2 and Eqs. (2.1)–(2.3)) can be applied to each continuous section. As a result, transmission conditions between the continuous sections must be formulated to determine the additional constants of integration, see Problem 2.3.

2.1 Cantilever rod with point loads

Given is a rod of length L and constant axial tensile stiffness EA as shown in Fig. 2.5. At the left-hand side there is a fixed support and the right-hand side is either elongated by a displacement u_0 (case a) or loaded by a single force F_0 (case b). Determine the analytical solution for the elongation $u_x(x)$, the strain $\varepsilon_x(x)$, and the stress $\sigma_x(x)$ along the rod axis. Sketch for both cases the corresponding distributions.

Fig. 2.6 Free-body diagram
of the rod with displacement
boundary condition

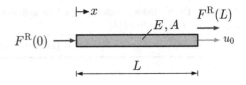

2.1 Solution

Case (a): Let us start the solution procedure by sketching the free-body diagram as
shown in Fig. 2.6. It should be noted here that the imposed displacement u_0 at the
right-hand boundary results in a reaction force $F^R(L)$.

The next step is to identify the boundary conditions of the problem. They can be
immediately stated as:

$$u_x(0) = 0, \tag{2.8}$$

$$u_x(L) = u_0. \tag{2.9}$$

Consideration of the first boundary condition in Eq. (2.1) results with $p_0 = 0$ directly
in $c_2 = 0$. Considering the second boundary condition in Eq. (2.1) gives then $c_1 = \frac{EAu_0}{L}$. Thus, the distributions of elongation, strain, and stress are obtained as:

$$u_x(x) = u_0 \frac{x}{L}, \tag{2.10}$$

$$\varepsilon_x(x) = \frac{du_x(x)}{dx} = \frac{u_0}{L}, \tag{2.11}$$

$$\sigma_x(x) = E\varepsilon_x(x) = \frac{u_0 E}{L}. \tag{2.12}$$

Case (b): Let us start the solution procedure by sketching the free-body diagram as
shown in Fig. 2.7.

The first boundary condition is again $u_x(0) = 0$ which results with Eq. (2.1)
directly in $c_2 = 0$. The second boundary condition might be not so obvious and
requires to consider of the force equilibrium for a small element at $x = L$, see Fig. 2.8.

The horizontal force equilibrium yields the second boundary condition as $N_x(L) = F_0$. Introducing this second condition into Eq. (2.2), the second constant of integration is obtained for $p_0 = 0$ as $c_1 = F_0$. Thus, the distributions of elongation, strain,
and stress are obtained as:

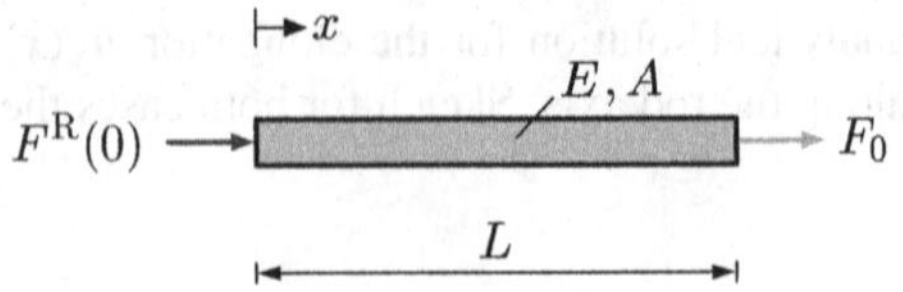

Fig. 2.7 Free-body diagram
of the rod with force
boundary condition

$$\mapsto x$$

$$E, A$$

$$F^R(0) \longrightarrow \boxed{} \longrightarrow F_0$$

$$\overleftarrow{\quad\quad\quad L \quad\quad\quad}$$

Fig. 2.8 Equilibrium
between internal normal
force N_x and external load
F_0

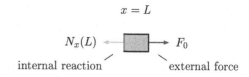

$x = L$

$N_x(L)$ F_0

internal reaction   external force

$$u_x(x) = \frac{F_0}{EA}x, \tag{2.13}$$

$$\varepsilon_x(x) = \frac{du_x(x)}{dx} = \frac{F_0}{EA}, \tag{2.14}$$

$$\sigma_x(x) = E\varepsilon_x(x) = \frac{F_0}{A}. \tag{2.15}$$

Equation (2.15) is the classical definition of engineering stress in the case of a uniaxial tensile test. The graphical representation of the field variables (displacement, strain, and stress) is shown in Fig. 2.9.

2.2 Cantilever rod with distributed load

Given is a rod of length L and constant axial tensile stiffness EA as shown in Fig. 2.10. At the left-hand side there is a fixed support and a constant distributed load p_0 is acting along the entire rod axis. Determine the analytical solution for the elongation $u_x(x)$, the strain $\varepsilon_x(x)$, and the stress $\sigma_x(x)$ along the rod axis.

2.2 Solution

Let us start the solution procedure by sketching the free-body diagram as shown in Fig. 2.11.

As outlined in the previous example, the boundary conditions can be stated as $u_x(0) = 0$ and $N_x(L) = 0$. However, we must consider now that a constant distributed load p_0 is acting and the evaluation of Eq. (2.1) based on the first boundary condition gives $c_2 = 0$. Application of the second boundary condition in Eq. (2.2) gives now $c_1 = p_0 L$. Thus, the distributions of elongation, strain, and stress are obtained as:

$$u_x(x) = \frac{p_0 L^2}{EA}\left(-\frac{1}{2}\left[\frac{x}{L}\right]^2 + \left[\frac{x}{L}\right]\right), \tag{2.16}$$

$$\varepsilon_x(x) = \frac{du_x(x)}{dx} = \frac{p_0 L}{EA}\left(-\left[\frac{x}{L}\right] + 1\right), \tag{2.17}$$

$$\sigma_x(x) = E\varepsilon_x(x) = \frac{p_0 L}{A}\left(-\left[\frac{x}{L}\right] + 1\right). \tag{2.18}$$

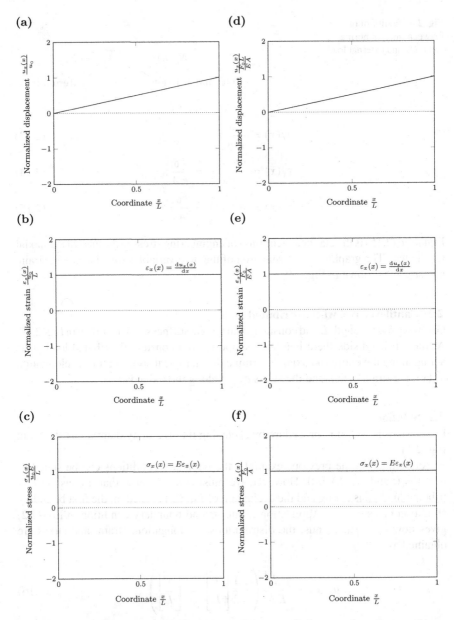

Fig. 2.9 Graphical representation of the field variables: **a–c** displacement boundary conditions (u_0), and **d–f** force boundary condition (F_0)

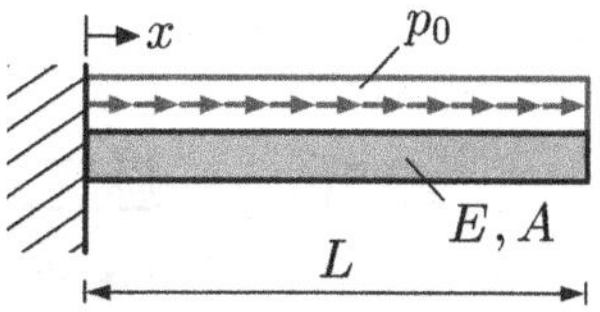

Fig. 2.10 Rod with distributed load

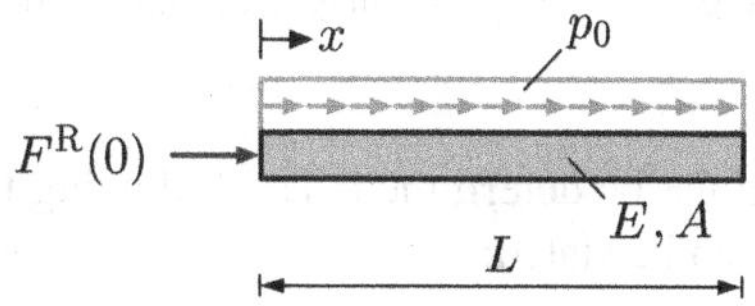

Fig. 2.11 Free-body diagram of the rod with distributed load

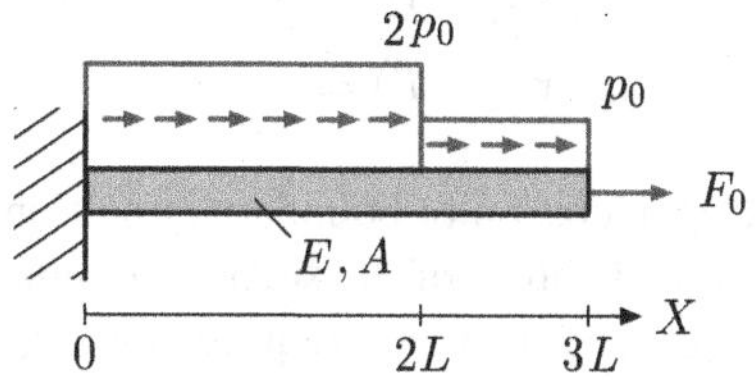

Fig. 2.12 Rod with different sections

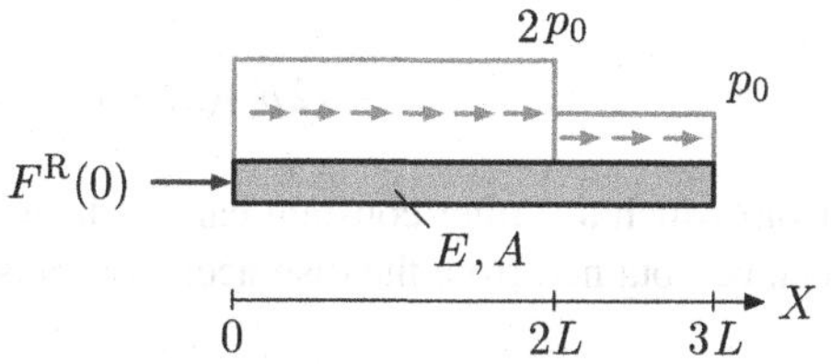

Fig. 2.13 Free-body diagram of the rod with different sections

2.3 Cantilever rod with different sections

Given is a rod of length $3L$ and constant axial tensile stiffness EA as shown in Fig. 2.12. At the left-hand side there is a fixed support and a constant distributed load $2p_0$ is acting in the range $0 \leq x \leq 2L$ whereas a load of p_0 is acting in the range $2L \leq x \leq 3L$. Determine the analytical solution for the elongation $u_x(x)$, the strain $\varepsilon_x(x)$, and the stress $\sigma_x(x)$ along the rod axis.

2.3 Solution

Let us start the solution procedure by sketching the free-body diagram as shown in Fig. 2.13.

The discontinuity in the distributed load can be handled by splitting the rod at $X = 2L$ into two parts, see Fig. 2.14. The left-hand part is now described by the local coordinate x_I with $0 \leq x_I \leq 2L$ while the right-hand part is described by the local coordinate x_{II} with $0 \leq x_{II} \leq L$.

Consideration of two parts means that Eqs. (2.1) and (2.2) must be applied to both sections and in total four integration constants, i.e. two for each section (here c_1 and c_2 for the left-hand section while c_3 and c_4 is assigned to the right-hand section),

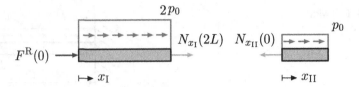

Fig. 2.14 Free-body diagram of the rod decomposed into two sections

must be determined. The following two boundary and two transmission conditions can be stated:

$$u_x(x_I = 0) = 0, \qquad\qquad N_x(x_{II} = L) = 0, \qquad\qquad (2.19)$$
$$u_x(x_I = 2L) = u_x(x_{II} = 0), \qquad N_x(x_I = 2L) = N_x(x_{II} = 0). \qquad (2.20)$$

Consideration of boundary condition $(2.19)_1$ in Eq. (2.1) gives immediately $c_2 = 0$. Consideration of the second boundary condition $(2.19)_2$ in Eq. (2.2) provides $c_3 = p_0 L$. The next step requires the application of the transmission conditions. Let us start with the transmission condition for the normal force $(2.20)_2$:

$$- (2p_0)(2L) + c_1 = c_3 \stackrel{\text{2nd BC}}{=} p_0 L, \qquad (2.21)$$

from which a further constant can be determined as $c_1 = 5 p_0 L$. The final constant can be obtained from the displacement transmission condition $(2.20)_1$:

$$- \frac{1}{2}(2p_0)(2L)^2 + c_1(2L) = c_4, \qquad (2.22)$$

which can be solved for the remaining constant: $c_4 = 6 p_0 L^2$. Thus, the distributions of elongation, strain, and stress are obtained as for each section as:

$$u_x(x_I) = \frac{p_0 L^2}{EA} \left(-\left[\frac{x_I}{L}\right]^2 + 5\left[\frac{x_I}{L}\right] \right), \qquad (2.23)$$

$$\varepsilon_x(x_I) = \frac{du_x(x_I)}{dx} = \frac{p_0 L}{EA} \left(-2\left[\frac{x_I}{L}\right] + 5 \right), \qquad (2.24)$$

$$\sigma_x(x_I) = E\varepsilon_x(x_I) = \frac{p_0 L}{A} \left(-2\left[\frac{x_I}{L}\right] + 5 \right), \qquad (2.25)$$

and

$$u_x(x_{\text{II}}) = \frac{p_0 L^2}{EA}\left(-\frac{1}{2}\left[\frac{x_{\text{II}}}{L}\right]^2 + \left[\frac{x_{\text{II}}}{L}\right] + 6\right), \tag{2.26}$$

$$\varepsilon_x(x_{\text{II}}) = \frac{du_x(x_{\text{II}})}{dx} = \frac{p_0 L}{EA}\left(-\left[\frac{x_{\text{II}}}{L}\right] + 1\right), \tag{2.27}$$

$$\sigma_x(x_{\text{II}}) = E\varepsilon_x(x_{\text{II}}) = \frac{p_0 L}{A}\left(-\left[\frac{x_{\text{II}}}{L}\right] + 1\right). \tag{2.28}$$

An alternative solution approach can be based on the MACAULAY brackets as outlined in Eq. (2.4). Based on this particular approach to express discontinuities, we can state the distribution of the distributed load in the global coordinate X as:

$$p_X(X) = 2p_0\left(\langle X\rangle^0 - \langle X - 2L\rangle^0\right) + p_0\left(\langle X - 2L\rangle^0\right). \tag{2.29}$$

This expression can be introduced in the second-order differential equation (see Table 2.2) as load function:

$$EA\frac{d^2 u_X(X)}{dX^2} = -2p_0\left(\langle X\rangle^0 - \langle X - 2L\rangle^0\right) - p_0\left(\langle X - 2L\rangle^0\right). \tag{2.30}$$

Integration twice gives:

$$EA\frac{d^1 u_X}{dX^1} = -2p_0\left(\langle X\rangle^1 - \langle X - 2L\rangle^1\right) - p_0\left(\langle X - 2L\rangle^1\right) + c_1, \tag{2.31}$$

$$EAu_X = -2p_0\left(\frac{1}{2}\langle X\rangle^2 - \frac{1}{2}\langle X - 2L\rangle^2\right) - p_0\left(\frac{1}{2}\langle X - 2L\rangle^2\right) + c_1 X + c_2. \tag{2.32}$$

The constants can be obtained based on the boundary conditions (2.19) as $c_2 = 0$ and $c_1 = 5p_0 L$. Thus, the distributions of elongation, strain, and stress are obtained in closed-form representation as:

$$u_X(X) = \frac{p_0}{EA}\left\{-\langle X\rangle^2 + \langle X - 2L\rangle^2 - \frac{1}{2}\langle X - 2L\rangle^2 + 5LX\right\}, \tag{2.33}$$

$$\varepsilon_X(X) = \frac{p_0}{EA}\left\{-2\langle X\rangle^1 + 2\langle X - 2L\rangle^1 - \langle X - 2L\rangle^1 + 5L\right\}, \tag{2.34}$$

$$\sigma_X(X) = \frac{p_0}{A}\left\{-2\langle X\rangle^1 + 2\langle X - 2L\rangle^1 - \langle X - 2L\rangle^1 + 5L\right\}. \tag{2.35}$$

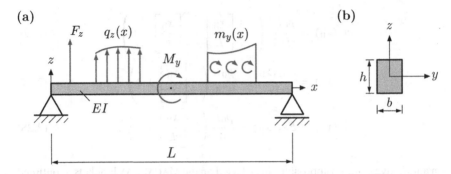

Fig. 2.15 General configuration for EULER–BERNOULLI beam problems: **a** example of boundary conditions and external loads; **b** cross-sectional area (bending occurs in the x-z plane)

2.2.2 Euler–Bernoulli Beams

A thin or EULER–BERNOULLI beam is defined as a long prismatic body whose axial dimension is much larger than its transverse dimensions [4, 11, 16, 27, 29]. This structural member is only loaded perpendicular to its longitudinal body axis by forces (single forces F_z or distributed loads q_z) or moments (single moments M_y or distributed moments m_y). Perpendicular means that the line of application of a force or the direction of a moment vector forms a right angle with the x-axis, see Fig. 2.15. As a result of this loading, the deformation occurs only perpendicular to its main axis.

Derivations are restricted many times to the following simplifications:

- only applying to straight beams,
- no elongation along the x-axis,
- no torsion around the x-axis,
- deformations in a single plane (here: x-z), i.e. symmetrical bending,
- infinitesimally small deformations and strains,
- simple cross sections, and
- the material is linear-elastic.

The three basic equations of continuum mechanics, i.e the kinematics relationship, the constitutive law and the equilibrium equation, as well as their combination to the describing partial differential equation are summarized in Table 2.4.

Under the assumption of constant material ($E = $ const.) and geometric ($I_y = $ const.) properties, the differential equation in Table 2.4 can be integrated four times for constant distributed load ($q_z = q_0 = $ const.) to obtain the general analytical solution of the problem:

$$
u_z(x) = \frac{1}{EI_y}\left(\frac{q_0 x^4}{24} + \frac{c_1 x^3}{6} + \frac{c_2 x^2}{2} + c_3 x + c_4\right), \tag{2.36}
$$

Table 2.4 Different formulations of the basic equations for a BERNOULLI beam (bending occurs in the x-z plane), with $\mathcal{L}_2(\dots) = \frac{d^2(\dots)}{dx^2}$

Specific formulation	General formulation
Kinematics	
$\varepsilon_x(x, z) = -z\dfrac{d^2 u_z(x)}{dx^2}$	$\varepsilon_x(x, z) = -z\mathcal{L}_2\left(u_z(x)\right)$
$\kappa = -\dfrac{d^2 u_z(x)}{dx^2}$	$\kappa = -\mathcal{L}_2\left(u_z(x)\right)$
Constitution	
$\sigma_x(x, z) = E\varepsilon_x(x, z)$	$\sigma_x(x, z) = C\varepsilon_x(x, z)$
$M_y(x) = EI_y\kappa(x)$	$M_y(x) = D\kappa(x)$
Equilibrium	
Force	
$\dfrac{dQ_z(x)}{dx} = -q_z(x)$	
Moment	
$\dfrac{dM_y(x)}{dx} = Q_z(x)$	
Combined	
$\dfrac{d^2 M_y(x)}{dx^2} + q_z(x) = 0$	$\mathcal{L}_2^{\mathrm{T}}\left(M_y(x)\right) + q_z(x) = 0$
PDE	
$\dfrac{d^2}{dx^2}\left(EI_y\dfrac{d^2 u_z(x)}{dx^2}\right) - q_z(x) = 0$	$\mathcal{L}_2^{\mathrm{T}}\left(D\mathcal{L}_2\left(u_z(x)\right)\right) - q_z(x) = 0$
$\dfrac{d}{dx}\left(EI_y\dfrac{d^2 u_z(x)}{dx^2}\right) = -Q_z(x)$	
$EI_y\dfrac{d^2 u_z(x)}{dx^2} = -M_y(x)$	

where the four constants of integration c_i $(i = 1, \dots, 4)$ must be determined based on the boundary conditions (see Table 2.5). The following equations for the shear force $Q_z(x)$, the bending moment $M_y(x)$, and the rotation $\varphi_y(x)$ were obtained based on one-, two- and three-times integration and might be useful to determine some of the constants of integration:

Table 2.5 Different boundary conditions and corresponding reactions for a continuum EULER–BERNOULLI beam (bending occurs in the x-z plane)

Case	Boundary Condition	Reaction
	$u_z(0) = 0, \varphi_y(0) = 0$	
	$u_z(0) = 0, M_y(0) = 0$	
	$u_z(0) = 0, M_y(0) = 0$	
	$\varphi_y(0) = 0, Q_z(0) = 0$	
	$u_z(L) = u_0, M_y(L) = 0$	
	$Q_z(L) = F_0, M_y(L) = 0$	
	$\varphi_y(L) = \varphi_0, Q_z(L) = 0$	
	$M_y(L) = M_0, Q_z(L) = 0$	
	$M_y(L) = 0, Q_z(L) = 0$	

Fig. 2.16 Internal reactions for a continuum EULER–BERNOULLI beam

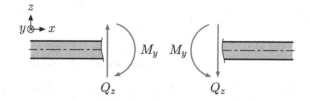

$$Q_z(x) = -q_0 x - c_1 , \tag{2.37}$$

$$M_y(x) = -\frac{q_0 x^2}{2} - c_1 x - c_2 , \tag{2.38}$$

$$\varphi_y(x) = -\frac{du_z(x)}{dx} = -\frac{1}{EI_y}\left(\frac{q_0 x^3}{6} + \frac{c_1 x^2}{2} + c_2 x + c_3 \right) . \tag{2.39}$$

The internal reactions in a beam become visible if one cuts—at an arbitrary location x—the member in two parts. As a result, two opposite oriented shear forces Q_z and bending moments M_y can be indicated. Summing up the internal reactions from both parts must result in zero. Their positive direction is connected with the positive coordinate directions at the positive face (outward surface normal vector parallel to the positive x-axis). This means that at a positive face the positive reactions have the same direction as the positive coordinate axes, see Fig. 2.16.

Once the internal bending moment M_y is known, the normal stress σ_x can be calculated:

$$\sigma_x(x, z) = \frac{M_y(x)}{I_y} z(x) , \tag{2.40}$$

whereas the shear force Q_z allows us to calculate the shear stress distribution. For a rectangular cross section (width b, height h, see Fig. 2.15) under the assumption that the shear stress is constant along the width, the following distribution is obtained [16]:

$$\tau_{xz}(x, z) = \frac{Q_z(x)}{2I_y}\left[\left(\frac{h}{2}\right)^2 - z^2 \right] . \tag{2.41}$$

Application of HOOKE's law (i.e., $\sigma_x = E\varepsilon_x$ and $\tau_{xz} = G\gamma_{xz}$) allows us to calculate the normal and shear strains. Typical distributions of the two stress components in a beam element are shown in Fig. 2.17. It can be seen that normal stress distribution is linear while the shear stress distribution is parabolic over the cross section.

Finally, it should be noted here that the one-dimensional EULER–BERNOULLI beam theory has its two-dimensional analogon in the form of KIRCHHOFF plates[2] [5–7, 12, 18, 30].

---

[2] Also called thin or shear-rigid plates.

(a)   (b)

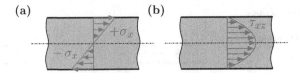

Fig. 2.17 Different stress distributions of an EULER–BERNOULLI beam with rectangular cross section and linear-elastic material behavior: **a** normal stress and **b** shear stress (bending occurs in the x-z plane)

2.4 Cantilever beam with different end loads and deformations

Calculate the analytical solutions for the deflection $u_z(x)$ and rotation $\varphi_y(x)$ of the cantilever beam shown in Fig. 2.18. Calculate in addition for all four cases the reactions at the fixed support and the distributions of the bending moment and shear force. It can be assumed for this exercise that the bending stiffness EI_y is constant.

2.4 Solution

Case (a): Let us start the solution procedure by sketching the free-body diagram as shown in Fig. 2.19a.

The consideration of the global force and moment equilibrium would allow to calculate the reactions at the fixed support, i.e., at $x = 0$:

$$\sum_i F_{z_i} = 0 \quad \Leftrightarrow \quad F_z^R(0) - F_0 = 0 \quad \Rightarrow \quad F_z^R(0) = F_0, \tag{2.42}$$

$$\sum_i M_{y_i} = 0 \quad \Leftrightarrow \quad M_y^R(0) + F_0 L = 0 \quad \Rightarrow \quad M_y^R(0) = -F_0 L. \tag{2.43}$$

The boundary conditions can be stated at the left-hand end as

$$u_z(0) = 0 \quad \text{and} \quad \varphi_y(0) = 0, \tag{2.44}$$

while the consideration of the force and moment equilibrium at the right-hand boundary (see Fig. 2.20) requires that

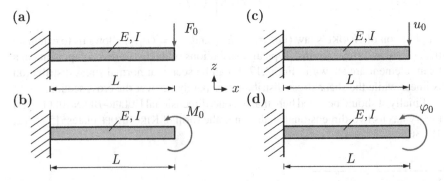

Fig. 2.18 Cantilever beam with different end loads and deformations: **a** single force; **b** single moment; **c** displacement; **d** rotation

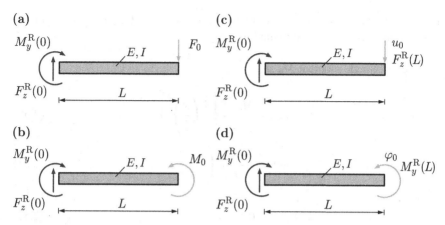

(a) $M_y^R(0)$ E,I F_0 $F_z^R(0)$ L

(b) $M_y^R(0)$ E,I M_0 $F_z^R(0)$ L

(c) $M_y^R(0)$ E,I u_0 $F_z^R(L)$ $F_z^R(0)$ L

(d) $M_y^R(0)$ E,I φ_0 $M_y^R(L)$ $F_z^R(0)$ L

Fig. 2.19 Free-body diagrams of the cantilever beams with different end loads and deformations: **a** single force; **b** single moment; **c** displacement; **d** rotation

Fig. 2.20 Equilibrium between internal reactions and external load at $x = L$

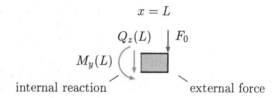

$$Q_z(L) = -F_0 \quad \text{and} \quad M_y(L) = 0 . \tag{2.45}$$

Consideration of the boundary condition $(2.44)_1$ in the general expression for the displacement distribution (2.36) gives the fourth constant of integration as: $c_4 = 0$. In a similar way, the third constant of integration can be obtained by considering the boundary condition $(2.44)_2$ in the general expression for the rotation distribution (2.39): $c_3 = 0$. Introducing the boundary conditions at the right-hand end, i.e. Eq. (2.45) in the expressions for the bending moment and shear force according to Eqs. (2.37) and (2.38), the remaining constants are obtained as: $c_1 = F_0$ and $c_2 = -F_0 L$. Thus, the distributions of deflection, rotational angle, bending moment, and shear force can be stated as:

$$u_z(x) = \frac{F_0 L^3}{EI} \left\{ \frac{1}{6} \left(\frac{x}{L} \right)^3 - \frac{1}{2} \left(\frac{x}{L} \right)^2 \right\} , \tag{2.46}$$

$$\varphi_y(x) = \frac{F_0 L^2}{EI} \left\{ -\frac{1}{2} \left(\frac{x}{L} \right)^2 + \left(\frac{x}{L} \right) \right\} , \tag{2.47}$$

$$M_y(x) = F_0 L \left\{ -\left(\frac{x}{L} \right) + 1 \right\} , \tag{2.48}$$

$$Q_z(x) = -F_0 . \tag{2.49}$$

The other three cases can be solved in a similar way and the final results for the distributions are summarized in the following:

Case (b): Single moment M_0 at $x = L$

$$u_z(x) = \frac{M_0 L^2}{EI} \left\{ \frac{1}{2} \left(\frac{x}{L} \right)^2 \right\} , \tag{2.50}$$

$$\varphi_y(x) = -\frac{M_0 L}{EI} \left(\frac{x}{L} \right) , \tag{2.51}$$

$$M_y(x) = -M_0 , \tag{2.52}$$

$$Q_z(x) = 0 . \tag{2.53}$$

Case (c): Displacement u_0 at $x = L$

$$u_z(x) = \left\{ \frac{1}{2} \left(\frac{x}{L} \right)^3 - \frac{3}{2} \left(\frac{x}{L} \right)^2 \right\} u_0 , \tag{2.54}$$

$$\varphi_y(x) = \left\{ -\frac{3}{2} \left(\frac{x}{L} \right)^2 + 3 \left(\frac{x}{L} \right) \right\} \frac{u_0}{L} , \tag{2.55}$$

$$M_y(x) = \frac{3EIu_0}{L^2} \left\{ -\left(\frac{x}{L} \right) + 1 \right\} , \tag{2.56}$$

$$Q_z(x) = -\frac{3EIu_0}{L^3} . \tag{2.57}$$

Case (d): Rotation φ_0 at $x = L$

$$u_z(x) = \frac{\varphi_0 L}{2} \left(\frac{x}{L} \right)^2 , \tag{2.58}$$

$$\varphi_y(x) = -\varphi_0 \left(\frac{x}{L} \right) , \tag{2.59}$$

$$M_y(x) = -\frac{\varphi_0 EI}{L} , \tag{2.60}$$

$$Q_z(x) = 0 . \tag{2.61}$$

2.5 Cantilever beam with distributed load

Given is a beam with different support conditions which is loaded by a constant distributed load q_0, see Fig. 2.21. It can be assumed for this exercise that the bending stiffness EI_y is constant. Calculate the analytical solution for the deflection $u_z(x)$,

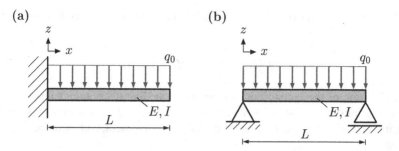

Fig. 2.21 Beam loaded under constant distributed load and different boundary conditions: **a** cantilever and **b** simply supported

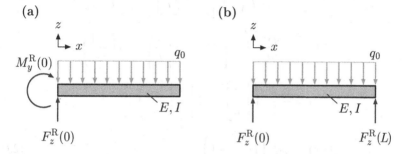

Fig. 2.22 Free-body diagrams of the beams loaded under constant distributed load and different boundary conditions: **a** cantilever and **b** simply supported

rotation $\varphi_y(x)$, the reactions at the supports as well as the distributions of the bending moment and shear force.

2.5 Solution

Case (a): Let us start the solution procedure by sketching the free-body diagram as shown in Fig. 2.22a.

The consideration of the global force and moment equilibrium would allow to calculate the reactions at the fixed support, i.e., at $x = 0$:

$$\sum_i F_{z_i} = 0 \;\Leftrightarrow\; F_z^{\mathrm{R}}(0) - q_0 L = 0 \;\Rightarrow\; F_z^{\mathrm{R}}(0) = q_0 L \,, \tag{2.62}$$

$$\sum_i M_{y_i} = 0 \;\Leftrightarrow\; M_y^{\mathrm{R}}(0) + \frac{q_0 L^2}{2} = 0 \;\Rightarrow\; M_y^{\mathrm{R}}(0) = -\frac{q_0 L^2}{2}. \tag{2.63}$$

The boundary conditions can be stated at the left-hand end as

$$u_z(0) = 0 \;\text{ and }\; \varphi_y(0) = 0 \,, \tag{2.64}$$

while the consideration of the force and moment equilibrium at the right-hand boundary (see Fig. 2.20) requires that

$$Q_z(L) = 0 \quad \text{and} \quad M_y(L) = 0. \tag{2.65}$$

Consideration of these conditions in the corresponding distributions results in the following constants of integration: $c_1 = q_0 L$, $c_2 = -\frac{1}{2} q_0 L^2$, $c_3 = c_4 = 0$. Thus, the distributions of deflection, rotational angle, bending moment, and shear force can be stated as:

$$u_z(x) = -\frac{q_0 L^4}{24 EI} \left(\left[\frac{x}{L} \right]^4 - 4 \left[\frac{x}{L} \right]^3 + 6 \left[\frac{x}{L} \right]^2 \right), \tag{2.66}$$

$$\varphi_y(x) = -\frac{q_0 L^3}{6 EI} \left(-\left[\frac{x}{L} \right]^3 + 3 \left[\frac{x}{L} \right]^2 - 3 \left[\frac{x}{L} \right] \right), \tag{2.67}$$

$$M_y(x) = \frac{q_0 L^2}{2} \left(\left[\frac{x}{L} \right]^2 - 2 \left[\frac{x}{L} \right] + 1 \right), \tag{2.68}$$

$$Q_z(x) = q_0 L \left(\left[\frac{x}{L} \right] - 1 \right). \tag{2.69}$$

Case (b): The set of boundary conditions is in this case given as

$$u_z(0) = 0, \qquad\qquad\qquad M_y(0) = 0, \tag{2.70}$$
$$u_z(L) = 0, \qquad\qquad\qquad M_y(L) = 0, \tag{2.71}$$

which results in the following constants of integration: $c_1 = \frac{q_0 L}{2}$, $c_2 = 0$, $c_3 = -\frac{q_0 L^3}{24}$, and $c_4 = 0$. Thus, the distributions of deflection, rotational angle, bending moment, and shear force can be stated as:

$$u_z(x) = -\frac{q_0 L^4}{24 EI} \left(\left[\frac{x}{L} \right]^4 - 2 \left[\frac{x}{L} \right]^3 + \left[\frac{x}{L} \right] \right), \tag{2.72}$$

$$\varphi_y(x) = -\frac{q_0 L^3}{24 EI} \left(-4 \left[\frac{x}{L} \right]^3 + 6 \left[\frac{x}{L} \right]^2 - 1 \right), \tag{2.73}$$

$$M_y(x) = \frac{q_0 L^2}{2} \left(\left[\frac{x}{L} \right]^2 - \left[\frac{x}{L} \right] \right), \tag{2.74}$$

$$Q_z(x) = \frac{q_0 L}{2} \left(2 \left[\frac{x}{L} \right] - 1 \right). \tag{2.75}$$

Fig. 2.23 Beam with different sections

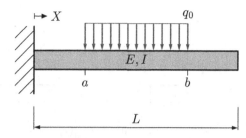

Fig. 2.24 Free-body diagram of the beam with different sections

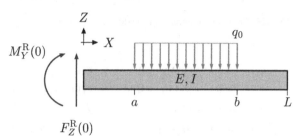

2.6 Cantilever beam with different sections

Given is a cantilever beam of length L and constant bending stiffness EI as shown in Fig. 2.23. At the left-hand side there is a fixed support and a constant distributed load p_0 is acting in the range $a \leq x \leq b$. Calculate the analytical solution for the deflection $u_z(x)$, rotation $\varphi_y(x)$, the reactions at the support as well as the distributions of the bending moment and shear force.

2.6 Solution

Let us start the solution procedure by sketching the free-body diagram of the entire structure as shown in Fig. 2.24.

The two discontinuities in regards to the load at $X = a$ and $X = b$ requires to split the structure in three parts as indicated in Fig. 2.25. The left-hand part is now described by the local coordinate x_I with $0 \leq x_I \leq a$, the middle part is described by the local coordinate x_{II} with $0 \leq x_{II} \leq b - a$ while the right-hand part is described by the local coordinate x_{III} with $0 \leq x_{III} \leq L - b$.

Consideration of three parts means that Eqs. (2.36)–(2.39) must be applied to all sections and in total 12 integration constants, i.e. four for each section (here $c_1 \ldots c_4$ for the left-hand, $c_5 \ldots c_8$ for the middle section while c_9 to c_{12} for the right-hand section), must be determined. The following four boundary and eight transmission conditions can be stated:

$$u_z(x_I = 0) = 0, \qquad\qquad M_y(x_{III} = L - b) = 0, \qquad (2.76)$$
$$\varphi_y(x_I = 0) = 0, \qquad\qquad Q_z(x_{III} = L - b) = 0, \qquad (2.77)$$

and

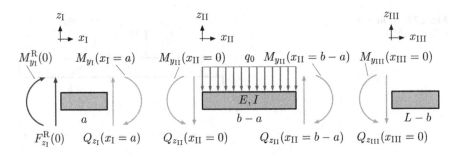

Fig. 2.25 Free-body diagrams of the different sections

$$u_z(x_I = a) = u_z(x_{II} = 0), \qquad u_z(x_{II} = b - a) = u_z(x_{III} = 0), \qquad (2.78)$$
$$\varphi_y(x_I = a) = \varphi_y(x_{II} = 0), \qquad \varphi_y(x_{II} = b - a) = \varphi_y(x_{III} = 0), \qquad (2.79)$$
$$Q_z(x_I = a) = Q_z(x_{II} = 0), \qquad Q_z(x_{II} = b - a) = Q_z(x_{III} = 0), \qquad (2.80)$$
$$M_y(x_I = a) = M_y(x_{II} = 0), \qquad M_y(x_{II} = b - a) = M_y(x_{III} = 0). \qquad (2.81)$$

The general solutions for the displacements, rotations, shear forces and bending moments, i.e., Eqs. (2.36)–(2.39), can be stated for the three sections as:

$$u_z(x_I) = \frac{1}{EI_y}\left(\frac{c_1 x_I^3}{6} + \frac{c_2 x_I^2}{2} + c_3 x_I + c_4\right), \qquad (2.82)$$

$$Q_z(x_I) = -c_1, \qquad (2.83)$$
$$M_y(x_I) = -c_1 x_I - c_2, \qquad (2.84)$$

$$\varphi_y(x_I) = -\frac{du_z(x_I)}{dx} = -\frac{1}{EI_y}\left(\frac{c_1 x_I^2}{2} + c_2 x_I + c_3\right), \qquad (2.85)$$

and for the second section

$$u_z(x_{II}) = \frac{1}{EI_y}\left(\frac{-q_0 x_{II}^4}{24} + \frac{c_5 x_{II}^3}{6} + \frac{c_6 x_{II}^2}{2} + c_7 x_{II} + c_8\right), \qquad (2.86)$$

$$Q_z(x_{II}) = +q_0 x_{II} - c_5, \qquad (2.87)$$

$$M_y(x_{II}) = +\frac{q_0 x_{II}^2}{2} - c_5 x_{II} - c_6, \qquad (2.88)$$

$$\varphi_y(x_{II}) = -\frac{du_z(x_{II})}{dx} = -\frac{1}{EI_y}\left(\frac{-q_0 x_{II}^3}{6} + \frac{c_5 x_{II}^2}{2} + c_6 x_{II} + c_7\right), \qquad (2.89)$$

and for the third section

$$u_z(x_{III}) = \frac{1}{EI_y}\left(\frac{c_9 x_{III}^3}{6} + \frac{c_{10} x_{III}^2}{2} + c_{11} x_{III} + c_{12}\right), \tag{2.90}$$

$$Q_z(x_{III}) = -c_9, \tag{2.91}$$

$$M_y(x_{III}) = -c_9 x_{III} - c_{10}, \tag{2.92}$$

$$\varphi_y(x_{III}) = -\frac{\mathrm{d}u_z(x_{III})}{\mathrm{d}x} = -\frac{1}{EI_y}\left(\frac{c_9 x_{III}^2}{2} + c_{10} x_{III} + c_{11}\right). \tag{2.93}$$

Consideration of the 12 boundary and transmissions conditions in this set of equations gives 12 conditions for the unknown constants of integration $c_1 \ldots c_{12}$ which can be expressed in matrix form as follows:

$$
\begin{bmatrix}
0 & 0 & 0 & 1 & 0 & 0 & 0 & 0 & 0 & 0 & 0 & 0 \\
0 & 0 & 1 & 0 & 0 & 0 & 0 & 0 & 0 & 0 & 0 & 0 \\
0 & 0 & 0 & 0 & 0 & 0 & 0 & 1 & 0 & 0 & 0 & 0 \\
\frac{a^3}{6} & \frac{a^2}{2} & 0 & 0 & 0 & 0 & 0 & -1 & 0 & 0 & 0 & 0 \\
\frac{a^2}{2} & a & 0 & 0 & 0 & 0 & -1 & 0 & 0 & 0 & 0 & 0 \\
-1 & 0 & 0 & 0 & 1 & 0 & 0 & 0 & 0 & 0 & 0 & 0 \\
-a & -1 & 0 & 0 & 0 & 1 & 0 & 0 & 0 & 0 & 0 & 0 \\
0 & 0 & 0 & 0 & \frac{(b-a)^3}{6} & \frac{(b-a)^2}{2} & (b-a) & 1 & 0 & 0 & 0 & -1 \\
0 & 0 & 0 & 0 & \frac{(b-a)^2}{2} & (b-a) & 1 & 0 & 0 & 0 & -1 & 0 \\
0 & 0 & 0 & 0 & -1 & 0 & 0 & 0 & 1 & 0 & 0 & 0 \\
0 & 0 & 0 & 0 & -(b-a) & -1 & 0 & 0 & 0 & 1 & 0 & 0 \\
0 & 0 & 0 & 0 & 0 & 0 & 0 & -(L-b) & -1 & 0 & 0 \\
\end{bmatrix}
\begin{bmatrix}
c_1 \\ c_2 \\ c_3 \\ c_4 \\ c_5 \\ c_6 \\ c_7 \\ c_8 \\ c_9 \\ c_{10} \\ c_{11} \\ c_{12}
\end{bmatrix}
=
\begin{bmatrix}
0 \\ 0 \\ 0 \\ 0 \\ 0 \\ 0 \\ 0 \\ \frac{q_0(b-a)^4}{24} \\ \frac{q_0(b-a)^3}{6} \\ -q_0(b-a) \\ \frac{-q_0(b-a)^2}{2} \\ 0
\end{bmatrix}.
\tag{2.94}
$$

Multiplication of the inversed coefficient matrix with the right-hand side allows us to determine the constants as:

$$c_1 = q_0(b-a), \qquad c_2 = -\tfrac{q_0}{2}(b^2 - a^2), \tag{2.95}$$

$$c_3 = 0, \qquad c_4 = 0, \tag{2.96}$$

$$c_5 = q_0(b-a), \qquad c_6 = -\tfrac{q_0}{2}(b-a)^2, \tag{2.97}$$

$$c_7 = -\tfrac{q_0 ab}{2}(b-a), \qquad c_8 = \tfrac{q_0 a^2}{12}(a^2 + 2ab - 3b^2), \tag{2.98}$$

$$c_9 = 0, \qquad c_{10} = 0, \tag{2.99}$$

$$c_{11} = -\tfrac{q_0}{6}(b^3 - a^3), \qquad c_{12} = -\tfrac{q_0}{24}(a^4 + 3b^4 - 4a^3 b). \tag{2.100}$$

Based on these constants of integration, the general expressions (2.82)–(2.93) for the distributions can be concretized as:

$$u_z(x_{\mathrm{I}}) = \frac{q_0}{EI}\left(\frac{b-a}{6}x_{\mathrm{I}}^3 - \frac{b^2-a^2}{4}x_{\mathrm{I}}^2\right), \tag{2.101}$$

$$u_z(x_{\mathrm{II}}) = \frac{q_0}{EI}\left(\frac{-x_{\mathrm{II}}^4}{24} + \frac{b-a}{6}x_{\mathrm{II}}^3 - \frac{(b-a)^2}{4}x_{\mathrm{II}}^2 - \frac{ab(b-a)}{2}x_{\mathrm{II}} + \right.$$
$$\left. + \frac{a^2(a^2+2ab-3b^2)}{12}\right), \tag{2.102}$$

$$u_z(x_{\mathrm{III}}) = \frac{q_0}{EI}\left(-\frac{b^3-a^3}{6}x_{\mathrm{III}} - \frac{a^4+3b^4-4a^3b}{24}\right), \tag{2.103}$$

and for the rotations

$$\varphi_y(x_{\mathrm{I}}) = -\frac{q_0}{2EI}\left((b-a)x_{\mathrm{I}}^2 - (b^2-a^2)x_{\mathrm{I}}\right), \tag{2.104}$$

$$\varphi_y(x_{\mathrm{II}}) = -\frac{q_0}{2EI}\left(-\frac{x_{\mathrm{II}}^3}{3} + (b-a)x_{\mathrm{II}}^2 - (b-a)^2x_{\mathrm{II}} - ab(b-a)\right), \tag{2.105}$$

$$\varphi_y(x_{\mathrm{III}}) = \frac{q_0}{6EI}\left(b^3-a^3\right), \tag{2.106}$$

and for the bending moments

$$M_y(x_{\mathrm{I}}) = q_0\left(-(b-a)x_{\mathrm{I}} + \frac{b^2-a^2}{2}\right), \tag{2.107}$$

$$M_y(x_{\mathrm{II}}) = \frac{q_0}{2}\left(x_{\mathrm{II}}^2 - 2(b-a)x_{\mathrm{II}} + (b-a)^2\right), \tag{2.108}$$

$$M_y(x_{\mathrm{III}}) = 0, \tag{2.109}$$

and for the shear forces

$$Q_z(x_{\mathrm{I}}) = -q_0(b-a), \tag{2.110}$$

$$Q_z(x_{\mathrm{II}}) = q_0(x_{\mathrm{II}} - (b-a)), \tag{2.111}$$

$$Q_z(x_{\mathrm{III}}) = 0. \tag{2.112}$$

An alternative solution approach can be based on the MACAULAY brackets as outlined in Eq. (2.4). Based on this particular approach to express discontinuities, we can state the distribution of the distributed load in the global coordinate X as:

$$q_Z(X) = -q_0\left(\langle X-a\rangle^0 - \langle X-b\rangle^0\right). \tag{2.113}$$

This expression can be introduced in the fourth-order differential equation (see Table 2.4) as load function:

$$EI\frac{d^4 u_z(X)}{dX^4} = q_Z(X) = -q_0 \left(\langle X - a \rangle^0 - \langle X - b \rangle^0 \right) . \tag{2.114}$$

Four times integration of the last equation gives:

$$EI\frac{d^3 u_z(X)}{dX^3} = -Q_Z(X) = -q_0 \left(\langle X - a \rangle^1 - \langle X - b \rangle^1 \right) + c_1 , \tag{2.115}$$

$$EI\frac{d^2 u_z(X)}{dX^2} = -M_Y(X) = -q_0 \left(\frac{1}{2}\langle X - a \rangle^2 - \frac{1}{2}\langle X - b \rangle^2 \right) + c_1 X + c_2 , \tag{2.116}$$

$$EI\frac{d^1 u_z(X)}{dX^1} = EI(-\varphi_Y(X)) = -q_0 \left(\frac{1}{6}\langle X - a \rangle^3 - \frac{1}{6}\langle X - b \rangle^3 \right) +$$

$$+ \frac{c_1}{2} X^2 + c_2 X + c_3 , \tag{2.117}$$

$$EI u_z(X) = -q_0 \left(\frac{1}{24}\langle X - a \rangle^4 - \frac{1}{24}\langle X - b \rangle^4 \right) + \frac{c_1}{6} X^3 + \frac{c_2}{2} X^2 +$$

$$+ c_3 X + C_4 . \tag{2.118}$$

The constants can be obtained based on the boundary conditions (2.76)–(2.77) as $c_1 = q_0(b - a)$, $c_2 = -\frac{q_0}{2}(b^2 - a^2)$, $c_3 = 0$, and $c_4 = 0$. Thus, the distribution of the deflection is obtained in closed-form representation as:

$$u_Z(X) = \frac{-q_0}{EI} \left(\frac{\langle X - a \rangle^4}{24} - \frac{\langle X - b \rangle^4}{24} - \frac{(b - a)}{6} X^3 + \frac{(b^2 - a^2)}{4} X^2 \right) . \tag{2.119}$$

It should be noted that the end deflection of the beam can be obtained for $X = L$ as:

$$u_Z(L) = \frac{-q_0}{EI} \left(-\frac{La^3}{6} + \frac{a^4}{24} + \frac{Lb^3}{6} - \frac{b^4}{24} \right) . \tag{2.120}$$

The special case that the distributed load extends over the entire beam, i.e., $a = 0$ and $b = L$, gives the classical result for the end deflection: $u_Z(L) = \frac{q_0 L^4}{8EI}$.

The distributions of the load q_Z, shear force Q_Z, and bending moment M_Y are shown in Fig. 2.26 and allows us to understand the dependency of these quantities.

Fig. 2.26 Beam with
different sections:
a distributed load, **b** shear
force, and **c** bending moment

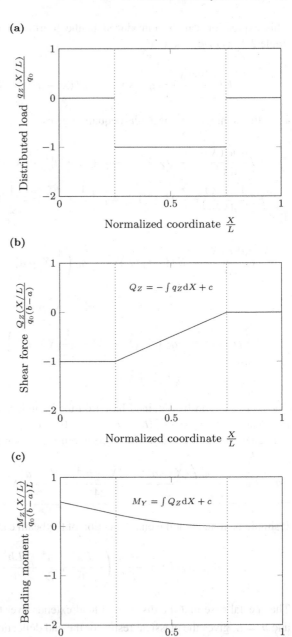

(a)

Distributed load $\frac{q_Z(X/L)}{q_0}$

Normalized coordinate $\frac{X}{L}$

(b)

Shear force $\frac{Q_Z(X/L)}{q_0(b-a)}$

$Q_Z = -\int q_Z \mathrm{d}X + c$

Normalized coordinate $\frac{X}{L}$

(c)

Bending moment $\frac{M_Z(X/L)}{q_0(b-a)L}$

$M_Y = \int Q_Z \mathrm{d}X + c$

Normalized coordinate $\frac{X}{L}$

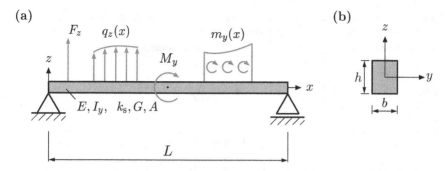

Fig. 2.27 General configuration for TIMOSHENKO beam problems: **a** example of boundary conditions and external loads; **b** cross-sectional area (bending occurs in the x-z plane)

2.2.3 Timoshenko Beams

A thick or TIMOSHENKO beam is defined as a long prismatic body whose axial dimension is much larger than its transverse dimensions [26, 32]. This structural member is only loaded perpendicular to its longitudinal body axis by forces (single forces F_z or distributed loads q_z) or moments (single moments M_y or distributed moments m_y). Perpendicular means that the line of application of a force or the direction of a moment vector forms a right angle with the x-axis, see Figs. 2.15 and 2.27. As a result of this loading, the deformation occurs only perpendicular to its main axis. The formulation is a shear-flexible theory which means that the shear forces contribute to the bending deformation.

Derivations are restricted many times to the following simplifications:

- only applying to straight beams,
- no elongation along the x-axis,
- no torsion around the x-axis,
- deformations in a single plane (here: x-z), i.e. symmetrical bending,
- infinitesimally small deformations and strains,
- simple cross sections, and
- the material is linear-elastic.

The three basic equations of continuum mechanics, i.e the kinematics relationship, the constitutive law and the equilibrium equation, as well as their combination to the describing partial differential equations are summarized in Table 2.6. It should be noted here that the deflection u_z and the rotation ϕ_y are now independent variables and both represented in the coupled differential equations.

Under the assumption of constant material (E, G) and geometric (I_y, A, k_s) properties, the system of differential equations in Table 2.6 can be solved for constant distributed loads ($q_z = q_0 = $ const. and $m_y = 0$) to obtain the general analytical solution of the problem [31, 32]:

Table 2.6 Different formulations of the basic equations for a TIMOSHENKO beam (bending in the x-z plane). e: generalized strains; s: generalized stresses

Specific formulation	General formulation
Kinematics	
$\begin{bmatrix} \frac{du_z}{dx} + \phi_y \\ \frac{d\phi_y}{dx} \end{bmatrix} = \begin{bmatrix} \frac{d}{dx} & 1 \\ 0 & \frac{d}{dx} \end{bmatrix} \begin{bmatrix} u_z \\ \phi_y \end{bmatrix}$	$e = \mathcal{L}_1 u$
Constitution	
$\begin{bmatrix} -Q_z \\ M_y \end{bmatrix} = \begin{bmatrix} -k_s AG & 0 \\ 0 & EI_y \end{bmatrix} \begin{bmatrix} \frac{du_z}{dx} + \phi_y \\ \frac{d\phi_y}{dx} \end{bmatrix}$	$s = De$
Equilibrium	
$\begin{bmatrix} \frac{d}{dx} & 0 \\ 1 & \frac{d}{dx} \end{bmatrix} \begin{bmatrix} -Q_z \\ M_y \end{bmatrix} + \begin{bmatrix} -q_z \\ +m_z \end{bmatrix} = \begin{bmatrix} 0 \\ 0 \end{bmatrix}$	$\mathcal{L}_1^T s + b = 0$
PDE	
$-\dfrac{d}{dx}\left[k_s GA\left(\dfrac{du_z}{dx} + \phi_y\right)\right] - q_z = 0,$ $\dfrac{d}{dx}\left(EI_y\dfrac{d\phi_y}{dx}\right) - k_s GA\left(\dfrac{du_z}{dx} + \phi_y\right) + m_y = 0$	$\mathcal{L}_1^T D\mathcal{L}_1 u + b = 0$

$$u_z(x) = \frac{1}{EI_y}\left(\frac{q_0 x^4}{24} + c_1\frac{x^3}{6} + c_2\frac{x^2}{2} + c_3 x + c_4\right), \qquad (2.121)$$

$$\phi_y(x) = -\frac{1}{EI_y}\left(\frac{q_0 x^3}{6} + c_1\frac{x^2}{2} + c_2 x + c_3\right) - \frac{q_0 x}{k_s AG} - \frac{c_1}{k_s AG}, \qquad (2.122)$$

$$M_y(x) = -\left(\frac{q_0 x^2}{2} + c_1 x + c_2\right) - \frac{q_0 EI_y}{k_s AG}, \qquad (2.123)$$

$$Q_z(x) = -(q_0 x + c_1), \qquad (2.124)$$

where the four constants of integration $c_i (i = 1, \ldots, 4)$ must be determined based on the boundary conditions, see Table 2.7.

The internal reactions in a beam become visible if one cuts—at an arbitrary location x—the member in two parts. As a result, two opposite oriented shear forces Q_z and bending moments M_y can be indicated. Summing up the internal reactions from both parts must result in zero. Their positive directions are connected with the positive coordinate directions at the positive face (outward surface normal vector parallel

Table 2.7 Different boundary conditions and their corresponding reactions for a continuum TIM-
OSHENKO beam (bending occurs in the x-z plane)

Case	Boundary Condition	Reaction
	$u_z(0) = 0, \phi_y(0) = 0$	
	$u_z(0) = 0, M_y(0) = 0$	
	$u_z(0) = 0, M_y(0) = 0$	
	$\phi_y(0) = 0, Q_z(0) = 0$	
	$u_z(L) = u_0, M_y(L) = 0$	
	$Q_z(L) = F_0, M_y(L) = 0$	
	$\phi_y(L) = \phi_0, Q_z(L) = 0$	
	$M_y(L) = M_0, Q_z(L) = 0$	
	$M_y(L) = 0, Q_z(L) = 0$	

Fig. 2.28 Internal reactions for a continuum TIMOSHENKO beam (bending occurs in the x-z plane)

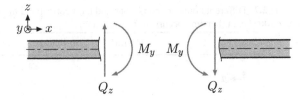

to the positive x-axis). This means that at a positive face the positive reactions have the same direction as the positive coordinate axes, see Fig. 2.28.

Once the internal bending moment M_y is known, the normal stress σ_x can be calculated:

$$\sigma_x(x, z) = \frac{M_y(x)}{I_y} z(x) = E \frac{\mathrm{d}\phi_y(x)}{\mathrm{d}x} z(x), \tag{2.125}$$

whereas the shear stress τ_{xz} is assumed constant over the cross section:

$$\tau_{xz} = \frac{Q_z(x)}{A_s} = \frac{Q_z(x)}{k_s A} = G\gamma_{xz}(x). \tag{2.126}$$

In the above equation, the relation between the shear area A_s and the actual cross-sectional area A is referred to as the shear correction factor k_s [9, 14]:

$$k_s = \frac{A_s}{A}. \tag{2.127}$$

The value of the shear correction factor is, for example, for a circular cross section equal to $\frac{9}{10}$ and for a square cross section equal to $\frac{5}{6}$, see [33]. The relationship between the YOUNG's and shear modulus (see Eqs. (2.125) and (2.126)) is given by [8]:

$$G = \frac{E}{2(1 + \nu)}, \tag{2.128}$$

where ν is POISSON's ratio. The graphical representations of the different stress components are shown in Fig. 2.29. The normal stress is, as in the case of the EULER–BERNOULLI beam, linearly distributed whereas the shear stress is now assumed to be constant.

If more realistic shear stress distributions are considered, one reaches so-called theories of higher-order [17, 22, 23]. Finally, it should be noted here that the one-dimensional TIMOSHENKO beam theory has its two-dimensional analogon in the form of REISSNER- MINDLIN plates[3] [5, 12, 19, 24, 28].

2.7 Beam under pure bending load

The cantilever TIMOSHENKO beam shown in Fig. 2.30 is loaded by a moment M_0

---

[3] Also called thick plates.

at the free right-hand end. The bending stiffness EI and the shear stiffness $k_s AG$ are constant and the total length of the beam is equal to L. Determine, based on the TIMOSHENKO beam theory, the bending line and compare the result with the EULER–BERNOULLI theory.

2.7 Solution

The set of equations for deflection, rotational angle, bending moment and shear force as given in Eqs. (2.121)–(2.124) reduces for $q_0 = 0$ to the following formulation:

$$u_z(x) = \frac{1}{EI_y}\left(c_1\frac{x^3}{6} + c_2\frac{x^2}{2} + c_3 x + c_4\right), \qquad (2.129)$$

$$\phi_y(x) = -\frac{1}{EI_y}\left(+c_1\frac{x^2}{2} + c_2 x + c_3\right) - \frac{c_1}{k_s AG}, \qquad (2.130)$$

$$M_y(x) = -(c_1 x + c_2), \qquad (2.131)$$

$$Q_z(x) = -(c_1). \qquad (2.132)$$

The boundary conditions for the case shown in Fig. 2.30 can be stated as

$$u_z(0) = 0, \qquad\qquad M_y(L) = -M_0, \qquad (2.133)$$

$$\varphi_y(0) = 0, \qquad\qquad Q_z(L) = 0, \qquad (2.134)$$

which allow to determine the constants of integration in Eqs. (2.129)–(2.132) as $c_1 = 0$, $c_2 = M_0$, $c_3 = 0$, and $c_4 = 0$. Thus, the bending line can be expressed as

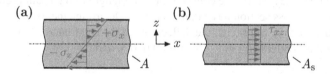

Fig. 2.29 Different stress distributions of a TIMOSHENKO beam with rectangular cross section and linear-elastic material behavior: **a** normal stress and **b** shear stress (bending occurs in the x-z plane)

Fig. 2.30 Beam loaded under pure bending moment

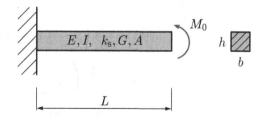

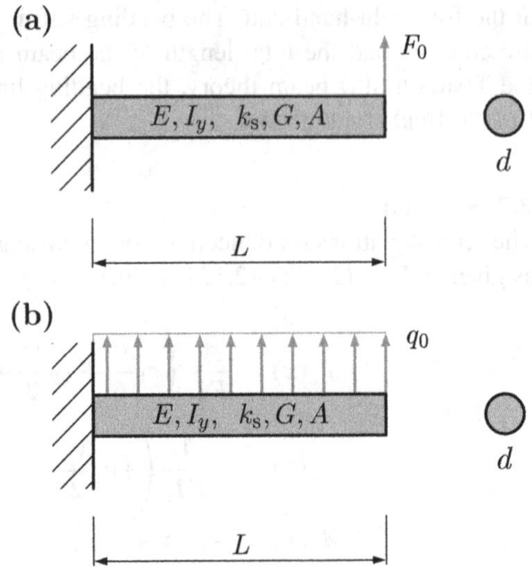

Fig. 2.31 Cantilever TIMOSHENKO beam: **a** single force case and **b** distributed load case

$$u_z(x) = \frac{M_0 x^2}{2EI}. \tag{2.135}$$

This result is identical with the solution according to the EULER–BERNOULLI beam theory.

2.8 Cantilever beam under the influence of a point or distributed load
The cantilever TIMOSHENKO beam shown in Fig. 2.31 is either loaded by a single force F_0 at its right-hand end or by a distributed load q_0. The bending stiffness EI and the shear stiffness $k_s AG$ are constant, the total length of the beam is equal to L, and the circular cross section has a diameter of d. Determine the expressions of the bending lines ($u_z(x)$) and sketch the deflections of the right-hand end ($x = L$) as a function of the slenderness ratio $\frac{d}{L}$ for $\nu = 0.0, 0.3$, and 0.5.

2.8 Solution
Case (a): The set of equations for deflection, rotational angle, bending moment and shear force as given in Eqs. (2.121)–(2.124) reduces for $q_0 = 0$ to the following formulation:

$$u_z(x) = \frac{1}{EI_y}\left(c_1 \frac{x^3}{6} + c_2 \frac{x^2}{2} + c_3 x + c_4\right), \tag{2.136}$$

$$\phi_y(x) = -\frac{1}{EI_y}\left(+c_1 \frac{x^2}{2} + c_2 x + c_3\right) - \frac{c_1}{k_s AG}, \tag{2.137}$$

$$M_y(x) = -(c_1 x + c_2), \tag{2.138}$$

$$Q_z(x) = -(c_1). \tag{2.139}$$

The boundary conditions for the case shown in Fig. 2.31a can be stated as

$$u_z(0) = 0, \qquad\qquad\qquad M_y(L) = 0, \qquad\qquad (2.140)$$
$$\varphi_y(0) = 0, \qquad\qquad\qquad Q_z(L) = F_0, \qquad\qquad (2.141)$$

which allow to determine the constants of integration in Eqs. (2.136)–(2.139) as $c_1 = -F_0, c_2 = F_0 L, c_3 = \frac{EIF_0}{k_s AG}$, and $c_4 = 0$. Thus, the bending line can be expressed as

$$u_z(x) = \frac{1}{EI}\left(-F_0\frac{x^3}{6} + F_0 L\frac{x^2}{2} + \frac{EIF_0}{k_s AG}x\right), \qquad (2.142)$$

or in normalized representation as:

$$\frac{u_z\left(\frac{x}{L}\right)}{\frac{F_0 L^3}{EI}} = -\frac{1}{6}\left(\frac{x}{L}\right)^3 + \frac{1}{2}\left(\frac{x}{L}\right)^2 + \frac{EI}{k_s AGL^2}\left(\frac{x}{L}\right). \qquad (2.143)$$

In the case of the considered circular cross section, one can use $k_s = \frac{9}{10}$, $A = \frac{\pi d^2}{4}$, and $I = \frac{\pi d^4}{64}$ to simplify Eq. (2.143):

$$\frac{u_z\left(\frac{x}{L}\right)}{\frac{F_0 L^3}{EI}} = -\frac{1}{6}\left(\frac{x}{L}\right)^3 + \frac{1}{2}\left(\frac{x}{L}\right)^2 + \frac{5}{36}(1+\nu)\left(\frac{x}{L}\right)\left(\frac{d}{L}\right)^2, \qquad (2.144)$$

or only at the right-hand end, i.e., $x = L$:

$$\frac{u_z\left(\frac{x}{L}=1\right)}{\frac{F_0 L^3}{EI}} = \frac{1}{3} + \frac{5}{36}(1+\nu)\left(\frac{d}{L}\right)^2. \qquad (2.145)$$

The graphical representation of the deflection at the right-hand end for different values of POISSON's ratio is given in Fig. 2.32.

Case (b): The set of equations for deflection, rotational angle, bending moment and shear force must be considered as given in Eqs. (2.121)–(2.124):

$$u_z(x) = \frac{1}{EI_y}\left(\frac{q_0 x^4}{24} + c_1\frac{x^3}{6} + c_2\frac{x^2}{2} + c_3 x + c_4\right), \qquad (2.146)$$

$$\phi_y(x) = -\frac{1}{EI_y}\left(\frac{q_0 x^3}{6} + c_1\frac{x^2}{2} + c_2 x + c_3\right) - \frac{q_0 x}{k_s AG} - \frac{c_1}{k_s AG}, \qquad (2.147)$$

$$M_y(x) = -\left(\frac{q_0 x^2}{2} + c_1 x + c_2\right) - \frac{q_0 EI_y}{k_s AG}, \qquad (2.148)$$

$$Q_z(x) = -(q_0 x + c_1). \qquad (2.149)$$

Fig. 2.32 Deflection of the
right-hand-end of a
TIMOSHENKO beam based on
analytical solutions for single
force loading **a** general view
and **b** magnification for
small slenderness ratios

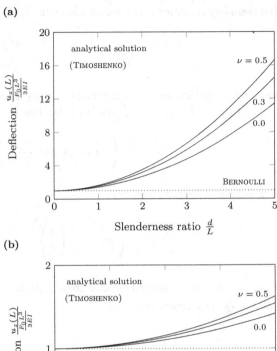

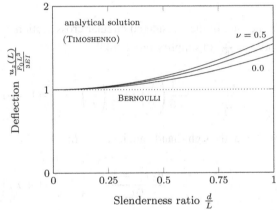

The boundary conditions for the case shown in Fig. 2.31b can be stated as

$$u_z(0) = 0, \qquad\qquad M_y(L) = 0, \qquad\qquad (2.150)$$

$$\varphi_y(0) = 0, \qquad\qquad Q_z(L) = 0, \qquad\qquad (2.151)$$

which allow to determine the constants of integration in Eqs. (2.146)–(2.149) as
$c_1 = -q_0 L$, $c_2 = \frac{q_0 L^2}{2} - \frac{q_0 E I}{k_s A G}$, $c_3 = \frac{q_0 L E I}{k_s A G}$, and $c_4 = 0$. Thus, the bending line can
be expressed as

$$u_z(x) = \frac{1}{EI}\left(\frac{q_0 x^4}{24} - \frac{q_0 L x^3}{6} + \left[\frac{q_0 L^2}{2} - \frac{q_0 E I}{k_s A G} \right]\frac{x^2}{2} + \frac{q_0 L E I}{k_s A G}x \right), \quad (2.152)$$

or in normalized representation as:

$$\frac{u_z\left(\frac{x}{L}\right)}{\frac{q_0 L^4}{EI}} = \frac{1}{24}\left(\frac{x}{L}\right)^4 - \frac{1}{6}\left(\frac{x}{L}\right)^3 + \frac{1}{2}\left[\frac{1}{2} - \frac{EI}{k_s A G L^2}\right]\left(\frac{x}{L}\right)^2 + \frac{EI}{k_s A G L^2}\left(\frac{x}{L}\right).$$
(2.153)

In the case of the considered circular cross section, one can use $k_s = \frac{9}{10}$, $A = \frac{\pi d^2}{4}$, and $I = \frac{\pi d^4}{64}$ to simplify Eq. (2.153):

$$\frac{u_z\left(\frac{x}{L}\right)}{\frac{q_0 L^4}{EI}} = \frac{1}{24}\left(\frac{x}{L}\right)^4 - \frac{1}{6}\left(\frac{x}{L}\right)^3 + \frac{1}{2}\left[\frac{1}{2} - \frac{5}{36}(1+\nu)\left(\frac{d}{L}\right)^2\right]\left(\frac{x}{L}\right)^2$$

$$+ \frac{5}{36}(1+\nu)\left(\frac{d}{L}\right)^2\left(\frac{x}{L}\right),$$
(2.154)

or only at the right-hand end, i.e., $x = L$:

$$\frac{u_z\left(\frac{x}{L}\right)}{\frac{q_0 L^4}{EI}} = \frac{1}{8} + \frac{5}{72}(1+\nu)\left(\frac{d}{L}\right)^2.$$
(2.155)

The graphical representation of the deflection at the right-hand end for different values of POISSON's ratio is given in Fig. 2.33.

2.9 Cantilever beam with two different sections

The cantilever TIMOSHENKO beam shown in Fig. 2.34 is composed of two sections, i.e., section one (I) with $0 \le X \le L_I$ and section two (II) with $L_I \le X \le L_{II}$. The beam is loaded by a single force F_I at $X = L_I$ and at its right-hand end by a single force F_{II}. The bending and the shear stiffnesses are $E I_I$ and $k_s A_I G$ in section I while $E I_{II}$ and $k_s A_{II} G$ holds for section II. This means that the beam is made of the same material and that the cross sections have similar shapes. Determine the expressions of the bending line.

2.9 Solution

The discontinuity in the cross section can be handled by splitting the beam at $X = L_I$ into two parts. The left-hand part is now described by the local coordinate x_I with $0 \le x_I \le L_I$ while the right-hand part is described by the local coordinate x_{II} with $0 \le x_{II} \le L_{II}$. Consideration of two parts means that Eqs. (2.121) and (2.124) must be applied to both sections and in total eight integration constants, i.e. four for each section (here $c_1, \ldots, c_4$ for the left-hand section while $c_5, \ldots, c_8$ is assigned to the right-hand section), must be determined:

Fig. 2.33 Deflection of the right-hand-end of a TIMOSHENKO beam based on analytical solutions for distributed force loading **a** general view and **b** magnification for small slenderness ratios

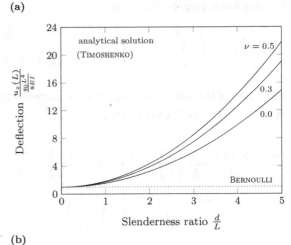

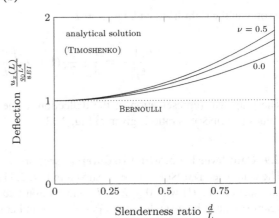

Fig. 2.34 Cantilever TIMOSHENKO beam with two different sections

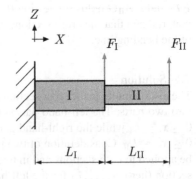

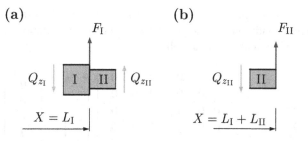

Fig. 2.35 Cantilever TIMOSHENKO beam with two different sections: **a** detail for transmission condition; **b** detail for boundary condition

$$u_z(x_\mathrm{I}) = \frac{1}{EI_y}\left(c_1\frac{x_\mathrm{I}^3}{6} + c_2\frac{x_\mathrm{I}^2}{2} + c_3 x_\mathrm{I} + c_4\right), \qquad (2.156)$$

$$\phi_y(x_\mathrm{I}) = -\frac{1}{EI_y}\left(c_1\frac{x_\mathrm{I}^2}{2} + c_2 x_\mathrm{I} + c_3\right) - \frac{c_1}{k_s AG}, \qquad (2.157)$$

$$M_y(x_\mathrm{I}) = -(c_1 x_\mathrm{I} + c_2), \qquad (2.158)$$

$$Q_z(x_\mathrm{I}) = -(c_1), \qquad (2.159)$$

and

$$u_z(x_\mathrm{II}) = \frac{1}{EI_y}\left(c_5\frac{x_\mathrm{II}^3}{6} + c_6\frac{x_\mathrm{II}^2}{2} + c_7 x_\mathrm{II} + c_8\right), \qquad (2.160)$$

$$\phi_y(x_\mathrm{II}) = -\frac{1}{EI_y}\left(c_5\frac{x_\mathrm{II}^2}{2} + c_6 x_\mathrm{II} + c_7\right) - \frac{c_5}{k_s AG}, \qquad (2.161)$$

$$M_y(x_\mathrm{II}) = -(c_5 x_\mathrm{II} + c_6), \qquad (2.162)$$

$$Q_z(x_\mathrm{II}) = -(c_5). \qquad (2.163)$$

The following four boundary and four transmission conditions can be stated (see Fig. 2.35):

$$u_z(x_\mathrm{I} = 0) = 0, \qquad\qquad M_y(x_\mathrm{II} = L_\mathrm{II}) = 0, \qquad (2.164)$$

$$\varphi_y(x_\mathrm{I} = 0) = 0, \qquad\qquad Q_z(x_\mathrm{II} = L_\mathrm{II}) = F_\mathrm{II}, \qquad (2.165)$$

and

$$u_z(x_\mathrm{I} = L_\mathrm{I}) = u_z(x_\mathrm{II} = 0), \qquad \varphi_y(x_\mathrm{I} = L_\mathrm{I}) = \varphi_y(x_\mathrm{II} = 0), \qquad (2.166)$$

$$Q_z(x_\mathrm{I} = L_\mathrm{I}) = F_\mathrm{I} + Q_z(x_\mathrm{II} = 0), \qquad M_y(x_\mathrm{I} = L_\mathrm{I}) = M_y(x_\mathrm{II} = 0). \qquad (2.167)$$

Consideration of the eight boundary and transmissions conditions in this set of equations gives eight conditions for the unknown constants of integration $c_1 \ldots c_8$ which can be expressed in matrix form as follows:

$$
\begin{bmatrix}
0 & 0 & 0 & 1 & 0 & 0 & 0 & 0 \\
\frac{1}{k_s A_I G} & 0 & \frac{1}{E I_I} & 0 & 0 & 0 & 0 & 0 \\
0 & 0 & 0 & 0 & -L_{\mathrm{II}} & -1 & 0 & 0 \\
0 & 0 & 0 & 0 & 1 & 0 & 0 & 0 \\
\frac{L_I^3}{6} & \frac{L_I^2}{2} & L_I & 1 & 0 & 0 & 0 & -\frac{I_I}{I_{\mathrm{II}}} \\
-\left(\frac{L_I^2}{2} + \frac{E I_I}{k_s A_I G}\right) & -L_I & -1 & 0 & \frac{E I_I}{k_s A_{\mathrm{II}} G} & 0 & \frac{I_I}{I_{\mathrm{II}}} & 0 \\
-1 & 0 & 0 & 0 & 1 & 0 & 0 & 0 \\
-L_I & -1 & 0 & 0 & 0 & 1 & 0 & 0
\end{bmatrix}
\begin{bmatrix} c_1 \\ c_2 \\ c_3 \\ c_4 \\ c_5 \\ c_6 \\ c_7 \\ c_8 \end{bmatrix}
=
\begin{bmatrix} 0 \\ 0 \\ 0 \\ -F_{\mathrm{II}} \\ 0 \\ 0 \\ F_I \\ 0 \end{bmatrix}. \qquad (2.168)
$$

Multiplication of the inverse coefficient matrix with the right-hand side allows us to determine the constants as:

$$c_1 = -(F_I + F_{\mathrm{II}}), \qquad c_2 = F_I L_I + F_{\mathrm{II}}(L_I + L_{\mathrm{LL}}), \qquad (2.169)$$

$$c_3 = \frac{E I_I (F_I + F_{\mathrm{II}})}{k_s A_I G}, \qquad c_4 = 0, \qquad (2.170)$$

$$c_5 = -F_{\mathrm{II}}, \qquad c_6 = F_{\mathrm{II}} L_{\mathrm{II}}, \qquad (2.171)$$

$$c_7 = \frac{1}{2} \frac{I_{\mathrm{II}}\left(k_s A_{\mathrm{II}} G \left[F_I L_I^2 + F_{\mathrm{II}} L_I^2 + 2 F_{\mathrm{II}} L_I L_{\mathrm{II}}\right] + 2 E I_I F_{\mathrm{II}}\right)}{I_I k_s A_{\mathrm{II}} G}, \qquad (2.172)$$

$$c_8 = \frac{1}{6} \frac{L_I I_{\mathrm{II}}\left(k_s A_I G \left[2 F_I L_I^2 + 2 F_{\mathrm{II}} L_I^2 + 3 F_{\mathrm{II}} L_I L_{\mathrm{II}}\right] + 6 E I_I (F_I + F_{\mathrm{II}})\right)}{I_I k_s A_I G}. \qquad (2.173)$$

Thus, based on these constants of integration, the bending lines given in Eqs. (2.156) and (2.160) are determined.

An interesting special case is obtained at the right-hand end for $A_I = A_{\mathrm{II}} = A$, $I_I = I_{\mathrm{II}} = I$, $L_I = L_{\mathrm{II}} = \frac{L}{2}$, and $F_I = 0$ and $F_{\mathrm{II}} = F$:

$$u_z(X = L) = \frac{F L^3}{3 E I} + \frac{F L}{k_s A G}. \qquad (2.174)$$

2.3   Energy-Based Approaches

As an alternative approach to the analytical solution procedures based on partial differential equations (see Sects. 2.2.1–2.2.3), the following section is related to energy approaches, in particular CASTIGLIANO's theorems, see [2, 13, 15, 16].

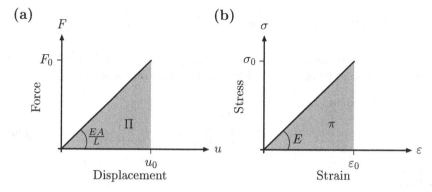

Fig. 2.36 Recorded data from a uniaxial tensile test: **a** force-displacement diagram; **b** stress-strain diagram

Let us first illustrate the energy which is stored in a material due to deformation, i.e. the so-called strain energy.[4] For an ideal uniaxial tensile test with linear-elastic material behavior, Fig. 2.36 illustrates schematic force-displacement and stress-strain diagrams.

The area under the force-displacement diagram (see Fig. 2.36a) represents the total strain energy (Π) and can be calculated as[5]:

$$\Pi = \frac{1}{2} F_0 u_0 , \qquad (2.175)$$

or in an integral approach:

$$\Pi = \int\limits_0^{u_0} F(u)\mathrm{d}u = \int\limits_0^{u_0} \frac{EA}{L} u \, \mathrm{d}u = \frac{EA}{2L} u_0^2 = \frac{F_0^2 L}{2EA} = \int\limits_0^L \frac{N_x(x)^2}{2EA}\mathrm{d}x . \qquad (2.176)$$

The transformations in the last equation used HOOKE's law and the equilibrium between the external load (F_0) and the internal reaction ($N_x(x)$). On the other hand, the area under the stress-strain diagram (see Fig. 2.36b) represents the volumetric strain energy ($\pi = \frac{\Pi}{V}$):

$$\pi = \int\limits_0^{\varepsilon_0} \sigma(\epsilon)\mathrm{d}\epsilon = \int\limits_0^{\varepsilon_0} E\varepsilon\mathrm{d}\varepsilon = \frac{E}{2}\varepsilon_0^2 = \frac{1}{2}\sigma_0\varepsilon_0 . \qquad (2.177)$$

The last equation can extended to the total strain energy in the following way:

---

[4]The total potential energy is the sum of the strain energy and the work done by the external loads.
[5]Confer the unit of energy: 1 J = 1 Nm = 1 Ws.

$$d\Pi = d\pi dV = \sigma d\varepsilon\, dV = E\varepsilon d\varepsilon\, dV = \frac{E\varepsilon^2}{2}dV = \frac{\sigma^2}{2E}(Adx) = \frac{N^2}{2EA}dx\,.$$
$$(2.178)$$

Similar derivations can be written for other simple modes of deformation and the following cases can be distinguished for linear-elastic material behavior:

- Tension or compression:

$$\Pi = \int\limits_0^L \frac{N_x(x)^2}{2EA}dx\,.\tag{2.179}$$

- Bending:

$$\Pi = \int\limits_0^L \frac{M_y(x)^2}{2EI_y}dx\,.\tag{2.180}$$

- Shear:

$$\Pi = \int\limits_0^L \frac{Q_z(x)^2}{2GA_s}dx = \int\limits_0^L \frac{Q_z(x)^2}{2k_sGA}dx\,.\tag{2.181}$$

- Torsion[6]:

$$\Pi = \int\limits_0^L \frac{M_x(x)^2}{2GI_p}dx\,.\tag{2.182}$$

Thus, the total strain energy in a rod/beam-like structural member can be expressed as

$$\Pi = \int\limits_0^L \frac{N_x(x)^2}{2EA}dx + \int\limits_0^L \frac{M_y(x)^2}{2EI_y}dx + \int\limits_0^L \frac{Q_z(x)^2}{2GA_s}dx + \int\limits_0^L \frac{M_x(x)^2}{2GI_p}dx\,,\tag{2.183}$$

where the N_x, M_y, Q_z, M_x represent the distributions of the internal reactions. Depending on the mode of deformation, the corresponding terms in Eq. (2.183) must be considered. Based on the following theorems which make use of the strain energy, different quantities can be determined:

CASTIGLIANO's first theorem:
The partial derivative of the total strain energy with respect to the generalized displacement (displacement or rotation) gives the generalized force (force or moment). In equations, this can be expressed as:

---

[6]Only shown for completeness and not further covered here.

$$\frac{\partial \Pi(x, u_i, \dots)}{\partial u_i} = F_i , \tag{2.184}$$

$$\frac{\partial \Pi(x, \varphi_i, \dots)}{\partial \varphi_i} = M_i . \tag{2.185}$$

CASTIGLIANO's second theorem:
The partial derivative of the total strain energy with respect to the generalized force (force or moment) gives the generalized displacement (displacement or rotation) in the direction of that generalized force. In equations, this can be expressed as:

$$\frac{\partial \Pi(x, F_i, \dots)}{\partial F_i} = u_i , \tag{2.186}$$

$$\frac{\partial \Pi(x, M_i, \dots)}{\partial M_i} = \varphi_i . \tag{2.187}$$

The procedure also allows us to determine deformations where no external generalized forces are acting. This can be handled by introducing an auxiliary generalized force (F_a or M_a) and setting the auxiliary quantity to zero in the final equation for the generalized displacement:

$$\left(\frac{\partial \Pi(x, F_{a_i}, \dots)}{\partial F_{a_i}} \right)_{F_{a_i}=0} = u_i , \tag{2.188}$$

$$\left(\frac{\partial \Pi(x, M_{a_i}, \dots)}{\partial M_{a_i}} \right)_{M_{a_i}=0} = \varphi_i . \tag{2.189}$$

Based on this procedure, it is even possible to calculate entire distributions if the auxiliary quantity is introduced at a variable position. For practical calculations with constant material and geometrical properties (EA, EI_y, $k_s GA$, GI_p), it might be useful to perform the partial derivative first and only after the integration. For example, the case of tension/compression can be written as:

$$u_{x,1} = \frac{\partial \Pi}{\partial F_1} = \frac{\partial}{\partial F_1} \left(\int_0^L \frac{N_x^2(x)}{2EA} dx \right) = \int_0^L \frac{N_x(x)}{EA} \frac{\partial N_x(x, F_1, \dots)}{\partial F_1} dx . \tag{2.190}$$

2.10 Cantilever rod with point loads (Alternative solution procedure of Problem 2.1)
Given is a rod of length L and constant axial tensile stiffness EA as shown in Fig. 2.37. At the left-hand side there is a fixed support and the right-hand side is either elongated by a displacement u_0 (case a) or loaded by a single force F_0 (case b). Determine based on CASTIGLIANO's theorems the solution for the elongation $u_x(x)$, the strain $\varepsilon_x(x)$, and the stress $\sigma_x(x)$ along the rod axis.

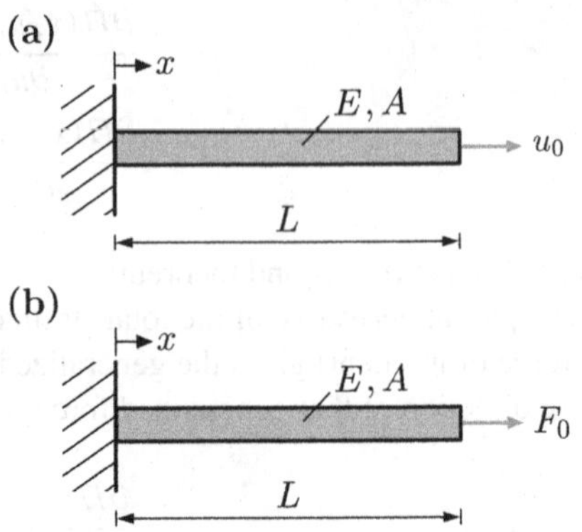

Fig. 2.37 Rod under different loading conditions: **a** displacement and **b** force

(a)

(b)

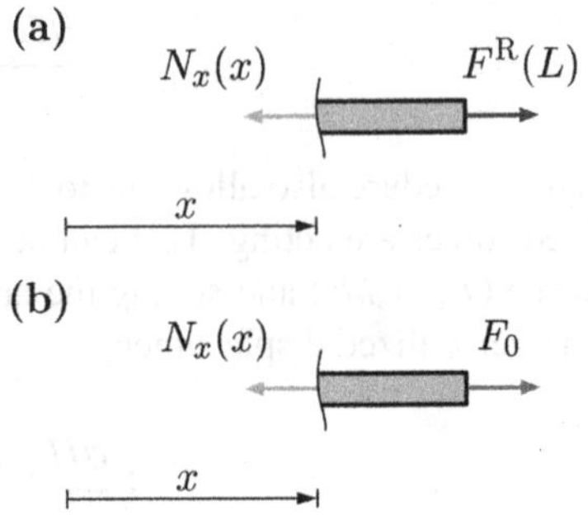

Fig. 2.38 Determination of the normal force distribution for the rod under different loading conditions: **a** displacement and **b** force

(a)

(b)

2.10 Solution

In case that only the reaction force $F^R(L)$ at $x = L$ (case a) or the displacement $u_x(L)$ at $x = L$ (case b) would be wanted, we could simply determine the normal force distributions as, see Fig. 2.38:

$$N_x(x) = F^R(L) \qquad \text{(case a)}, \qquad (2.191)$$
$$N_x(x) = F_0 \qquad \text{(case b)}. \qquad (2.192)$$

For case (a), we can state based on CASTIGLIANO's second theorem that

$$u_0 = \frac{\partial \Pi}{\partial F^R(L)} = \frac{\partial}{\partial F^R(L)} \int_0^L \frac{N_x^2(x)}{2EA}\,\mathrm{d}x = \int_0^L \frac{N_x(x)}{EA}\frac{\partial N_x(x, F^R)}{\partial F^R(L)}\,\mathrm{d}x$$

$$= \int_0^L \frac{F^R(L)}{EA} \times 1\mathrm{d}x = \frac{F^R(L)}{EA}[x]_0^L = \frac{F^R(L)L}{EA}, \qquad (2.193)$$

or solved for the unknown reaction force at $x = L$:

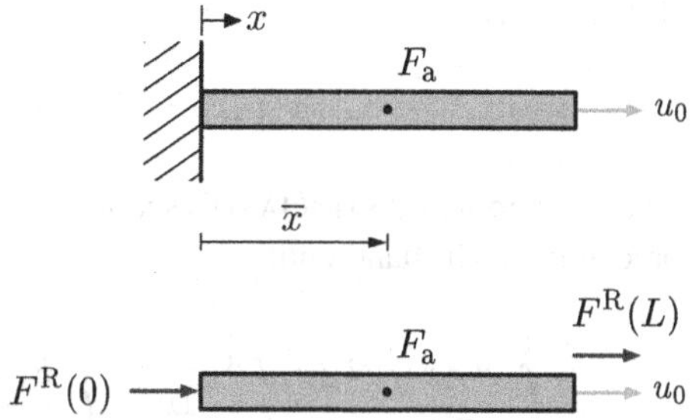

Fig. 2.39 Rod with displacement boundary condition and auxiliary force

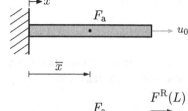

$$F^{R}(L) = \frac{E A u_0}{L}. \qquad (2.194)$$

For case (b), we can state based on CASTIGLIANO's second theorem that

$$u_x(L) = \frac{\partial \Pi}{\partial F_0} = \int_0^L \frac{N_x(x)}{E A} \frac{\partial N_x(x, F_0)}{\partial F_0} \, dx = \int_0^L \frac{F_0}{E A} \times 1 dx = \frac{F_0 L}{E A}. \qquad (2.195)$$

However, if we need to find the distributions of displacement ($u_x = u_x(x)$), stress ($\sigma_x = \sigma_x(x)$), and strain ($\varepsilon_x = \varepsilon_x(x)$), we need to follow a slightly different approach. For this purpose, an auxiliary force F_a is introduced at an arbitrary position $\bar{x}$. This is shown for case (a) in Fig. 2.39 together with the corresponding free-body diagram.

From the horizontal force equilibrium, we can conclude that

$$+ F^{R}(0) + F_a + F^{R}(L) = 0 \quad \text{or} \quad F^{R}(0) = -F_a - F^{R}(L). \qquad (2.196)$$

Since we have now at $x = \bar{x}$ a discontinuity, we must determine the normal force distribution for two sections, see Fig. 2.40.

For the section $x < \bar{x}$, the internal normal force can be expressed as

$$N_x(x) = -F^{R}(0) = F_a + F^{R}(L), \qquad (2.197)$$

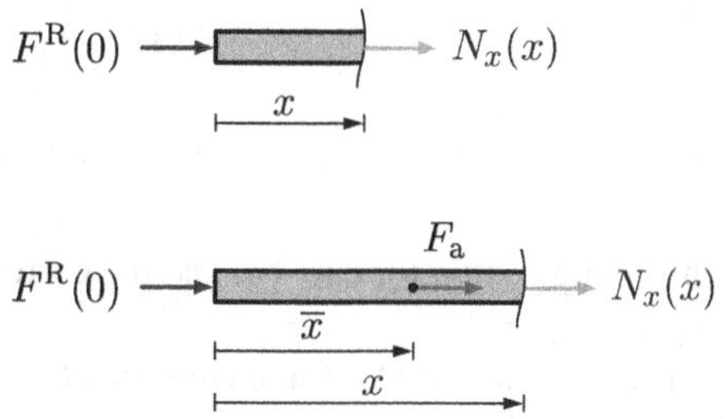

Fig. 2.40 Rod with displacement boundary condition and auxiliary force: different sections for normal force determination

while the section $x \geq \overline{x}$ gives:

$$N_x(x) = -F^R(0) - F_a = F^R(L). \qquad (2.198)$$

Let us first apply CASTIGLIANO's second theorem to determine the unknown reaction force at the right-hand end:

$$
u_0 = \int_0^L \frac{N_x(x)\, \partial N_x(x, F^R)}{EA\, \partial F^R(L)}\, dx = \int_0^{\overline{x}} \frac{F_a + F^R(L)}{EA} \times 1 dx + \int_{\overline{x}}^L \frac{F^R(L)}{EA} \times 1 dx
$$

$$
= \frac{F_a + F^R(L)}{EA}\overline{x} + \frac{F^R(L)}{EA}(L - \overline{x}) = \frac{F_a\overline{x}}{EA} + \frac{F^R(L)L}{EA}. \qquad (2.199)
$$

With $F_a \to 0$ (and $\overline{x} \to x$), one obtains the reactions force as:

$$F^R(L) = \frac{EAu_0}{L}. \qquad (2.200)$$

The next application of CASTIGLIANO's second theorem allows us to determine the distribution of the displacement field:

$$
u_x(x) = \int_0^L \frac{N_x(x)\, \partial N_x(x, F_a)}{EA\, \partial F_a}\, dx = \int_0^{\overline{x}} \frac{F_a + F^R(L)}{EA} \times 1 dx + \int_{\overline{x}}^L \frac{F^R(L)}{EA} \times 0 dx
$$

$$
= \frac{F_a + F^R(L)}{EA}\overline{x}. \qquad (2.201)
$$

With $F_a \to 0$ and $\overline{x} \to x$, one obtains the displacement field as:

$$u_x(x) = \frac{F^R(L)x}{EA} = u_0\frac{x}{L}. \qquad (2.202)$$

The application of the kinematics and constitutive relationship (see Table 2.2) gives immediately the strain and stress distributions:

$$\varepsilon_x(x) = \frac{\partial u_x(x)}{\partial x} = \frac{u_0}{L}, \qquad (2.203)$$

$$\sigma_x(x) = E\varepsilon_x(x) = \frac{u_0 E}{L}. \qquad (2.204)$$

The configuration for case (b) and the corresponding free-body diagram is shown in Fig. 2.41.

From the horizontal force equilibrium, we can calculate the reaction force at the left-hand end:

Fig. 2.41 Rod with force boundary condition and auxiliary force

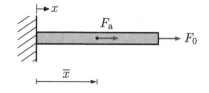

Fig. 2.42 Rod with force boundary condition and auxiliary force: different sections for normal force determination

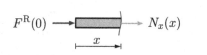

$$+ F^R(0) + F_a + F_0 = 0 \quad \text{or} \quad F^R(0) = -F_0 - F_a. \tag{2.205}$$

Due to the discontinuity, the normal force distribution is required for two sections, see Fig. 2.42.

For the section $x < \overline{x}$, the internal normal force can be expressed as

$$N_x(x) = -F^R(0) = F_0 + F_a, \tag{2.206}$$

while the section $x \geq \overline{x}$ gives:

$$N_x(x) = F_0. \tag{2.207}$$

Application of CASTIGLIANO's second theorem allows us to determine the distribution of the displacement field:

$$u_x(x) = \int_0^L \frac{N_x(x) \partial N_x(x, F_a)}{EA \quad \partial F_a} dx = \int_0^{\overline{x}} \frac{F_0 + F_a}{EA} \times 1 dx + \int_{\overline{x}}^L \frac{F_0}{EA} \times 0 dx$$

$$= \frac{F_0 + F_a}{EA} \overline{x}. \tag{2.208}$$

With $F_a \to 0$ and $\overline{x} \to x$, one obtains the displacement field as:

$$u_x(x) = \frac{F_0 x}{EA}. \tag{2.209}$$

Fig. 2.43 Rod with
distributed load

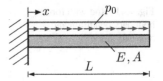

The application of the kinematics and constitutive relationships (see Table 2.2) gives
immediately the strain and stress distributions:

$$\varepsilon_x(x) = \frac{\partial u_x(x)}{\partial x} = \frac{F_0}{EA}, \tag{2.210}$$

$$\sigma_x(x) = E\varepsilon_x(x) = \frac{F_0}{A}. \tag{2.211}$$

2.11 Cantilever rod with distributed load (Alternative solution procedure of Problem 2.2)

Given is a rod of length L and constant axial tensile stiffness EA as shown in Fig. 2.43.
At the left-hand side there is a fixed support and a constant distributed load p_0 is
acting along the entire rod axis. Determine based on CASTIGLIANO's theorems the
analytical solution for the elongation $u_x(x)$, the strain $\varepsilon_x(x)$, and the stress $\sigma_x(x)$
along the rod axis.

2.11 Solution

The determination of the distributions of displacement ($u_x = u_x(x)$), stress ($\sigma_x =
\sigma_x(x)$), and strain ($\varepsilon_x = \varepsilon_x(x)$) requires that an auxiliary force F_a is introduced at
an arbitrary position $\overline{x}$. This is shown in Fig. 2.44 together with the corresponding
free-body diagram.

From the horizontal force equilibrium, we can conclude that

$$+ F^R(0) + F_a + p_0 L = 0 \quad \text{or} \quad F^R(0) = -F_a - p_0 L. \tag{2.212}$$

Fig. 2.44 Rod with
distributed load: introduction
of auxiliary force

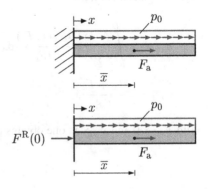

Fig. 2.45 Rod with distributed load: different sections for normal force determination

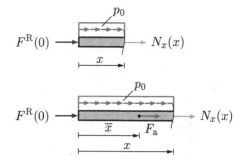

Since we have now at $x = \overline{x}$ a discontinuity, we must determine the normal force distribution for two sections, see Fig. 2.45.

For the section $x < \overline{x}$, the internal normal force can be expressed as

$$N_x(x) = F_a + p_0(L - x),\qquad(2.213)$$

while the section $x \geq \overline{x}$ gives:

$$N_x(x) = p_0(L - x).\qquad(2.214)$$

Application of CASTIGLIANO's second theorem allows us to determine the distribution of the displacement field:

$$
\begin{aligned}
u_x(x) &= \int_0^L \frac{N_x(x)}{EA}\frac{\partial N_x(x, F_a)}{\partial F_a}\,\mathrm{d}x \\
&= \int_0^{\overline{x}} \frac{F_a + p_0(L - x)}{EA}\times 1\mathrm{d}x + \int_{\overline{x}}^L \frac{p_0(L - x)}{EA}\times 0\mathrm{d}x \\
&= \frac{1}{EA}\left(F_a\overline{x} + p_0 L\overline{x} - \frac{p_0\overline{x}^2}{2} \right).
\end{aligned}\qquad(2.215)
$$

With $F_a \to 0$ and $\overline{x} \to x$, one obtains the displacement field as:

$$u_x(x) = \frac{p_0}{EA}\left(Lx - \frac{x^2}{2} \right) = \frac{p_0 L^2}{EA}\left(-\frac{1}{2}\left[\frac{x}{L}\right]^2 + \left[\frac{x}{L}\right] \right).\qquad(2.216)$$

The application of the kinematics and constitutive relationships (see Table 2.2) gives immediately the strain and stress distributions:

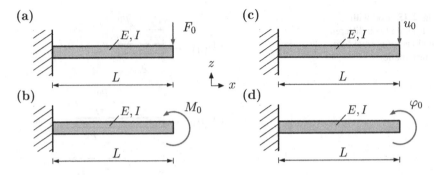

Fig. 2.46 Cantilever beam with different end loads and deformations: **a** single force; **b** single moment; **c** displacement; **d** rotation

$$\varepsilon_x(x) = \frac{\partial u_x(x)}{\partial x} = \frac{p_0 L}{EA}\left(-\left[\frac{x}{L}\right] + 1\right),\tag{2.217}$$

$$\sigma_x(x) = E\varepsilon_x(x) = \frac{p_0 L}{A}\left(-\left[\frac{x}{L}\right] + 1\right).\tag{2.218}$$

2.12 Cantilever beam with different end loads and deformations (Alternative solution procedure of Problem 2.4)

Calculate based on CASTIGLIANO's theorems the analytical solutions for the deflection $u_z(x)$ and rotation $\varphi_y(x)$ of the cantilever beam shown in Fig. 2.46. Calculate in addition for all four cases the reactions at the fixed support and the distributions of the bending moment and shear force. It can be assumed for this exercise that the bending stiffness EI_y is constant.

2.12 Solution

Case (a): The determination of the distributions of deflection ($u_z = u_z(x)$) and rotation ($\varphi_y = \varphi_y(x)$) requires that an auxiliary force F_a is introduced at an arbitrary position $\bar{x}$. This is shown in Fig. 2.47 together with the corresponding free-body diagram.

From the vertical force and moment equilibrium, we can conclude that

$$+F_z^R(0) - F_a - F_0 = 0 \quad \text{or} \quad F_z^R(0) = F_a + F_0,\tag{2.219}$$

$$+M_y^R(0) + F_a\bar{x} + F_0 L = 0 \quad \text{or} \quad M_y^R(0) = -F_a\bar{x} - F_0 L.\tag{2.220}$$

Since we have now at $x = \bar{x}$ a discontinuity, we must determine the bending moment (the shear force distribution is only required if the shear contribution on the deformation should be considered) distribution for two sections, see Fig. 2.48.

For the section $x < \bar{x}$, the internal bending moment can be expressed as

$$M_y(x) = -F_z^R(0)x - M_y^R(0) = F_a(\bar{x} - x) + F_0(L - x),\tag{2.221}$$

Fig. 2.47 Cantilever beam
with force boundary
condition: introduction of
auxiliary force

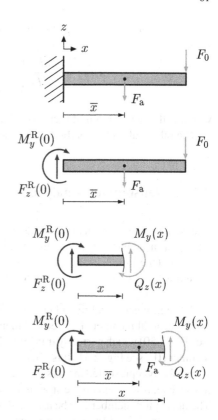

Fig. 2.48 Cantilever beam
with force boundary
condition: different sections
for internal reactions

while the section $x \geq \overline{x}$ gives:

$$M_y(x) = -F_z^R(0)x - M_y^R(0) + F_a(x - \overline{x}) = F_0(L - x). \qquad (2.222)$$

Application of CASTIGLIANO's second theorem allows us to determine the distribution
of the displacement field:

$$u_z(x) = \int_0^L \frac{M_y(x)}{EI} \frac{\partial M_y(x, F_a)}{\partial F_a} \, \mathrm{d}x$$

$$= \int_0^{\overline{x}} \frac{M_y(x)}{EI} \frac{\partial M_y(x, F_a)}{\partial F_a} \, \mathrm{d}x + \int_{\overline{x}}^L \frac{M_y(x)}{EI} \frac{\partial M_y(x, F_a)}{\partial F_a} \, \mathrm{d}x$$

$$= \int_0^{\overline{x}} \frac{F_a(\overline{x} - x) + F_0(L - x)}{EI} \times (x - \overline{x}) \, \mathrm{d}x + \int_0^{\overline{x}} \frac{F_a(L - x)}{EI} \times 0 \, \mathrm{d}x .$$

The evaluation of these integrals gives finally under consideration of $F_a \to 0$ and $\bar{x} \to x$:

$$u_z(x) = \frac{F_0 L^3}{EI}\left(-\frac{1}{6}\left[\frac{x}{L}\right]^3 + \frac{1}{2}\left[\frac{x}{L}\right]^2\right), \qquad (2.223)$$

which is the deflection in direction of F_a.

The other subproblems (b–d) can be solved in a similar manner.

2.4 Extensometer Analysis

The extensometer shown and illustrated in Figs. 1.1 and 1.2 can be modeled in a first attempt as a ⊔-shaped frame with different properties for the horizontal and vertical members (see Fig. 2.49a). Looking at this mechanical model, it is obvious that the structure is symmetric with respect to a vertical axis and can be reduced as indicated in Fig. 2.49b.

A rough mechanical model can be obtained by splitting the frame into a vertical and horizontal member. The vertical member (I) is assumed to be a cantilever beam (see Fig. 2.50) which perfectly transmits the reaction moment and force to the vertical member (II), see Fig. 2.51. It is obvious that the small rotation in the frame corner is not perfectly represented in this simple model. However, it allows us to derive a simple design equation based on the straight beam equations provided in Sects. 2.2.1–2.2.2. The horizontal member can be assumed to be a simply supported beam of length L_{II} as shown in Fig. 2.51a or as a cantilever of length $\frac{L_{II}}{2}$ as shown in Fig. 2.51b.

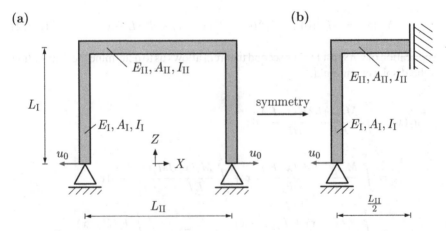

Fig. 2.49 Mechanical model of the extensometer: **a** entire sensor and **b** consideration of symmetry

(a)   (b)   (c)

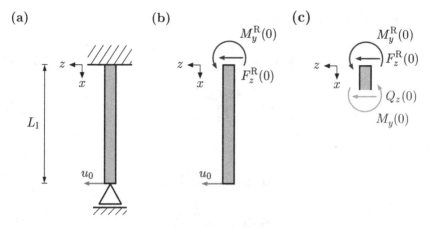

Fig. 2.50 Simplified approach for vertical members: **a** approximation as cantilever beam; **b** free-body diagram; **c** infinitesimal element at $x = 0$

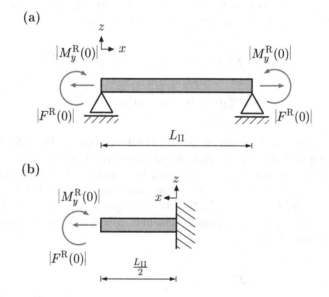

Fig. 2.51 Simplified approach for horizontal member: **a** approximation as simply supported beam and **b** consideration of symmetry

Let us have a closer look at the vertical member as shown in Fig. 2.50. From Eqs. (2.36) and (2.39) we can conclude[7] with $c_1 = -\frac{3E_I I_I u_0}{L_I^3}$ and $c_2 = \frac{33E_I I_I u_0}{L_I^2}$ the

---

[7]An alternative way in the elastic material range would be as follows: consider the maximum deflection of a cantilever beam which is loaded by a tip load F_0. Rearrange this equation for the force, which is now understood as the reaction force, gives $F^R = \frac{3E_I I_I u_0}{L_I^3}$. This allows to calculate the shear force and the bending moment at $x = 0$.

bending moment and shear force distributions to be:

$$M_y(x) = \frac{3E_I I_1 u_0}{L_I^2}\left(\frac{x}{L} - 1\right), \tag{2.224}$$

$$Q_z(x) = \frac{3E_I I_1 u_0}{L_I^3}, \tag{2.225}$$

or at $x = 0$:

$$M_y(0) = -\frac{3E_I I_1 u_0}{L_I^2}, \tag{2.226}$$

$$Q_z(0) = \frac{3E_I I_1 u_0}{L_I^3}. \tag{2.227}$$

These internal reactions must be balanced at $x = 0$ by the reactions of the fixed support. The force and moment equilibrium at $x = L$ reads:

$$+Q_z(0) + F_z^R(0) = 0 \quad \Rightarrow \quad F_z^R(0) = -\frac{3E_I I_1 u_0}{L_I^3}, \tag{2.228}$$

$$+M_y(0) + M_y^R(0) = 0 \quad \Rightarrow \quad M_y^R(0) = \frac{3E_I I_1 u_0}{L_I^2}. \tag{2.229}$$

These reactions are now applied at the horizontal member, see Fig. 2.51. To avoid any confusion with the sign of these quantities, it is advised to simply take the absolute values and consider the correct directions as indicated in the figure. This configuration relates to the case that the base sample is under tensile load.

Let us first consider the case that only the bending moment is acting, i.e. the case of pure bending. The internal bending moment distribution for both cases shown is Fig. 2.51 is obtained as:

$$M_y(x) = -\frac{3E_I I_1 u_0}{L_I^2} = \text{const.} \tag{2.230}$$

Equation (2.40) together with HOOKE's law allows us to express the normal strain in the horizontal members (II) as:

$$\varepsilon_{x,\text{II}}(z) = \frac{M_y(x)}{E_\text{II} I_\text{II}} z(x) = -\frac{3E_I I_1 u_0}{E_\text{II} I_\text{II} L_I^2} z(x). \tag{2.231}$$

In the next step, we can express the displacement u_0 by the strain in the specimen ε_{sp}, i.e.,

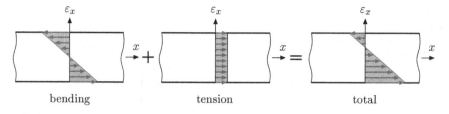

Fig. 2.52 Strain distributions in the horizontal member of the extensometer: **a** pure bending; **b** pure tension; **c** superposition of both cases

$$\varepsilon_{\mathrm{sp}} = \frac{2u_0}{L_{\mathrm{II}}} = \frac{u_0}{\frac{L_{\mathrm{II}}}{2}}, \tag{2.232}$$

which allows us to express the strain in the horizontal member of the extensometer as:

$$\varepsilon_{x,\mathrm{II}}(z) = -\frac{3}{2} \times \frac{E_{\mathrm{I}} I_{\mathrm{I}} L_{\mathrm{II}}}{E_{\mathrm{II}} I_{\mathrm{II}} L_{\mathrm{I}}^2} \times \varepsilon_{\mathrm{sp}} z(x). \tag{2.233}$$

The strain distribution under pure bending is sketched in Fig. 2.52a where a linear distribution can be observed. Furthermore, the distribution is symmetric with a compressive regime for $z > 0$ and a tensile regime for $z < 0$.

Reviewing again Fig. 2.50b, we can identify a shear force $F^{\mathrm{R}}(0)$ which acts on the horizontal member as a tensile force, see Fig. 2.51. This 'tensile' force results in the following tensile strain:

$$\varepsilon_{x,\mathrm{II}} = \frac{|F^{\mathrm{R}}|}{E_{\mathrm{II}} A_{\mathrm{II}}} = \frac{3 E_{\mathrm{I}} I_{\mathrm{I}} u_0}{L_{\mathrm{I}}^3 E_{\mathrm{II}} A_{\mathrm{II}}} = \frac{3}{2} \times \frac{E_{\mathrm{I}} I_{\mathrm{I}} L_{\mathrm{II}}}{E_{\mathrm{II}} A_{\mathrm{II}} L_{\mathrm{I}}^3} \times \varepsilon_{\mathrm{sp}}. \tag{2.234}$$

The strain components given in Eqs. (2.233) and (2.234) can be superposed to obtain the total axial strain (see Fig. 2.52c) in the horizontal member of the extensometer as:

$$\varepsilon_{x,\mathrm{II}} = \frac{3}{2}\left(-\frac{E_{\mathrm{I}} I_{\mathrm{I}} L_{\mathrm{II}}}{E_{\mathrm{II}} I_{\mathrm{II}} L_{\mathrm{I}}^2} \times z(x) + \frac{E_{\mathrm{I}} I_{\mathrm{I}} L_{\mathrm{II}}}{E_{\mathrm{II}} A_{\mathrm{II}} L_{\mathrm{I}}^3}\right) \varepsilon_{\mathrm{sp}}. \tag{2.235}$$

Let us assume in the following a rectangular cross section (with width b_{II} and height h_{II}) for the horizontal member. Based on the relationship $A_{\mathrm{II}} = \frac{12 I_{\mathrm{II}}}{h_{\mathrm{II}}^2}$, the total strain can be expressed as:

$$\varepsilon_{x,\mathrm{II}} = \frac{3}{2} \times \frac{E_{\mathrm{I}} I_{\mathrm{I}} L_{\mathrm{II}}}{E_{\mathrm{II}} I_{\mathrm{II}} L_{\mathrm{I}}}\left(-\frac{z(x)}{L_{\mathrm{I}}} + \frac{1}{12}\left(\frac{h_{\mathrm{II}}}{L_{\mathrm{I}}}\right)^2\right) \varepsilon_{\mathrm{sp}}. \tag{2.236}$$

Fig. 2.53 Schematic
representation of the
calibration curve for the
extensometer

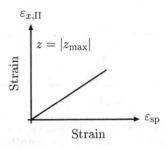

The extreme values at the free surfaces, i.e. $z = +\frac{h_{\mathrm{II}}}{2}$ and $z = -\frac{h_{\mathrm{II}}}{2}$, are obtained as follows:

$$\varepsilon_{x,\mathrm{II}}\Big|_{z=+\frac{h_{\mathrm{II}}}{2}} = \frac{3}{2} \times \frac{E_{\mathrm{I}} I_{\mathrm{I}} L_{\mathrm{II}}}{E_{\mathrm{II}} I_{\mathrm{II}} L_{\mathrm{I}}} \times \frac{h_{\mathrm{II}}}{L_{\mathrm{I}}} \times \left(-\frac{1}{2} + \frac{1}{12}\left(\frac{h_{\mathrm{II}}}{L_{\mathrm{I}}}\right)\right) \varepsilon_{\mathrm{sp}}, \quad (2.237)$$

$$\varepsilon_{x,\mathrm{II}}\Big|_{z=-\frac{h_{\mathrm{II}}}{2}} = \frac{3}{2} \times \frac{E_{\mathrm{I}} I_{\mathrm{I}} L_{\mathrm{II}}}{E_{\mathrm{II}} I_{\mathrm{II}} L_{\mathrm{I}}} \times \frac{h_{\mathrm{II}}}{L_{\mathrm{I}}} \times \left(+\frac{1}{2} + \frac{1}{12}\left(\frac{h_{\mathrm{II}}}{L_{\mathrm{I}}}\right)\right) \varepsilon_{\mathrm{sp}}. \quad (2.238)$$

Based on Eqs. (2.237) and/or (2.238), it is now possible to calculate and draw the calibration curve for the extensometer, i.e. the relation between the measured strain in the extensometer ($\varepsilon_{x,\mathrm{II}}$) and the strain in the specimen ($\varepsilon_{\mathrm{sp}}$), see Fig. 2.53. From a practical point of view, one could measure the strain on the top, or the bottom (larger signal since two positive strain components are summed up) of the beam or even average both signals (under consideration that the distribution is no longer symmetric).

Both Eqs. (2.237) and (2.238) can be generally written as

$$\varepsilon_{x,\mathrm{II}} = \varepsilon_{\mathrm{sp}}(E_{\mathrm{I}}, E_{\mathrm{II}}, I_{\mathrm{I}}, I_{\mathrm{II}}, L_{\mathrm{I}}, L_{\mathrm{II}}), \quad (2.239)$$

which allows us to design the extensometer in the boundaries of minimum strain (sensitivity) and maximum strain (failure of the strain gage).

Let us look in the following at a solution procedure which is based on the strain energy as outlined in Sect. 2.3. This allows us to consider the entire frame (see Fig. 2.49) without the strong simplification in regards to the connection of the vertical and horizontal members.

The horizontal force and moment equilibrium (see Fig. 2.54) gives the internal reactions of the vertical member as follows:

$$Q_z(x_{\mathrm{I}}) = F_0^{\mathrm{R}}, \quad (2.240)$$

$$M_y(x_{\mathrm{I}}) = -F_0^{\mathrm{R}}(L_{\mathrm{I}} - x_{\mathrm{I}}). \quad (2.241)$$

Fig. 2.54 Vertical beam section for the determination of the internal reactions

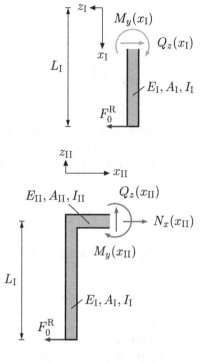

Fig. 2.55 Section of the frame structure for the determination of the internal reactions in the horizontal member

For the internal reactions of the horizontal member, it is advantageous to consider the left-hand half as shown in Fig. 2.55. Horizontal and vertical force as well as the moment equilibrium give the following internal reactions in the horizontal member:

$$N_x(x_{\text{II}}) = F_0^{\text{R}}\,, \tag{2.242}$$

$$Q_z(x_{\text{II}}) = 0\,, \tag{2.243}$$

$$M_y(x_{\text{II}}) = -F_0^{\text{R}} L_{\text{I}}\,. \tag{2.244}$$

It should be noted here that the reaction force F_0^{R} is still unknown. Based on CAS-TIGLIANO's second theorem, it is possible to express the horizontal displacement as[8]:

$$u_0 = \int_0^{L_{\text{I}}} \frac{M_y(x_{\text{I}})}{E_{\text{I}} I_{\text{I}}} \frac{\partial M_y(x_{\text{I}}, F_0^{\text{R}})}{\partial F_0^{\text{R}}} \mathrm{d}x_{\text{I}} + \int_0^{\frac{L_{\text{II}}}{2}} \frac{M_y(x_{\text{II}})}{E_{\text{II}} I_{\text{II}}} \frac{\partial M_y(x_{\text{II}}, F_0^{\text{R}})}{\partial F_0^{\text{R}}} \mathrm{d}x_{\text{II}}+$$

---

[8]It is assumed that the beams are thin and that the shear force is not contributing to the bending deformation modes.

$$+ \int_{0}^{\frac{L_{II}}{2}} \frac{N_x(x_{II})}{E_{II}A_{II}} \frac{\partial N_x(x_{II}, F_0^R)}{\partial F_0^R} dx_{II}$$

$$= F_0^R \left(\frac{1}{3} \frac{L_I^3}{E_I I_I} + \frac{1}{2} \frac{L_I^2 L_{II}}{E_{II} I_{II}} + \frac{1}{2} \frac{L_{II}}{E_{II} A_{II}} \right). \tag{2.245}$$

If we assume a square cross section for the horizontal member (width b_{II} and height h_{II}), we can relate the cross sectional area to the second moment of area, i.e. $A_{II} = \frac{12 I_{II}}{h_{II}^2}$, and Eq. (2.245) can be expressed as:

$$u_0 = F_0^R \left(\frac{1}{3} \frac{L_I^3}{E_I I_I} + \frac{1}{2} \frac{L_I^2 L_{II}}{E_{II} I_{II}} + \frac{1}{24} \frac{L_{II} h_{II}^2}{E_{II} I_{II}} \right), \tag{2.246}$$

or rearranged for the unknown reaction force:

$$F_0^R = \frac{u_0}{\frac{1}{3} \frac{L_I^3}{E_I I_I} + \frac{1}{2} \frac{L_I^2 L_{II}}{E_{II} I_{II}} + \frac{1}{24} \frac{L_{II} h_{II}^2}{E_{II} I_{II}}}. \tag{2.247}$$

Based on this reaction force, the internal reactions are known in both members, see Eqs. (2.240)–(2.244). Let us now calculate the total strain in the horizontal member of the extensometer. The axial strain due to the bending deformation can be expressed as:

$$\varepsilon_{x,II} = \frac{1}{E_{II}} \frac{M_y(x_{II})}{I_{II}} z(x_{II}) = \frac{1}{E_{II} I_{II}} \left(-\frac{u_0 L_I}{\frac{1}{3} \frac{L_I^3}{E_I I_I} + \frac{1}{2} \frac{L_I^2 L_{II}}{E_{II} I_{II}} + \frac{1}{24} \frac{L_{II} h_{II}^2}{E_{II} I_{II}}} \right) z(x_{II}), \tag{2.248}$$

which can be rearranged under consideration of $u_0 = \frac{L_{II}}{2} \varepsilon_{sp}$ to:

$$\varepsilon_{x,II} = -\frac{\varepsilon_{sp}}{\frac{2}{3} \frac{E_{II} I_{II} L_I}{E_I I_I L_{II}} + 1 + \frac{1}{12} \frac{h_{II}^2}{L_I^2}} \left(\frac{z_{II}}{L_I} \right). \tag{2.249}$$

On the other hand, the axial strain due to the tensile deformation can be expressed as:

$$\varepsilon_{x,II} = \frac{1}{E_{II}} \frac{F_0^R}{A_{II}} = \frac{1}{E_{II} A_{II}} \left(\frac{u_0}{\frac{1}{3} \frac{L_I^3}{E_I I_I} + \frac{1}{2} \frac{L_I^2 L_{II}}{E_{II} I_{II}} + \frac{1}{2} \frac{L_{II}}{E_{II} A_{II}}} \right), \tag{2.250}$$

which can be rearranged under consideration of $A_{II} = \frac{12 I_{II}}{h_{II}^2}$ and $u_0 = \frac{L_{II}}{2} \varepsilon_{sp}$ to:

$$\varepsilon_{x,\mathrm{II}} = \frac{1}{12}\frac{h_{\mathrm{II}}^2}{L_{\mathrm{I}}^2}\left(\frac{\varepsilon_{sp}}{\frac{2}{3}\frac{E_{\mathrm{II}}I_{\mathrm{II}}L_{\mathrm{I}}}{E_{\mathrm{I}}I_{\mathrm{I}}L_{\mathrm{II}}}+1+\frac{1}{12}\frac{h_{\mathrm{II}}^2}{L_{\mathrm{I}}^2}}\right). \tag{2.251}$$

Thus, the total strain resulting from bending and tension is obtained as:

$$\varepsilon_{x,\mathrm{II}} = \frac{1}{\frac{2}{3}\frac{E_{\mathrm{II}}I_{\mathrm{II}}L_{\mathrm{I}}}{E_{\mathrm{I}}I_{\mathrm{I}}L_{\mathrm{II}}}+1+\frac{1}{12}\frac{h_{\mathrm{II}}^2}{L_{\mathrm{I}}^2}}\left(-\frac{z_{\mathrm{II}}}{L_{\mathrm{I}}}+\frac{1}{12}\left(\frac{h_{\mathrm{II}}}{L_{\mathrm{I}}}\right)^2\right)\varepsilon_{sp}. \tag{2.252}$$

The extreme values at the free surfaces, i.e. $z = +\frac{h_{\mathrm{II}}}{2}$ and $z = -\frac{h_{\mathrm{II}}}{2}$, are obtained as follows:

$$\varepsilon_{x,\mathrm{II}}\Big|_{z=+\frac{h_{\mathrm{II}}}{2}} = \frac{1}{\frac{2}{3}\frac{E_{\mathrm{II}}I_{\mathrm{II}}L_{\mathrm{I}}}{E_{\mathrm{I}}I_{\mathrm{I}}L_{\mathrm{II}}}+1+\frac{1}{12}\frac{h_{\mathrm{II}}^2}{L_{\mathrm{I}}^2}}\times\frac{h_{\mathrm{II}}}{L_{\mathrm{I}}}\times\left(-\frac{1}{2}+\frac{1}{12}\left(\frac{h_{\mathrm{II}}}{L_{\mathrm{I}}}\right)\right)\varepsilon_{sp}, \tag{2.253}$$

$$\varepsilon_{x,\mathrm{II}}\Big|_{z=-\frac{h_{\mathrm{II}}}{2}} = \frac{1}{\frac{2}{3}\frac{E_{\mathrm{II}}I_{\mathrm{II}}L_{\mathrm{I}}}{E_{\mathrm{I}}I_{\mathrm{I}}L_{\mathrm{II}}}+1+\frac{1}{12}\frac{h_{\mathrm{II}}^2}{L_{\mathrm{I}}^2}}\times\frac{h_{\mathrm{II}}}{L_{\mathrm{I}}}\times\left(+\frac{1}{2}+\frac{1}{12}\left(\frac{h_{\mathrm{II}}}{L_{\mathrm{I}}}\right)\right)\varepsilon_{sp}. \tag{2.254}$$

Let us do a simple calculation at $z = -\frac{h_{\mathrm{II}}}{2}$ for the special case $E_{\mathrm{I}} = E_{\mathrm{II}}$, $I_{\mathrm{I}} = I_{\mathrm{II}}$, $L_{\mathrm{I}} = L_{\mathrm{II}}$, and $h_{\mathrm{II}} = \frac{L_{\mathrm{I}}}{10}$. From Eq. (2.254), we get

$$\varepsilon_{x,\mathrm{II}}\Big|_{z=-\frac{h_{\mathrm{II}}}{2}} = \frac{61}{2001}\varepsilon_{sp} = 0.0305\varepsilon_{sp}, \tag{2.255}$$

while the simplified model according to Eq. (2.238) gives:

$$\varepsilon_{x,\mathrm{II}}\Big|_{z=-\frac{h_{\mathrm{II}}}{2}} = \frac{61}{800}\varepsilon_{sp} = 0.07625\varepsilon_{sp}. \tag{2.256}$$

Obviously there is quite a significant difference between both approaches but the results have at least the same order of magnitude.

The derivation of the equation for the displacement u_0 as given in Eqs. (2.245) and (2.246) was based on the assumption that the shear force is not contributing to the deformation of the frame. The results for the shear force in Eqs. (2.240) and (2.243) indicate that only the vertical frame part is loaded by a shear force. In the case that this member is designed as a short beam, i.e. the application of the thick beam might be more appropriate, CASTIGLIANO's statement can be modified to the following expression:

$$u_0 = \int_0^{L_I} \frac{M_y(x_I)}{E_I I_I} \frac{\partial M_y(x_I, F_0^R)}{\partial F_0^R} dx_I + \underbrace{\int_0^{L_I} \frac{Q_z(x_I)}{k_{s,I} G_I A_I} \frac{\partial Q_z(x_I, F_0^R)}{\partial F_0^R} dx_I +}_{\text{shear contribution}}$$

$$+ \int_0^{\frac{L_{II}}{2}} \frac{M_y(x_{II})}{E_{II} I_{II}} \frac{\partial M_y(x_{II}, F_0^R)}{\partial F_0^R} dx_{II} + \int_0^{\frac{L_{II}}{2}} \frac{N_x(x_{II})}{E_{II} A_{II}} \frac{\partial N_x(x_{II}, F_0^R)}{\partial F_0^R} dx_{II}$$

$$= F_0^R \left(\frac{1}{3} \frac{L_I^3}{E_I I_I} + \frac{1}{2} \frac{L_I^2 L_{II}}{E_{II} I_{II}} + \frac{1}{2} \frac{L_{II}}{E_{II} A_{II}} + \frac{L_I}{k_{s,I} G_I A_I} \right). \tag{2.257}$$

The last equation can be rearranged for the unknown reaction force:

$$F_0^R = \frac{u_0}{\dfrac{1}{3} \dfrac{L_I^3}{E_I I_I} + \dfrac{1}{2} \dfrac{L_I^2 L_{II}}{E_{II} I_{II}} + \dfrac{1}{2} \dfrac{L_{II}}{E_{II} A_{II}} + \dfrac{L_I}{k_{s,I} G_I A_I}}. \tag{2.258}$$

Under the consideration of a square cross section, i.e. $k_{s,I} = \frac{5}{6}$, $G_I = \frac{E_I}{2(1+\nu_I)}$, $A_i = \frac{12 I_i}{h_i^2}$, $I_i = \frac{b_i h_i^3}{12}$, and that the width is the same, i.e. $b_I = b_{II}$, one can easily derive the following normalized expression:

$$\frac{F_0^R}{\frac{3 E_{II} I_{II} u_0}{L_{II}^3}} = \frac{\frac{2}{3}}{\frac{2}{3} \frac{E_{II}}{E_I} \left(\frac{h_{II}}{L_{II}} \right)^3 \left(\frac{L_I}{h_I} \right)^3 + \left(\frac{L_I}{L_{II}} \right)^3 + \frac{1}{12} \left(\frac{h_{II}}{L_{II}} \right)^2 + \frac{2}{5} (1 + \nu_I) \frac{E_{II}}{E_I} \left(\frac{h_{II}}{L_{II}} \right)^3 \left(\frac{L_I}{h_I} \right)}. \tag{2.259}$$

It should be noted that the last expression in the denominator (which contains POISSON's ratio) stems from the consideration of the shear contribution. Let us now do some simple estimates to predict the significance of the different contributions. The different fractions in the denominator are evaluated as a function of $\frac{h_I}{L_I}$ in Table 2.8 for the special case $E_{II} = E_I$, $L_{II} = L_I$, $h_{II} = \frac{L_{II}}{10}$, and $\nu_I = 0.3$.

We can conclude from Table 2.8 that the dominant mode of deformation is bending in member II. Increasing the height (h_I) of member I with respect to its length (L_I) increases the shear contribution in this members compared to the bending fraction. However, both contributions reduce their share in the total deformation. Thus, we can justify from this investigation that we do not need to consider the contribution of the shear force on the deformation in this particular case.

Let us mention here that the presented approach allows also to estimate the influence of the deadweight, see Fig. 2.56. The distributed bending loads in the horizontal frame elements are given by $q_g = \frac{dF_g}{dx} = \varrho_I A_I g$ whereas the distributed axial load in the vertical frame element is given by $p_g = \frac{dF_g}{dx} = \varrho_{II} A_{II} g$.

Table 2.8 Sensitivity of different deformation modes on the normalized reaction force as a function of the slenderness ratio $\frac{h_I}{L_I}$

	Member I		Member II	
	Bending	Shear	Bending	Tension
	Equation			
$\frac{h_I}{L_I}$	$\frac{2}{3}\frac{E_{II}}{E_I}\left(\frac{h_{II}}{L_{II}}\right)^3\left(\frac{L_I}{h_I}\right)^3$	$\frac{2}{5}(1+\nu_I)\frac{E_{II}}{E_I}\left(\frac{h_{II}}{L_{II}}\right)^3\left(\frac{L_I}{h_I}\right)$	$\left(\frac{L_I}{L_{II}}\right)^3$	$\frac{1}{12}\left(\frac{h_{II}}{L_{II}}\right)^2$
$\frac{1}{10}$	0.6666666667	0.0052000000	1.0	0.0008333333
$\frac{1}{9}$	0.4860000000	0.0046800000	1.0	0.0008333333
$\frac{1}{8}$	0.3413333333	0.0041600000	1.0	0.0008333333
$\frac{1}{7}$	0.2286666667	0.0036400000	1.0	0.0008333333
$\frac{1}{6}$	0.1440000000	0.0031200000	1.0	0.0008333333
$\frac{1}{5}$	0.0833333333	0.0026000000	1.0	0.0008333333
$\frac{1}{4}$	0.0426666667	0.0020800000	1.0	0.0008333333
$\frac{1}{3}$	0.0180000000	0.0015600000	1.0	0.0008333333
$\frac{1}{2}$	0.0053333333	0.0010400000	1.0	0.0008333333
1	0.0006666667	0.0005200000	1.0	0.0008333333

Fig. 2.56 Extensometer under consideration of the deadweight

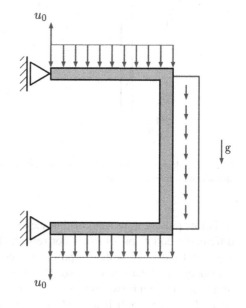

(a)

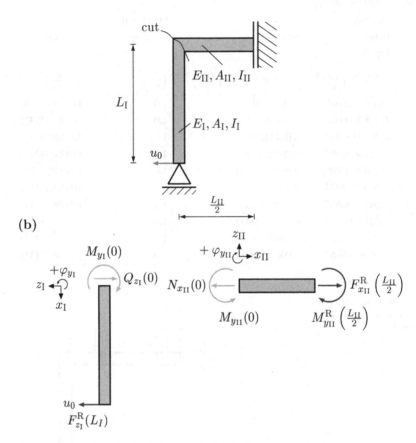

(b)

Fig. 2.57 Mechanical model for refined analysis based on differential equations: **a** symmetry consideration **b** single members

Let us present at the end of this section a refined analytical approach based on differential equations which is not restricted to the simplifications as outlined in Figs. 2.50 and 2.51. For this purpose, let us consider again the ⊔-shaped frame under consideration of the symmetry, see Fig. 2.57a. In the next step, we 'cut' the frame in its joining corner into two parts, i.e., the vertical beam I and the horizontal generalized beam II, see Fig. 2.57b. It should be noted that the reactions at $x_I = 0$ (negative face) and $x_{II} = 0$ (negative face) are now the internal reactions with their sign convention[9] as shown in Figs. 2.2 and 2.16.

---

[9]For the simplified analytical approach as shown in Fig. 2.50, the reactions at $x_I = 0$ are the support reactions from the assumed fixed support. Thus, arbitrary directions can be assumed.

The next step is to state the general solution for the deformations and internal reactions for both members as outlined in Sects. 2.2.2 and 2.2.1:

Member I (pure beam):

$$u_{z_I}(x) = \frac{1}{E_I I_{y_I}}\left(\frac{c_1 x_I^3}{6} + \frac{c_2 x_I^2}{2} + c_3 x_I + c_4\right), \tag{2.260}$$

$$\varphi_{y_I}(x) = -\frac{1}{E_I I_{y_I}}\left(\frac{c_1 x_I^2}{2} + c_2 x_I + c_3\right), \tag{2.261}$$

$$M_{y_I}(x) = -c_1 x_I - c_2, \tag{2.262}$$

$$Q_{z_I}(x) = -c_1. \tag{2.263}$$

Member II (generalized beam):

$$u_{z_{II}}(x) = \frac{1}{E_{II} I_{y_{II}}}\left(\frac{c_5 x_{II}^3}{6} + \frac{c_6 x_{II}^2}{2} + c_7 x_{II} + c_8\right), \tag{2.264}$$

$$\varphi_{y_{II}}(x) = -\frac{1}{E_{II} I_{y_{II}}}\left(\frac{c_5 x_{II}^2}{2} + c_6 x_{II} + c_7\right), \tag{2.265}$$

$$M_{y_{II}}(x) = -c_5 x_{II} - c_6, \tag{2.266}$$

$$Q_{z_{II}}(x) = -c_5, \tag{2.267}$$

$$u_x(x) = \frac{1}{E_{II} A_{II}}(c_9 x + c_{10}), \tag{2.268}$$

$$N_{x_{II}}(x) = c_9. \tag{2.269}$$

As can be seen from the above general equations, there are 10 constants of integration which require 10 conditions for their determination. Looking again on Fig. 2.57, we can state the following boundary conditions for member I:

$$u_{z_I}(x_I = L_I) = u_0, \tag{2.270}$$

$$M_{y_I}(x_I = L_I) = 0. \tag{2.271}$$

Member II is subjected to the following boundary conditions:

$$u_{x_{II}}(x_{II} = \tfrac{L_{II}}{2}) = 0, \tag{2.272}$$

$$\varphi_{y_{II}}(x_{II} = \tfrac{L_{II}}{2}) = 0, \tag{2.273}$$

$$Q_{z_{II}}(x_{II} = \tfrac{L_{II}}{2}) = 0, \tag{2.274}$$

$$u_{z_{II}}(x_{II} = 0) = 0. \tag{2.275}$$

The remaining four conditions can be specified as transmission conditions:

$$u_{z_I}(x_I = 0) = -u_{x_{II}}(x_{II} = 0), \qquad (2.276)$$
$$\varphi_{y_I}(x_I = 0) = -\varphi_{y_{II}}(x_{II} = 0), \qquad (2.277)$$
$$M_{y_I}(x_I = 0) = M_{y_{II}}(x_{II} = 0), \qquad (2.278)$$
$$Q_{z_I}(x_I = 0) = N_{x_{II}}(x_{II} = 0). \qquad (2.279)$$

Let us add a few comments on the above stated transmission conditions: In case that a positive horizontal displacement $(u_{z_I}(x_I))$ is obtained at the upper end of the vertical beam, the horizontal beam must move at its left-hand end in the negative x_{II}-direction to ensure compatibility. In the same sense for the rotation: In case that a positive rotation $(\varphi_{y_I}(x_I))$ is obtained at the upper end of the vertical beam, the horizontal beam must rotate at its left-hand end in negative sense (expressed in its own coordinate system) to ensure compatibility. In case that a positive internal bending moment $(M_{y_I}(x_I))$ is obtained at the upper end of the vertical beam, the horizontal beam must have at its left-hand end a positive internal bending moment (expressed in its own coordinate system). Pay attention to that fact that these two internal bending moments have opposite directions of rotations and cancel each other in sum. In case that a positive internal shear force $(Q_{z_I}(x_I))$ is obtained at the upper end of the vertical beam, the horizontal beam must have at its left-hand end a positive internal normal force (expressed in its own coordinate system). Pay attention to that fact that these two internal forces have opposite directions and cancel each other in sum.

Evaluation of the ten conditions given in Eqs. (2.270)–(2.279) results in ten equations for the unknown constants of integration $c_1, \ldots, c_{10}$. These ten equations can be written in matrix form as follows:

$$
\begin{bmatrix}
\frac{L_I^3}{6} & \frac{L_I^2}{2} & L_I & 1 & 0 & 0 & 0 & 0 & 0 & 0 \\
-L_I & -1 & 0 & 0 & 0 & 0 & 0 & 0 & 0 & 0 \\
0 & 0 & 0 & 0 & 0 & 0 & 0 & 0 & \frac{L_{II}}{2} & 1 \\
0 & 0 & 0 & 0 & 0 & 0 & 0 & 1 & 0 & 0 \\
0 & 0 & 0 & 0 & -1 & 0 & 0 & 0 & 0 & 0 \\
0 & 0 & 0 & 0 & \frac{L_{II}^2}{8} & \frac{L_{II}}{2} & 1 & 0 & 0 & 0 \\
0 & 0 & 0 & \frac{1}{E_I I_{y_I}} & 0 & 0 & 0 & 0 & 0 & \frac{1}{E_{II} A_{II}} \\
0 & 0 & -\frac{1}{E_I I_{y_I}} & 0 & 0 & 0 & -\frac{1}{E_{II} I_{y_{II}}} & 0 & 0 & 0 \\
0 & -1 & 0 & 0 & 0 & 1 & 0 & 0 & 0 & 0 \\
-1 & 0 & 0 & 0 & 0 & 0 & 0 & 0 & -1 & 0
\end{bmatrix}
\begin{bmatrix}
c_1 \\ c_2 \\ c_3 \\ c_4 \\ c_5 \\ c_6 \\ c_7 \\ c_8 \\ c_9 \\ c_{10}
\end{bmatrix}
=
\begin{bmatrix}
E_I I_{y_I} u_0 \\ 0 \\ 0 \\ 0 \\ 0 \\ 0 \\ 0 \\ 0 \\ 0 \\ 0
\end{bmatrix}.
$$

$$(2.280)$$

Multiplication of the inversed coefficient matrix with the right-hand side allows us to determine the constants of integration as:

$$
\begin{bmatrix} c_1 \\ c_2 \\ c_3 \\ c_4 \\ c_5 \\ c_6 \\ c_7 \\ c_8 \\ c_9 \\ c_{10} \end{bmatrix}
=
\begin{bmatrix}
-\dfrac{6 A_{II} E_I E_{II} I_I I_{II} u_0}{\left(3 A_{II} E_I I_I L_I^2 + 3 E_I I_I I_{II}\right) L_{II} + 2 A_{II} E_{II} I_{II} L_I^3} \\[2ex]
\dfrac{6 A_{II} E_I E_{II} I_I I_{II} L_{II} u_0}{\left(3 A_{II} E_I I_I L_I^2 + 3 E_I I_I I_{II}\right) L_{II} + 2 A_{II} E_{II} I_{II} L_I^3} \\[2ex]
\dfrac{3 A_{II} E_I^2 I_I^2 L_I L_{II} u_0}{\left(3 A_{II} E_I I_I L_I^2 + 3 E_I I_I I_2\right) L_I + 2 A_{II} E_{II} I_{II} L_I^3} \\[2ex]
\dfrac{3 E_I^2 I_I^2 I_{II} L_{II} u_0}{\left(3 A_{II} E_I I_I L_I^2 + 3 E_I I_I I_{II}\right) L_{II} + 2 A_{II} E_{II} I_{II} L_I^3} \\[2ex]
0 \\[2ex]
\dfrac{6 A_{II} E_I E_{II} I_I I_{II} L_I u_0}{\left(3 A_{II} E_I I_I L_I^2 + 3 E_I I_I I_{II}\right) L_{II} + 2 A_{II} E_{II} I_{II} L_I^3} \\[2ex]
-\dfrac{3 A_{II} E_I E_{II} I_I I_{II} L_I L_{II} u_0}{\left(3 A_{II} E_I I_I L_I^2 + 3 E_I I_I I_{II}\right) L_{II} + 2 A_{II} E_{II} I_{II} L_I^3} \\[2ex]
0 \\[2ex]
\dfrac{6 A_{II} E_I E_{II} I_I I_{II} u_0}{\left(3 A_{II} E_I I_I L_I^2 + 3 E_I I_I I_{II}\right) L_{II} + 2 A_{II} E_{II} I_{II} L_I^3} \\[2ex]
-\dfrac{3 A_{II} E_I E_{II} I_I I_{II} L_{II} u_0}{\left(3 A_{II} E_I I_I L_I^2 + 3 E_I I_I I_{II}\right) L_{II} + 2 A_{II} E_{II} I_{II} L_I^3}
\end{bmatrix}
. \tag{2.281}
$$

The constant of integration c_1 can be re-arranged to:

$$
c_1 = -\frac{u_0}{\dfrac{L_I^2 L_{II}}{2 E_{II} I_{y_{II}}} + \dfrac{L_{II}}{2 A_{II} E_{II}} + \dfrac{L_I^3}{3 E_I I_{y_I}}}. \tag{2.282}
$$

This gives with the relation $F_{z_I}^R(L_I) = Q_{z_I}(L_I) = -c_1$ the same exact reaction force as in the case of the energy approach, see Eq. 2.245.

Let us look on a different configuration, which is commonly used for high temperature applications, see Fig. 2.58.

Ceramic extensions are fitted to the classical extensometer (represented by section III in Fig. 2.59) and these extensions are inserted into a heat chamber or furnace to measure the axial strain of a specimen.

The solution procedure is again based on the strain energy approach, as outlined in Sect. 2.3, which allows us to consider the entire frame under consideration of the symmetry (see Fig. 2.59b). The horizontal force and the moment equilibrium (see Fig. 2.60) gives the internal reactions of the vertical member I as follows:

$$
Q_z(x_I) = F_0^R, \tag{2.283}
$$

$$
M_y(x_I) = -F_0^R(L_I + L_{III} - x_I). \tag{2.284}
$$

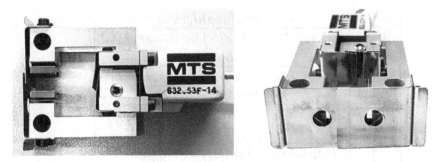

Fig. 2.58 High-temperature axial extensometer (MTS Systems Corporation, USA). To be used with precision ground ceramic extension rods (not shown here)

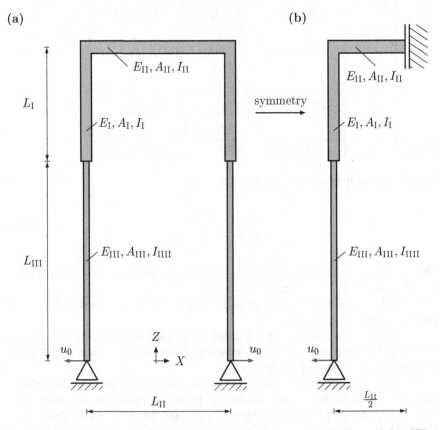

Fig. 2.59 Mechanical model of an extensometer with additional extensions (see index 'III'): **a** entire sensor and **b** consideration of symmetry

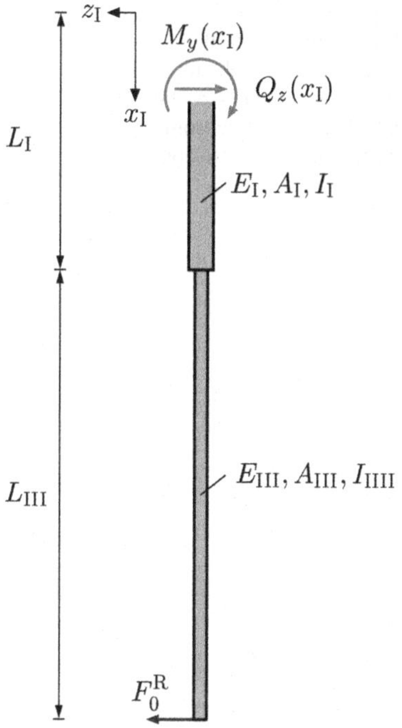

Fig. 2.60 Vertical beam section I for the determination of the internal reactions of the extensometer with additional extensions

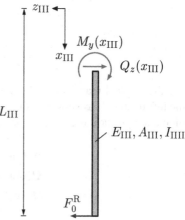

Fig. 2.61 Vertical beam section III for the determination of the internal reactions of the extensometer with additional extensions

In the same way, we can state the horizontal force and the moment equilibrium (see Fig. 2.61) to obtain the internal reactions of the vertical member III as follows:

$$Q_z(x_{\mathrm{III}}) = F_0^{\mathrm{R}}, \tag{2.285}$$

$$M_y(x_{\mathrm{III}}) = -F_0^{\mathrm{R}}(L_{\mathrm{III}} - x_{\mathrm{III}})\,. \tag{2.286}$$

Fig. 2.62 Section II of the
frame structure for the
determination of the internal
reactions in the horizontal
member of the extensometer
with additional extensions

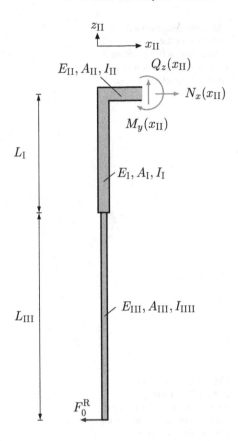

For the internal reactions of the horizontal member, it is advantageous to consider
the left-hand half as shown in Fig. 2.62. Horizontal and vertical force as well as the
moment equilibrium give the following internal reactions in the horizontal member:

$$N_x(x_{II}) = F_0^R, \tag{2.287}$$

$$Q_z(x_{II}) = 0, \tag{2.288}$$

$$M_y(x_{II}) = -F_0^R(L_I + L_{III}). \tag{2.289}$$

It should be noted here that the reaction force F_0^R is still unknown. Based on CAS-
TIGLIANO's second theorem, it is possible to express the horizontal displacement
as[10]:

---

[10]It is again assumed that the beams are thin and that the shear force is not contributing to the
bending deformation modes.

$$u_0 = \int_0^{L_I} \frac{M_y(x_I)}{E_I I_I} \frac{\partial M_y(x_I, F_0^R)}{\partial F_0^R} dx_I + \int_0^{L_{III}} \frac{M_y(x_{III})}{E_{III} I_{III}} \frac{\partial M_y(x_{III}, F_0^R)}{\partial F_0^R} dx_{III}$$

$$+ \int_0^{\frac{L_{II}}{2}} \frac{M_y(x_{II})}{E_{II} I_{II}} \frac{\partial M_y(x_{II}, F_0^R)}{\partial F_0^R} dx_{II} + \int_0^{\frac{L_{II}}{2}} \frac{N_x(x_{II})}{E_{II} A_{II}} \frac{\partial N_x(x_{II}, F_0^R)}{\partial F_0^R} dx_{II}$$

$$= F_0^R \left(\frac{L_I(\frac{1}{3}L_I^2 + L_I L_{III} + L_{III}^2)}{E_I I_I} + \frac{L_{II}(L_I + L_{III})^2}{2E_{II} I_{II}} + \frac{L_{II}}{2E_{II} A_{II}} + \frac{L_{III}^3}{3E_{III} I_{III}} \right) .$$

$$(2.290)$$

It should be noted here that the last equation contains the special case $L_{III} = 0$ as given in Eq. (2.245). Equation (2.290) can be rearranged for the unknown reaction force:

$$F_0^R = \frac{u_0}{\left(\frac{L_I(\frac{1}{3}L_I^2 + L_I L_{III} + L_{III}^2)}{E_I I_I} + \frac{L_{II}(L_I + L_{III})^2}{2E_{II} I_{II}} + \frac{L_{II}}{2E_{II} A_{II}} + \frac{L_{III}^3}{3E_{III} I_{III}} \right)} , \qquad (2.291)$$

or under the assumption of a square cross section for the horizontal member (width b_{II} and height h_{II}), we can relate the cross sectional area to the second moment of area, i.e. $A_{II} = \frac{12 I_{II}}{h_{II}^2}$, and Eq. (2.291) can be expressed as:

$$F_0^R = \frac{u_0}{\left(\frac{L_I(\frac{1}{3}L_I^2 + L_I L_{III} + L_{III}^2)}{E_I I_I} + \frac{L_{II}(L_I + L_{III})^2}{2E_{II} I_{II}} + \frac{L_{II} h_{II}^2}{24 E_{II} I_{II}} + \frac{L_{III}^3}{3E_{III} I_{III}} \right)} . \qquad (2.292)$$

Based on this reaction force, the internal reactions are known in both members, see Eqs. (2.287)–(2.289). Let us now calculate the total strain in the horizontal member of the extensometer. The axial strain due to the bending deformation can be expressed as:

$$\varepsilon_{x,II} = \frac{1}{E_{II}} \frac{M_y(x_{II})}{I_{II}} z(x_{II})$$

$$= \frac{1}{E_{II} I_{II}} \left(-\frac{u_0(L_I + L_{II})}{\left(\frac{L_I(\frac{1}{3}L_I^2 + L_I L_{III} + L_{III}^2)}{E_I I_I} + \frac{L_{II}(L_I + L_{III})^2}{2E_{II} I_{II}} + \frac{L_{II} h_{II}^2}{24 E_{II} I_{II}} + \frac{L_{III}^3}{3E_{III} I_{III}} \right)} \right) z(x_{II}) ,$$

$$(2.293)$$

which can be rearranged under consideration of $u_0 = \frac{L_{II}}{2} \varepsilon_{sp}$ to:

$$\varepsilon_{x,\mathrm{II}} = -\frac{\varepsilon_{sp} \times z_{\mathrm{II}} \left(L_{\mathrm{I}} + L_{\mathrm{III}}\right)}{\frac{E_{\mathrm{II}} I_{\mathrm{II}}}{E_{\mathrm{I}} I_{\mathrm{I}}} \left(\frac{2L_{\mathrm{I}}^3}{3L_{\mathrm{II}}} + \frac{L_{\mathrm{I}}^2 L_{\mathrm{III}}}{L_{\mathrm{II}}} + \frac{L_{\mathrm{I}} L_{\mathrm{III}}^2}{L_{\mathrm{II}}}\right) + \left(L_{\mathrm{I}} + L_{\mathrm{III}}\right)^2 + \frac{h_{\mathrm{II}}^2}{12} + \frac{2E_{\mathrm{II}} I_{\mathrm{II}} L_{\mathrm{III}}^3}{3E_{\mathrm{III}} I_{\mathrm{III}} L_{\mathrm{II}}}}. \tag{2.294}$$

On the other hand, the axial strain due to the tensile deformation can be expressed under the assumption $A_{\mathrm{II}} = \frac{12 I_{\mathrm{II}}}{h_{\mathrm{II}}^2}$ as:

$$\varepsilon_{x,\mathrm{II}} = \frac{1}{E_{\mathrm{II}} A_{\mathrm{II}}} \frac{F_0^R}{}$$

$$= \frac{1}{E_{\mathrm{II}} A_{\mathrm{II}}} \left(\frac{u_0}{\left(\frac{L_{\mathrm{I}}\left(\frac{1}{3}L_{\mathrm{I}}^2 + L_{\mathrm{I}} L_{\mathrm{III}} + L_{\mathrm{III}}^2\right)}{E_{\mathrm{I}} I_{\mathrm{I}}} + \frac{L_{\mathrm{II}}\left(L_{\mathrm{I}} + L_{\mathrm{III}}\right)^2}{2E_{\mathrm{II}} I_{\mathrm{II}}} + \frac{L_{\mathrm{II}} h_{\mathrm{II}}^2}{24 E_{\mathrm{II}} I_{\mathrm{II}}} + \frac{L_{\mathrm{III}}^3}{3E_{\mathrm{III}} I_{\mathrm{III}}}\right)}\right)$$

$$= \frac{\varepsilon_{sp} \times \frac{h_{\mathrm{II}}^2}{12}}{\frac{E_{\mathrm{II}} I_{\mathrm{II}}}{E_{\mathrm{I}} I_{\mathrm{I}}} \left(\frac{2L_{\mathrm{I}}^3}{3L_{\mathrm{II}}} + \frac{L_{\mathrm{I}}^2 L_{\mathrm{III}}}{L_{\mathrm{II}}} + \frac{L_{\mathrm{I}} L_{\mathrm{III}}^2}{L_{\mathrm{II}}}\right) + \left(L_{\mathrm{I}} + L_{\mathrm{III}}\right)^2 + \frac{h_{\mathrm{II}}^2}{12} + \frac{2E_{\mathrm{II}} I_{\mathrm{II}} L_{\mathrm{III}}^3}{3E_{\mathrm{III}} I_{\mathrm{III}} L_{\mathrm{II}}}}. \tag{2.295}$$

Thus, the total strain resulting from bending and tension is obtained as:

$$\varepsilon_{x,\mathrm{II}} =$$

$$= \frac{-\frac{z_{\mathrm{II}}}{L_{\mathrm{I}}} \left(1 + \frac{L_{\mathrm{III}}}{L_{\mathrm{I}}}\right) + \frac{1}{12}\left(\frac{h_{\mathrm{II}}}{L_{\mathrm{I}}}\right)^2}{\frac{E_{\mathrm{II}} I_{\mathrm{II}}}{E_{\mathrm{I}} I_{\mathrm{I}}} \left(\frac{2L_{\mathrm{I}}}{3L_{\mathrm{II}}} + \frac{L_{\mathrm{III}}}{L_{\mathrm{II}}} + \frac{L_{\mathrm{III}}^2}{L_{\mathrm{I}} L_{\mathrm{II}}}\right) + \left(1 + \frac{L_{\mathrm{III}}}{L_{\mathrm{I}}}\right)^2 + \frac{1}{12}\left(\frac{h_{\mathrm{II}}}{L_{\mathrm{I}}}\right)^2 + \frac{2E_{\mathrm{II}} I_{\mathrm{II}} L_{\mathrm{III}}^3}{3E_{\mathrm{III}} I_{\mathrm{III}} L_{\mathrm{I}}^2 L_{\mathrm{II}}}} \times \varepsilon_{sp}. \tag{2.296}$$

The extreme values at the free surfaces, i.e. $z = +\frac{h_{\mathrm{II}}}{2}$ and $z = -\frac{h_{\mathrm{II}}}{2}$, are obtained as follows:

$$\varepsilon_{x,\mathrm{II}}\Big|_{z=+\frac{h_{\mathrm{II}}}{2}}$$

$$= \frac{\frac{h_{\mathrm{II}}}{L_{\mathrm{I}}} \left(-\frac{1}{2} + \frac{1}{12}\left(\frac{h_{\mathrm{II}}}{L_{\mathrm{I}}}\right)\right)}{\frac{E_{\mathrm{II}} I_{\mathrm{II}}}{E_{\mathrm{I}} I_{\mathrm{I}}} \left(\frac{2L_{\mathrm{I}}}{3L_{\mathrm{II}}} + \frac{L_{\mathrm{III}}}{L_{\mathrm{II}}} + \frac{L_{\mathrm{III}}^2}{L_{\mathrm{I}} L_{\mathrm{II}}}\right) + \left(1 + \frac{L_{\mathrm{III}}}{L_{\mathrm{I}}}\right)^2 + \frac{1}{12}\left(\frac{h_{\mathrm{II}}}{L_{\mathrm{I}}}\right)^2 + \frac{2E_{\mathrm{II}} I_{\mathrm{II}} L_{\mathrm{III}}^3}{3E_{\mathrm{III}} I_{\mathrm{III}} L_{\mathrm{I}}^2 L_{\mathrm{II}}}} \times \varepsilon_{sp}, \tag{2.297}$$

$$\varepsilon_{x,\mathrm{II}}\Big|_{z=-\frac{h_{\mathrm{II}}}{2}}$$

$$= \frac{\frac{h_{\mathrm{II}}}{L_{\mathrm{I}}} \left(+\frac{1}{2} + \frac{1}{12}\left(\frac{h_{\mathrm{II}}}{L_{\mathrm{I}}}\right)\right)}{\frac{E_{\mathrm{II}} I_{\mathrm{II}}}{E_{\mathrm{I}} I_{\mathrm{I}}} \left(\frac{2L_{\mathrm{I}}}{3L_{\mathrm{II}}} + \frac{L_{\mathrm{III}}}{L_{\mathrm{II}}} + \frac{L_{\mathrm{III}}^2}{L_{\mathrm{I}} L_{\mathrm{II}}}\right) + \left(1 + \frac{L_{\mathrm{III}}}{L_{\mathrm{I}}}\right)^2 + \frac{1}{12}\left(\frac{h_{\mathrm{II}}}{L_{\mathrm{I}}}\right)^2 + \frac{2E_{\mathrm{II}} I_{\mathrm{II}} L_{\mathrm{III}}^3}{3E_{\mathrm{III}} I_{\mathrm{III}} L_{\mathrm{I}}^2 L_{\mathrm{II}}}} \times \varepsilon_{sp}. \tag{2.298}$$

(a)   (b)

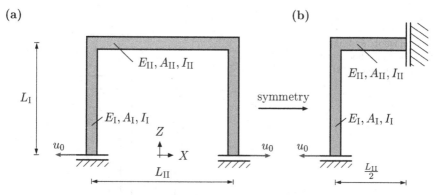

Fig. 2.63 Mechanical model of the extensometer (no rotation at specimen): **a** entire sensor and **b** consideration of symmetry

Fig. 2.64 Vertical beam section for the determination of the internal reactions (no rotation at specimen)

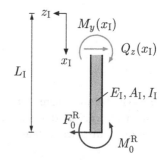

Let us investigate in the following a configuration which does not allow any rotation of the extensometer on the specimen, see Fig. 2.63. Thus, the extensometer can be imagined as bolted on the specimen.[11]

The horizontal force and moment equilibrium (see Fig. 2.64) gives the internal reactions of the vertical member as follows:

$$Q_z(x_I) = F_0^R,$$ (2.299)

$$M_y(x_I) = -F_0^R(L_I - x_I) + M_0^R.$$ (2.300)

For the internal reactions of the horizontal member, it is advantageous to consider the left-hand half as shown in Fig. 2.65. Horizontal and vertical force as well as the moment equilibrium give the following internal reactions in the horizontal member:

---

[11] An alternative configuration uses a pair of knife edges according to ASTM E-399 [1] standard for free rotation of the points of contact between the gage and the specimen. These edges are prestressed between two mating edges, which are bolted to the specimen, see [10] for details. This configuration is many times used on compact tension (CT) specimens for determining the J-integral in the context of experimental fracture mechanics.

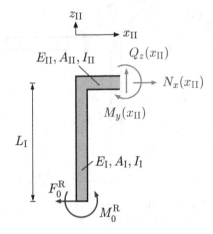

Fig. 2.65 Section of the frame structure for the determination of the internal reactions in the horizontal member (no rotation at specimen)

$$N_x(x_{\mathrm{II}}) = F_0^{\mathrm{R}}, \tag{2.301}$$

$$Q_z(x_{\mathrm{II}}) = 0, \tag{2.302}$$

$$M_y(x_{\mathrm{II}}) = -F_0^{\mathrm{R}} L_{\mathrm{I}} + M_0^{\mathrm{R}}. \tag{2.303}$$

It should be noted here that the reaction force F_0^{R} and the reaction moment M_0^{R} are still unknown. Based on CASTIGLIANO's second theorem, it is possible to express the horizontal displacement as[12]:

$$
\begin{aligned}
u_0 = & \int_0^{L_{\mathrm{I}}} \frac{M_y(x_{\mathrm{I}}) \, \partial M_y(x_{\mathrm{I}}, F_0^{\mathrm{R}}, M_0^{\mathrm{R}})}{E_{\mathrm{I}} I_{\mathrm{I}}} \, dx_{\mathrm{I}} + \\
& + \int_0^{\frac{L_{\mathrm{II}}}{2}} \frac{M_y(x_{\mathrm{II}}) \, \partial M_y(x_{\mathrm{II}}, F_0^{\mathrm{R}}, M_0^{\mathrm{R}})}{E_{\mathrm{II}} I_{\mathrm{II}}} \, dx_{\mathrm{II}} + \\
& + \int_0^{\frac{L_{\mathrm{II}}}{2}} \frac{N_x(x_{\mathrm{II}}) \, \partial N_x(x_{\mathrm{II}}, F_0^{\mathrm{R}})}{E_{\mathrm{II}} A_{\mathrm{II}}} \, dx_{\mathrm{II}},
\end{aligned}
\tag{2.304}
$$

or after performing the integrations:

$$u_0(F_0^{\mathrm{R}}, M_0^{\mathrm{R}}) = \frac{1}{E_{\mathrm{I}} I_{\mathrm{I}}} \left(\frac{F_0^{\mathrm{R}} L_{\mathrm{I}}^3}{3} - \frac{M_0^{\mathrm{R}} L_{\mathrm{I}}^2}{2} \right) + \frac{(F_0^{\mathrm{R}} L_{\mathrm{I}}^2 - M_0^{\mathrm{R}} L_{\mathrm{I}}) \frac{L_{\mathrm{II}}}{2}}{E_{\mathrm{II}} I_{\mathrm{II}}}$$

---

[12]It is assumed that the beams are thin and that the shear force is not contributing to the bending deformation modes.

$$+ \frac{F_0^R \frac{L_{II}}{2}}{E_{II} A_{II}} \tag{2.305}$$

$$= \left(\frac{L_I^3}{3 E_I I_I} + \frac{L_I^2 L_{II}}{2 E_{II} I_{II}} + \frac{L_{II}}{2 E_{II} A_{II}} \right) F_0^R + \left(-\frac{L_I^2}{2 E_I I_I} - \frac{L_I L_{II}}{2 E_{II} I_{II}} \right) M_0^R . \tag{2.306}$$

It should be noted that the last equation still contains two unknowns, i.e. F_0^R and M_0^R. A second application of CASTIGLIANO's second theorem with respect to the rotation at the contact point of extensometer and specimens reads:

$$0 = \varphi_0 = \int_0^{L_I} \frac{M_y(x_I)}{E_I I_I} \frac{\partial M_y(x_I, F_0^R, M_0^R)}{\partial M_0^R} dx_I +$$

$$+ \int_0^{\frac{L_{II}}{2}} \frac{M_y(x_{II})}{E_{II} I_{II}} \frac{\partial M_y(x_{II}, F_0^R, M_0^R)}{\partial M_0^R} dx_{II} + \tag{2.307}$$

$$+ \int_0^{\frac{L_{II}}{2}} \frac{N_x(x_{II})}{E_{II} A_{II}} \frac{\partial N_x(x_{II}, F_0^R)}{\partial M_0^R} dx_{II} ,$$

or after performing the integrations:

$$0 = \frac{M_0^R L_I}{E_I I_I} - \frac{F_0^R L_I^2}{2 E_I I_I} - \frac{F_0^R L_I L_{II}}{2 E_{II} I_{II}} + \frac{M_0^R L_{II}}{2 E_{II} I_{II}}, \tag{2.308}$$

or rearranged for the reaction moment:

$$M_0^R = \frac{F_0^R L_I}{2} \times \frac{1 + \frac{E_I I_I}{E_{II} I_{II}} \frac{L_{II}}{L_I}}{1 + \frac{1}{2} \frac{E_I I_I}{E_{II} I_{II}} \frac{L_{II}}{L_I}} . \tag{2.309}$$

The last equation can be introduced into Eq. (2.306) to give the expression for the reaction force:

$$F_0^R = \frac{u_0}{\frac{L_I^3}{3 E_I I_I} + \frac{L_I^2 L_{II}}{2 E_{II} I_{II}} + \frac{L_{II} h_{II}^2}{24 E_{II} I_{II}} - \frac{L_I^3}{4 E_I I_I} \frac{\left(1 + \frac{E_I I_I}{E_{II} I_{II}} \frac{L_{II}}{L_I} \right)^2}{1 + \frac{1}{2} \frac{E_I I_I}{E_{II} I_{II}} \frac{L_{II}}{L_I}}} . \tag{2.310}$$

Back-substitution of the last equation into Eq. (2.309) gives the corresponding expression for the reaction moment:

$$M_0^R = \cfrac{u_0 \cfrac{L_I}{2} \cfrac{1 + \frac{E_I I_I}{E_{II} I_{II}} \frac{L_{II}}{L_I}}{1 + \frac{1}{2} \frac{E_I I_I}{E_{II} I_{II}} \frac{L_{II}}{L_I}}}{\frac{L_I^3}{3E_I I_I} + \frac{L_I^2 L_{II}}{2E_{II} I_{II}} + \frac{L_{II} h_{II}^2}{24 E_{II} I_{II}} - \frac{L_I^3}{4E_I I_I} \cfrac{\left(1 + \frac{E_I I_I}{E_{II} I_{II}} \frac{L_{II}}{L_I}\right)^2}{1 + \frac{1}{2} \frac{E_I I_I}{E_{II} I_{II}} \frac{L_{II}}{L_I}}} . \tag{2.311}$$

Let us now calculate the total strain in the horizontal member of the extensometer. The axial strain due to the bending deformation can be expressed as:

$$\varepsilon_{x,II} = \frac{1}{E_{II}} \frac{M_y(x_{II})}{I_{II}} z(x_{II})$$

$$= \frac{1}{E_{II} I_{II}} \left(\cfrac{u_0 \left(-L_I + \frac{L_I}{2} \cfrac{1 + \frac{E_I I_I}{E_{II} I_{II}} \frac{L_{II}}{L_I}}{1 + \frac{1}{2} \frac{E_I I_I}{E_{II} I_{II}} \frac{L_{II}}{L_I}}\right)}{\frac{L_I^3}{3E_I I_I} + \frac{L_I^2 L_{II}}{2E_{II} I_{II}} + \frac{L_{II} h_{II}^2}{24 E_{II} I_{II}} - \frac{L_I^3}{4E_I I_I} \cfrac{\left(1 + \frac{E_I I_I}{E_{II} I_{II}} \frac{L_{II}}{L_I}\right)^2}{1 + \frac{1}{2} \frac{E_I I_I}{E_{II} I_{II}} \frac{L_{II}}{L_I}}} \right) z(x_{II}), \tag{2.312}$$

which can be rearranged under consideration of $u_0 = \frac{L_{II}}{2} \varepsilon_{sp}$ to:

$$\varepsilon_{x,II} = \cfrac{\left(-1 + \frac{1}{2} \cfrac{1 + \frac{E_I I_I}{E_{II} I_{II}} \frac{L_{II}}{L_I}}{1 + \frac{1}{2} \frac{E_I I_I}{E_{II} I_{II}} \frac{L_{II}}{L_I}}\right) \varepsilon_{sp}}{\frac{2}{3} \frac{E_{II} I_{II}}{E_I I_I} \frac{L_I}{L_{II}} + 1 + \frac{1}{12} \frac{h_{II}^2}{L_I^2} - \frac{1}{2} \frac{E_{II} I_{II}}{E_I I_I} \frac{L_I}{L_{II}} \cfrac{\left(1 + \frac{E_I I_I}{E_{II} I_{II}} \frac{L_{II}}{L_I}\right)^2}{1 + \frac{1}{2} \frac{E_I I_I}{E_{II} I_{II}} \frac{L_{II}}{L_I}}} \left(\frac{z_{II}}{L_I}\right). \tag{2.313}$$

On the other hand, the axial strain due to the tensile deformation can be expressed as:

$$\varepsilon_{x,II} = \frac{1}{E_{II}} \frac{F_0^R}{A_{II}}$$

$$= \frac{1}{E_{II} A_{II}} \left(\cfrac{u_0}{\frac{L_I^3}{3E_I I_I} + \frac{L_I^2 L_{II}}{2E_{II} I_{II}} + \frac{L_{II} h_{II}^2}{24 E_{II} I_{II}} - \frac{L_I^3}{4E_I I_I} \cfrac{\left(1 + \frac{E_I I_I}{E_{II} I_{II}} \frac{L_{II}}{L_I}\right)^2}{1 + \frac{1}{2} \frac{E_I I_I}{E_{II} I_{II}} \frac{L_{II}}{L_I}}} \right), \tag{2.314}$$

which can be rearranged under consideration of $A_{II} = \frac{12 I_{II}}{h_{II}^2}$ and $u_0 = \frac{L_{II}}{2} \varepsilon_{sp}$ to:

$$\varepsilon_{x,\mathrm{II}} = \frac{h_{\mathrm{II}}^2}{12L_{\mathrm{I}}^2} \times \frac{\varepsilon_{\mathrm{sp}}}{\frac{2}{3}\frac{E_{\mathrm{II}}I_{\mathrm{II}}}{E_{\mathrm{I}}I_{\mathrm{I}}}\frac{L_{\mathrm{I}}}{L_{\mathrm{II}}} + 1 + \frac{1}{12}\frac{h_{\mathrm{II}}^2}{L_{\mathrm{I}}^2} - \frac{1}{2}\frac{E_{\mathrm{II}}I_{\mathrm{II}}}{E_{\mathrm{I}}I_{\mathrm{I}}}\frac{L_{\mathrm{I}}}{L_{\mathrm{II}}}\frac{\left(1+\frac{E_{\mathrm{I}}I_{\mathrm{I}}}{E_{\mathrm{II}}I_{\mathrm{II}}}\frac{L_{\mathrm{II}}}{L_{\mathrm{I}}}\right)^2}{1+\frac{1}{2}\frac{E_{\mathrm{I}}I_{\mathrm{I}}}{E_{\mathrm{II}}I_{\mathrm{II}}}\frac{L_{\mathrm{II}}}{L_{\mathrm{I}}}}} . \tag{2.315}$$

Thus, the total strain resulting from bending and tension is obtained as:

$$\varepsilon_{x,\mathrm{II}} = \frac{\left(-1+\frac{1}{2}\dfrac{1+\frac{E_{\mathrm{I}}I_{\mathrm{I}}}{E_{\mathrm{II}}I_{\mathrm{II}}}\frac{L_{\mathrm{II}}}{L_{\mathrm{I}}}}{1+\frac{1}{2}\frac{E_{\mathrm{I}}I_{\mathrm{I}}}{E_{\mathrm{II}}I_{\mathrm{II}}}\frac{L_{\mathrm{II}}}{L_{\mathrm{I}}}}\right)\left(\frac{z_{\mathrm{II}}}{L_{\mathrm{I}}}\right) + \frac{1}{12}\left(\frac{h_{\mathrm{II}}}{L_{\mathrm{I}}}\right)^2}{\frac{2}{3}\frac{E_{\mathrm{II}}I_{\mathrm{II}}}{E_{\mathrm{I}}I_{\mathrm{I}}}\frac{L_{\mathrm{I}}}{L_{\mathrm{II}}} + 1 + \frac{1}{12}\frac{h_{\mathrm{II}}^2}{L_{\mathrm{I}}^2} - \frac{1}{2}\frac{E_{\mathrm{II}}I_{\mathrm{II}}}{E_{\mathrm{I}}I_{\mathrm{I}}}\frac{L_{\mathrm{I}}}{L_{\mathrm{II}}}\frac{\left(1+\frac{E_{\mathrm{I}}I_{\mathrm{I}}}{E_{\mathrm{II}}I_{\mathrm{II}}}\frac{L_{\mathrm{II}}}{L_{\mathrm{I}}}\right)^2}{1+\frac{1}{2}\frac{E_{\mathrm{I}}I_{\mathrm{I}}}{E_{\mathrm{II}}I_{\mathrm{II}}}\frac{L_{\mathrm{II}}}{L_{\mathrm{I}}}}} \times \varepsilon_{\mathrm{sp}} . \tag{2.316}$$

The extreme values at the free surfaces, i.e. $z = +\frac{h_{\mathrm{II}}}{2}$ and $z = -\frac{h_{\mathrm{II}}}{2}$, are obtained as follows:

$$\varepsilon_{x,\mathrm{II}}\Big|_{z=\pm\frac{h_{\mathrm{II}}}{2}} = \pm\frac{h_{\mathrm{II}}}{L_{\mathrm{I}}}\times$$
$$\frac{\left(-1+\frac{1}{2}\dfrac{1+\frac{E_{\mathrm{I}}I_{\mathrm{I}}}{E_{\mathrm{II}}I_{\mathrm{II}}}\frac{L_{\mathrm{II}}}{L_{\mathrm{I}}}}{1+\frac{1}{2}\frac{E_{\mathrm{I}}I_{\mathrm{I}}}{E_{\mathrm{II}}I_{\mathrm{II}}}\frac{L_{\mathrm{II}}}{L_{\mathrm{I}}}}\right)\left(\frac{1}{2}\right) + \frac{1}{12}\left(\frac{h_{\mathrm{II}}}{L_{\mathrm{I}}}\right)}{\frac{2}{3}\frac{E_{\mathrm{II}}I_{\mathrm{II}}}{E_{\mathrm{I}}I_{\mathrm{I}}}\frac{L_{\mathrm{I}}}{L_{\mathrm{II}}} + 1 + \frac{1}{12}\frac{h_{\mathrm{II}}^2}{L_{\mathrm{I}}^2} - \frac{1}{2}\frac{E_{\mathrm{II}}I_{\mathrm{II}}}{E_{\mathrm{I}}I_{\mathrm{I}}}\frac{L_{\mathrm{I}}}{L_{\mathrm{II}}}\frac{\left(1+\frac{E_{\mathrm{I}}I_{\mathrm{I}}}{E_{\mathrm{II}}I_{\mathrm{II}}}\frac{L_{\mathrm{II}}}{L_{\mathrm{I}}}\right)^2}{1+\frac{1}{2}\frac{E_{\mathrm{I}}I_{\mathrm{I}}}{E_{\mathrm{II}}I_{\mathrm{II}}}\frac{L_{\mathrm{II}}}{L_{\mathrm{I}}}}} \times \varepsilon_{\mathrm{sp}} . \tag{2.317}$$

Let us do a simple calculation at $z = -\frac{h_{\mathrm{II}}}{2}$ for the special case $E_{\mathrm{I}} = E_{\mathrm{II}}$, $I_{\mathrm{I}} = I_{\mathrm{II}}$, $L_{\mathrm{I}} = L_{\mathrm{II}}$, and $h_{\mathrm{II}} = \frac{L_{\mathrm{I}}}{10}$. From Eq. (2.317), we get

$$\varepsilon_{x,\mathrm{II}}\Big|_{z=-\frac{h_{\mathrm{II}}}{2}} = \frac{61}{2000}\varepsilon_{\mathrm{sp}} = 0.0474\varepsilon_{\mathrm{sp}} , \tag{2.318}$$

while the model according to Eq. (2.254), i.e. for free rotation at the contact point of extensometer and specimen, gave:

$$\varepsilon_{x,\mathrm{II}}\Big|_{z=-\frac{h_{\mathrm{II}}}{2}} = \frac{61}{2000}\varepsilon_{\mathrm{sp}} = 0.0305\varepsilon_{\mathrm{sp}} . \tag{2.319}$$

Thus, the configuration without rotation provides a higher calibration factor for the strain in the specimen.

Let us present now the analytical approach based on differential equations. For this purpose, let us consider again the ⊔-shaped frame under consideration of the symmetry, see Fig. 2.66a. In the next step, we 'cut' the frame in its joining corner into two parts, i.e., the vertical beam I and the horizontal generalized beam II, see

(a)

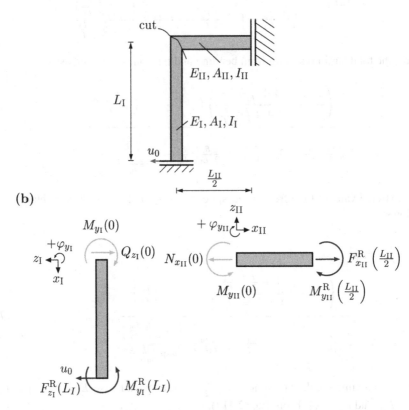

(b)

Fig. 2.66 Mechanical model for analysis based on differential equations (no rotation at specimen): a symmetry consideration b single members

Fig. 2.66b. It should be noted that the reactions at $x_I = 0$ (negative face) and $x_{II} = 0$ (negative face) are now the internal reactions with their sign convention as shown in Figs. 2.2 and 2.16.

The next step is to state the general solution for the deformations and internal reactions for both members as outlined in Sects. 2.2.2 and 2.2.1:

Member I (pure beam):

$$u_{z_I}(x) = \frac{1}{E_I I_{y_I}} \left(\frac{c_1 x_I^3}{6} + \frac{c_2 x_I^2}{2} + c_3 x_I + c_4 \right), \tag{2.320}$$

$$\varphi_{y_I}(x) = -\frac{1}{E_I I_{y_I}} \left(\frac{c_1 x_I^2}{2} + c_2 x_I + c_3 \right), \tag{2.321}$$

$$M_{y_{\mathrm{I}}}(x) = -c_1 x_{\mathrm{I}} - c_2 \, , \tag{2.322}$$

$$Q_{z_{\mathrm{I}}}(x) = -c_1 \, . \tag{2.323}$$

Member II (generalized beam):

$$u_{z_{\mathrm{II}}}(x) = \frac{1}{E_{\mathrm{II}} I_{y_{\mathrm{II}}}} \left(\frac{c_5 x_{\mathrm{II}}^3}{6} + \frac{c_6 x_{\mathrm{II}}^2}{2} + c_7 x_{\mathrm{II}} + c_8 \right) \, , \tag{2.324}$$

$$\varphi_{y_{\mathrm{II}}}(x) = -\frac{1}{E_{\mathrm{II}} I_{y_{\mathrm{II}}}} \left(\frac{c_5 x_{\mathrm{II}}^2}{2} + c_6 x_{\mathrm{II}} + c_7 \right) \, , \tag{2.325}$$

$$M_{y_{\mathrm{II}}}(x) = -c_5 x_{\mathrm{II}} - c_6 \, , \tag{2.326}$$

$$Q_{z_{\mathrm{II}}}(x) = -c_5 \, , \tag{2.327}$$

$$u_x(x) = \frac{1}{E_{\mathrm{II}} A_{\mathrm{II}}} (c_9 x + c_{10}) \, , \tag{2.328}$$

$$N_{x_{\mathrm{II}}}(x) = c_9 \, . \tag{2.329}$$

As can be seen from the above general equations, there are 10 constants of integration which require 10 conditions for their determination. Looking again on Fig. 2.66, we can state the following boundary conditions for member I:

$$u_{z_{\mathrm{I}}}(x_{\mathrm{I}} = L_{\mathrm{I}}) = u_0 \, , \tag{2.330}$$

$$\varphi_{y_{\mathrm{I}}}(x_{\mathrm{I}} = L_{\mathrm{I}}) = 0 \, . \tag{2.331}$$

Member II is subjected to the following boundary conditions:

$$u_{x_{\mathrm{II}}}(x_{\mathrm{II}} = \tfrac{L_{\mathrm{II}}}{2}) = 0 \, , \tag{2.332}$$

$$\varphi_{y_{\mathrm{II}}}(x_{\mathrm{II}} = \tfrac{L_{\mathrm{II}}}{2}) = 0 \, , \tag{2.333}$$

$$Q_{z_{\mathrm{II}}}(x_{\mathrm{II}} = \tfrac{L_{\mathrm{II}}}{2}) = 0 \, , \tag{2.334}$$

$$u_{z_{\mathrm{II}}}(x_{\mathrm{II}} = 0) = 0 \, . \tag{2.335}$$

The remaining four conditions can be specified as transmission conditions:

$$u_{z_{\mathrm{I}}}(x_{\mathrm{I}} = 0) = -u_{x_{\mathrm{II}}}(x_{\mathrm{II}} = 0) \, , \tag{2.336}$$

$$\varphi_{y_{\mathrm{I}}}(x_{\mathrm{I}} = 0) = -\varphi_{y_{\mathrm{II}}}(x_{\mathrm{II}} = 0) \, , \tag{2.337}$$

$$M_{y_{\mathrm{I}}}(x_{\mathrm{I}} = 0) = M_{y_{\mathrm{II}}}(x_{\mathrm{II}} = 0) \, , \tag{2.338}$$

$$Q_{z_{\mathrm{I}}}(x_{\mathrm{I}} = 0) = N_{x_{\mathrm{II}}}(x_{\mathrm{II}} = 0) \, . \tag{2.339}$$

Evaluation of the ten conditions given in Eqs. (2.330)–(2.339) results in ten equations for the unknown constants of integration $c_1, \ldots, c_{10}$. These ten equations can be written in matrix form as follows:

$$
\begin{bmatrix}
\frac{L_I^3}{6} & \frac{L_I^2}{2} & L_I & 1 & 0 & 0 & 0 & 0 & 0 & 0 \\
\frac{-L_I^2}{2} & L_I & 1 & 0 & 0 & 0 & 0 & 0 & 0 & 0 \\
0 & 0 & 0 & 0 & 0 & 0 & 0 & 0 & \frac{L_{II}}{2} & 1 \\
0 & 0 & 0 & 0 & \frac{L_{II}^2}{8} & \frac{L_{II}}{2} & 1 & 0 & 0 & 0 \\
0 & 0 & 0 & 0 & -1 & 0 & 0 & 0 & 0 & 0 \\
0 & 0 & 0 & 0 & 0 & 0 & 0 & 1 & 0 & 0 \\
0 & 0 & 0 & \frac{1}{E_I I_{y_I}} & 0 & 0 & 0 & 0 & 0 & \frac{1}{E_{II} A_{II}} \\
0 & 0 & -\frac{1}{E_I I_{y_I}} & 0 & 0 & 0 & -\frac{1}{E_{II} I_{y_{II}}} & 0 & 0 & 0 \\
0 & -1 & 0 & 0 & 0 & 1 & 0 & 0 & 0 & 0 \\
-1 & 0 & 0 & 0 & 0 & 0 & 0 & 0 & -1 & 0
\end{bmatrix}
\begin{bmatrix} c_1 \\ c_2 \\ c_3 \\ c_4 \\ c_5 \\ c_6 \\ c_7 \\ c_8 \\ c_9 \\ c_{10} \end{bmatrix}
=
\begin{bmatrix} E_I I_{y_I} u_0 \\ 0 \\ 0 \\ 0 \\ 0 \\ 0 \\ 0 \\ 0 \\ 0 \\ 0 \end{bmatrix}.
$$

$$\tag{2.340}$$

Multiplication of the inversed coefficient matrix with the right-hand side allows us to determine the constants of integration as:

$$
\begin{bmatrix} c_1 \\ c_2 \\ c_3 \\ c_4 \\ c_5 \\ c_6 \\ c_7 \\ c_8 \\ c_9 \\ c_{10} \end{bmatrix}
=
\begin{bmatrix}
-\dfrac{E_I I_I \left(6 A_{II} E_I E_{II} I_I L_{II} + 12 A_{II} E_{II}^2 I_{II} L_I\right) u_0}{3 E_I^2 I_I^2 L_{II}^2 + \left(2 A_{II} E_I E_{II} I_I L_I^3 + 6 E_I E_{II} I_I I_{II} L_I\right) L_{II} + A_{II} E_{II}^2 I_{II} L_I^4} \\[2mm]
\dfrac{6 A_{II} E_I E_{II}^2 I_I I_{II} L_I^2 u_0}{3 E_I^2 I_I^2 L_{II}^2 + \left(2 A_{II} E_I E_{II} I_I L_I^3 + 6 E_I E_{II} I_I I_{II} L_I\right) L_{II} + A_{II} E_{II}^2 I_{II} L_I^4} \\[2mm]
\dfrac{3 A_{II} E_I^2 E_{II} I_I^2 L_I^2 L_{II} u_0}{3 E_I^2 I_I^2 L_{II}^2 + \left(2 A_{II} E_I E_{II} I_I L_I^3 + 6 E_I E_{II} I_I I_{II} L_I\right) L_{II} + A_{II} E_{II}^2 I_{II} L_I^4} \\[2mm]
\dfrac{E_I I_I \left(3 E_I^2 I_I^2 L_{II}^2 + 6 E_I E_{II} I_I I_{II} L_I L_{II}\right) u_0}{3 E_I^2 I_I^2 L_{II}^2 + \left(2 A_{II} E_I E_{II} I_I L_I^3 + 6 E_I E_{II} I_I I_{II} L_I\right) L_{II} + A_{II} E_{II}^2 I_{II} L_I^4} \\[2mm]
0 \\[2mm]
\dfrac{6 A_{II} E_I E_{II}^2 I_I I_{II} L_I^2 u_0}{3 E_I^2 I_I^2 L_{II}^2 + \left(2 A_{II} E_I E_{II} I_I L_I^3 + 6 E_I E_{II} I_I I_{II} L_I\right) L_{II} + A_{II} E_{II}^2 I_{II} L_I^4} \\[2mm]
-\dfrac{3 A_{II} E_I E_{II}^2 I_I I_{II} L_I^2 L_{II} u_0}{3 E_I^2 I_I^2 L_{II}^2 + \left(2 A_{II} E_I E_{II} I_I L_I^3 + 6 E_I E_{II} I_I I_{II} L_I\right) L_{II} + A_{II} E_{II}^2 I_{II} L_I^4} \\[2mm]
0 \\[2mm]
\dfrac{E_I I_I \left(6 A_{II} E_I E_{II} I_I L_{II} + 12 A_{II} E_{II}^2 I_{II} L_I\right) u_0}{3 E_I^2 I_I^2 L_{II}^2 + \left(2 A_{II} E_I E_{II} I_I L_I^3 + 6 E_I E_{II} I_I I_{II} L_I\right) L_{II} + A_{II} E_{II}^2 I_{II} L_I^4} \\[2mm]
-\dfrac{E_I I_I \left(3 A_{II} E_I E_{II} I_I L_{II}^2 + 6 A_{II} E_{II}^2 I_{II} L_I L_{II}\right) u_0}{3 E_I^2 I_I^2 L_{II}^2 + \left(2 A_{II} E_I E_{II} I_I L_I^3 + 6 E_I E_{II} I_I I_{II} L_I\right) L_{II} + A_{II} E_{II}^2 I_{II} L_I^4}
\end{bmatrix}.
$$

$$\tag{2.341}$$

The constants of integration c_1 and c_2 can be re-arranged to:

$$
c_1 = \frac{-6 \left(1 + 2 \frac{E_{II} I_{II} L_I}{E_I I_I L_{II}}\right) u_0}{3 \frac{L_{II}}{A_{II} E_{II}} + 2 \frac{L_I^3}{E_I I_I} + 6 \frac{I_{II} L_I}{A_{II} E_I I_I} + \frac{E_{II} I_{II} L_I^4}{E_I^2 I_I^2 L_{II}}}
\tag{2.342}
$$

$$c_2 = \frac{6u_0}{3\frac{E_I I_I L_{II}^2}{A_{II} E_{II}^2 I_{II} L_I^2} + 2\frac{L_I L_{II}}{E_{II} I_{II}} + 6\frac{L_{II}}{A_{II} E_{II} L_I} + \frac{L_I^2}{E_I I_I}}. \tag{2.343}$$

Based on Eqs. (2.322) and (2.323) and the equilibrium between the inner reactions and the reactions at the lower support, the following relationships can be deducted:

$$F_{z_I^R} = Q_{z_I(L_I)} = -c_1 \tag{2.344}$$

$$M_{y_I^R} M_{y_I(L_I)} = -c_1 L_I - c_2, \tag{2.345}$$

which gives after some transformations the same representations as in Eqs. (2.310) and (2.311).

2.5 Supplementary Problems

2.13 Rod loaded by a single force in its middle

Given is a rod of length $2L$ and axial tensile stiffness EA which is fixed at both ends, see Fig. 2.67. A single force F_0 is acting in the middle ($X = L$) in positive direction. Determine the expression for the displacement $u_X(X)$ and the normal fore $N_X(X)$ based on the consideration of two sections or alternatively based on the application of a discontinuous function. Sketch both distributions.

2.14 Rod loaded by a distributed load

Given is a rod of length $2L$ and axial tensile stiffness EA which is fixed at both ends, see Fig. 2.68. A distributed load p_0 is acting in the range $L \leq X \leq 2L$ in positive direction. Determine the expression for the displacement $u_X(X)$ and the normal fore $N_X(X)$ based on the consideration of two sections or alternatively based on the application of a discontinuous function. Sketch both distributions.

Fig. 2.67 Rod loaded by a single force in its middle

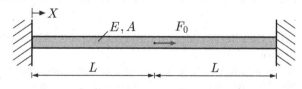

Fig. 2.68 Rod loaded by a distributed load

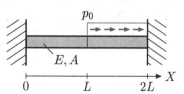

Fig. 2.69 Cantilever
TIMOSHENKO beam: **a** single
force case and **b** distributed
load case

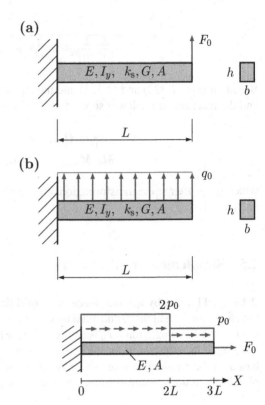

Fig. 2.70 Rod with different
sections

2.15 Cantilever beam under the influence of a point or distributed load—rectangular cross section

The cantilever TIMOSHENKO beam shown in Fig. 2.69 is either loaded by a single force F_0 at its right-hand end or by a distributed load q_0. The bending stiffness EI and the shear stiffness $k_s AG$ are constant, the total length of the beam is equal to L, and the rectangular cross section has the dimensions of $b \times h$. Determine the expressions of the bending lines ($u_z(x)$) and sketch the deflections of the right-hand end ($x = L$) as a function of the slenderness ratio $\frac{h}{L}$ for $\nu = 0.0, 0.3$, and 0.5.

2.16 Cantilever rod with different sections (Alternative solution procedure of Problem 2.3)

Given is a rod of length $3L$ and constant axial tensile stiffness EA as shown in Fig. 2.70. At the left-hand side there is a fixed support and a constant distributed load $2p_0$ is acting in the range $0 \leq x \leq 2L$ whereas a load of p_0 is acting in the range $2L \leq x \leq 3L$. Determine based on CASTIGLIANO's theorems the analytical solution for the elongation $u_x(x)$, the strain $\varepsilon_x(x)$, and the stress $\sigma_x(x)$ along the rod axis.

Fig. 2.71 Beam-like
structure

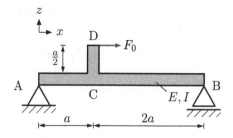

2.17 Beam-like structure: energy approach

Given is a beam-like structure of length $3a$ and bending stiffness EI which is simply supported, see Fig. 2.71. A single force F_0 is acting at a vertical extension in positive x-direction. Determine based on CASTIGLIANO's theorems the horizontal displacement of the load application point (D) and the vertical displacement of point C.

2.18 Calibration factor for the standard extensometer

The calibration factor for the standard extensometer has been obtained in Eq. (2.238) based on the simplified approach as:

$$\varepsilon_{x,\text{II}}\Big|_{z=-\frac{h_\text{II}}{2}} = \frac{3}{2} \times \frac{E_\text{I} I_\text{I} L_\text{II}}{E_\text{II} I_\text{II} L_\text{I}} \times \frac{h_\text{II}}{L_\text{I}} \times \left(+\frac{1}{2} + \frac{1}{12}\left(\frac{h_\text{II}}{L_\text{I}}\right)\right) \varepsilon_{\text{sp}}. \qquad (2.346)$$

Check if a particular design version allows to obtain a calibration factor of 1, i.e. $\varepsilon_{x,\text{II}} = 1 \times \varepsilon_{\text{sp}}$.

2.19 Elastic limit of the standard extensometer

The calibration factor for the standard extensometer has been obtained in Eq. (2.238) under the assumption of pure linear-elastic material behavior. Identify the critical location for initial yielding and derive an equation for the maximum elongation $u_{0,\text{max}}$ to avoid plastic material behavior.

2.20 Stiffening effect of the extensometer

The reaction force based on the refined energy approach was obtained in Eq. (2.247) as:

$$F_0^\text{R} = \frac{u_0}{\frac{1}{3}\frac{L_\text{I}^3}{E_\text{I} I_\text{I}} + \frac{1}{2}\frac{L_\text{I}^2 L_\text{II}}{E_\text{II} I_\text{II}} + \frac{1}{24}\frac{L_\text{II} h_\text{II}^2}{E_\text{II} I_\text{II}}}. \qquad (2.347)$$

Derive a general equation for the ratio between this reaction force (F_0^R) and the tensile force (F_{sp}) in the specimen. This ratio can be used to characterize the stiffening of the specimen by the extensometer. Calculate in addition numerical values for the following standard specimen with $A_{\text{sp}} = 40$ mm^2 and $E_{\text{sp}} = 70000$ MPa. Assume an extensometer with the dimensions $L_\text{I} = L_\text{II} = 25$ mm, $E_\text{I} = E_\text{II} = 210000$ MPa. Vary the cross-sectional dimensions of the extensometer as follows: $h_I = h_{II} = 2, 3, 4$ mm and $b_I = b_{II} = 5, 6, 7$ mm.

2.21 Calibration factor of the extensometer and corresponding strains

The calibration factor for the extensometer shown in Fig. 2.49 was obtained based on the refined energy approach in Eqs. (2.253)–(2.254) as:

$$\frac{\varepsilon_{x,II}\big|_{z=+\frac{h_{II}}{2}}}{\varepsilon_{sp}} = \frac{1}{\frac{2}{3}\frac{E_{II}I_{II}L_I}{E_I I_I L_{II}} + 1 + \frac{1}{12}\frac{h_{II}^2}{L_I^2}} \times \frac{h_{II}}{L_I} \times \left(-\frac{1}{2} + \frac{1}{12}\left(\frac{h_{II}}{L_I}\right)\right), \qquad (2.348)$$

$$\frac{\varepsilon_{x,II}\big|_{z=-\frac{h_{II}}{2}}}{\varepsilon_{sp}} = \frac{1}{\frac{2}{3}\frac{E_{II}I_{II}L_I}{E_I I_I L_{II}} + 1 + \frac{1}{12}\frac{h_{II}^2}{L_I^2}} \times \frac{h_{II}}{L_I} \times \left(+\frac{1}{2} + \frac{1}{12}\left(\frac{h_{II}}{L_I}\right)\right). \qquad (2.349)$$

Calculate the numerical values of the calibration factor for a typical extensometer with $L_{II} = 25$ mm, $L_I = 40.5$ mm, and $h_I = h_{II} = 1$ mm, see [25]. Furthermore, assume $E_I = E_{II} = 210000$ MPa and vary the cross-sectional dimensions $h_I = h_{II}$ as 1, 2, 3 mm. Calculate for a maximum measuring displacement $u_{0,\max} = 2.5$ mm the maximum strains in the member II.

2.22 Simple design of a double-cantilever displacement gage

The design of a displacement gage for crack-extension measurement is outlined in [10]. The basic configuration, i.e., a double-cantilever design, is shown in Fig. 2.72. It is suggested to use a titanium allow with a Young's modulus of $E = 103421.4$ MPa (15×10^6 psi) and an initial yield stress of $R_p = 827.37$ MPa (12×10^4 psi).[13] To guarantee an operation in the elastic range, it is suggested to limit the maximum strain to 75% of the elastic range.

Consider two different configurations with $L = 31.75$ mm and $L = 38.10$ mm (i.e., 1.25 in and 1.50 in, respectively). The strain gages are mounted as close as possible to the support and their length equals 6.35 mm (0.25 in). Derive and plot the following design curves as a function of the beam height h:

- maximum beam slope φ_{lim},
- maximum force per beam width F_{lim}/b,
- sensitivity $d\varepsilon_G/du_0$,
- useful range $2u_0$.

Finally take $L = 31.75$ mm and $h = 1.016$ mm (0.04 in) and calculate the characteristic values of the displacement gage.

---

[13]All units were converted to the SI system.

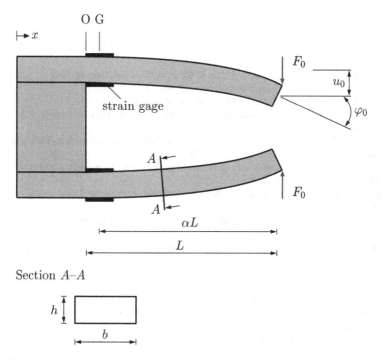

Fig. 2.72 Double-cantilever displacement gage for crack-extension measurement. Adapted from [10]

References

1. ASTM E299–12 (2013) Standard test method for linear-elastic plane-strain fracture toughness K_{Ic} of metallic materials. American Society for Testing and Materials, West Conshohocken
2. Altenbach H (2016) Holzmann/Meyer/Schumpich Technische Mechanik Festigkeitslehre. Springer Vieweg, Wiesbaden
3. Altenbach H, Öchsner A (eds) (2020) Encyclopedia of continuum mechanics. Springer, Berlin
4. Beer FP, Johnston ER Jr, DeWolf JT, Mazurek DF (2009) Mechanics of materials. McGraw-Hill, New York
5. Blaauwendraad J (2010) Plates and FEM: surprises and pitfalls. Springer, Dordrecht
6. Boresi AP, Schmidt RJ (2003) Advanced mechanics of materials. Wiley, New York
7. Budynas RG (1999) Advanced strength and applied stress analysis. McGraw-Hill Book, Singapore
8. Chen WF, Saleeb AF (1982) Constitutive equations for engineering materials. Volume 1: Elasticity and modelling. Wiley, New York
9. Cowper GR (1966) The shear coefficient in Timoshenko's beam theory. J Appl Mech 33:335–340
10. Fisher DM, Bubsey RT, Srawley JE (1966) Design and use of a displacement gage for crack extension measurements. NASA TN-D-3724. https://ntrs.nasa.gov/archive/nasa/casi.ntrs.nasa.gov/19670001426.pdf. Cited 8 Mai 2020
11. Gere JM, Timoshenko SP (1991) Mechanics of materials. PWS-KENT Publishing Company, Boston
12. Gould PL (1988) Analysis of shells and plates. Springer, New York

13. Gross D, Hauger W, Schröder J, Wall WA, Bonet J (2011) Engineering mechanics 2: mechanics of materials. Springer, Berlin
14. Gruttmann F, Wagner W (2001) Shear correction factors in Timoshenko's beam theory for arbitrary shaped cross-sections. Comput Mech 27:199–207
15. Hearn EJ (1997) Mechanics of materials 1. Butterworth-Heinemann, Oxford
16. Hibbeler RC (2008) Mechanics of materials. Prentice Hall, Singapore
17. Levinson M (1981) A new rectangular beam theory. J Sound Vib 74:81–87
18. Melosh RJ (1961) A stiffness matrix for the analysis of thin plates in bending. J Aerosp Sci 1:34–64
19. Mindlin RD (1951) Influence of rotary inertia and shear on flexural motions isotropic, elastic plates. J Appl Mech-T ASME 18:1031–1036
20. Öchsner A (2014) Elasto-plasticity of frame structure elements: modeling and simulation of rods and beams. Springer, Berlin
21. Öchsner A (2020) Partial differential equations of classical structural members: a consistent approach. Springer, Cham
22. Reddy JN (1997) Mechanics of laminated composite plates: theory and analysis. CRC Press, Boca Raton
23. Reddy JN (1997) On locking-free shear deformable beam finite elements. Comput Method Appl M 149:113–132
24. Reissner E (1945) The effect of transverse shear deformation on the bending of elastic plates. J Appl Mech-T ASME 12:A68–A77
25. Sandner Messtechnik GmbH (2020) Instruction Manual: High precision extensometer (strain gauge transducers). https://sandner-messtechnik.com/pdf/em/Bedienungsanleitung_Extensometer_E.pdf. Cited 12 Juli 2020
26. Timoshenko SP (1921) On the correction for shear of the differential equation for transverse vibrations of prismatic bars. Philos Mag 41:744–746
27. Timoshenko S (1940) Strength of materials - part I elementary theory and problems. D. Van Nostrand Company, New York
28. Timoshenko S, Woinowsky-Krieger S (1959) Theory of plates and shells. McGraw-Hill Book Company, New York
29. Timoshenko SP, Goodier JN (1970) Theory of elasticity. McGraw-Hill, New York
30. Ventsel E, Krauthammer T (2001) Thin plates and shells: theory, analysis, and applications. Marcel Dekker, New York
31. Wang CM (1995) Timoshenko beam-bending solutions in terms of Euler-Bernoulli solutions. J Eng Mech-ASCE 121:763–765
32. Wang CM, Reddy JN, Lee KH (2000) Shear deformable beams and plates: relationships with classical solution. Elsevier, Oxford
33. Weaver W Jr, Gere JM (1980) Matrix analysis of framed structures. Van Nostrand Reinhold Company, New York

Chapter 3
Finite Element Method

Abstract This chapter treats one-dimensional finite elements with two nodes. Rods for tensile deformation and thin and thick beams for bending deformation are introduced based on their elemental finite element equation and the corresponding relationships for post-processing. Both element types are superposed to the generalized beam element which can elongate and bend. In a further step, the elements are arranged in a single plane to form truss or frame structures. The provided elements are finally applied to the extensometer design problem.

3.1 General Idea of the Method

The general idea of the finite element method is illustrated in Fig. 3.1. The solution of the differential equation (see Table 2.2) of the continuum rod gives the displacement field $u_x(x)$, i.e., the displacement at any location x of the considered domain $0 \leq x \leq L$, see Fig. 3.1a.

It is easy to accept that such a detailed description of the problem is quite difficult or even impossible for complex structures. Thus, the major idea of the finite element method is to limit the description to a finite number of points, the so-called nodes, and to reduce the complexity of the problem, see Fig. 3.1b. Following this idea, the displacement is only calculated at these nodes. These nodes also define the boundary[1] of so-called elements, which subdivide the considered domain in smaller parts (so-called discretization). Furthermore, the nodal displacements are interpolated between these nodal values within an element. To distinguish the node and element numbering, we use Arabic numerals $(1, 2, \ldots)$ for the nodes and Roman numerals $(\mathrm{I}, \mathrm{II}, \ldots)$ for the elements.

The same idea is adopted for beams. The solution of the differential equation (see Table 2.4) for a thin continuum beam provides the deflection $u_z(x)$ at any location of the beam, see Fig. 3.2a. In the case of a thick beam (see Table 2.6), the rotation

---

[1]There are also more advanced elements with inner nodes. However, this is not treated here. For further details, see [7].

A. Öchsner, *A Project-Based Introduction to Computational Statics*,
https://doi.org/10.1007/978-3-030-58771-0_3

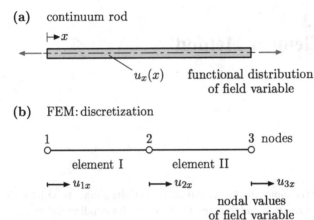

(a)   continuum rod

$\mapsto x$

$u_x(x)$ functional distribution
                  of field variable

(b)   FEM: discretization

1   2   3 nodes

element I   element II

$\longmapsto u_{1x}$ $\longmapsto u_{2x}$ $\longmapsto u_{3x}$

nodal values
of field variable

Fig. 3.1   a Continuum rod and **b** discretization with two finite elements

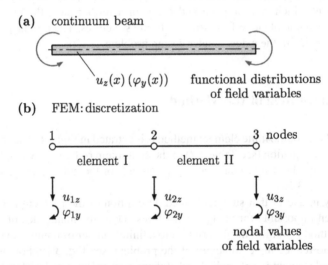

(a)   continuum beam

$u_z(x)\,(\varphi_y(x))$ functional distributions
                                of field variables

(b)   FEM: discretization

1   2   3 nodes

element I   element II

u_{1z} u_{2z} u_{3z}
φ_{1y} φ_{2y} φ_{3y}

nodal values
of field variables

Fig. 3.2   a Continuum beam and **b** discretization with two finite elements

would be a second independent field variable. For the finite element approach, each
node of a beam element[2] has two independent degrees of freedom, i.e., the deflection
u_z and the rotation φ_y.

---

[2]If we consider bending in a single plane (here: xz).

Fig. 3.3 Definition of the
one-dimensional linear rod
element: **a** deformations;
b external loads. The nodes
are symbolized by the two
circles at the ends (○)

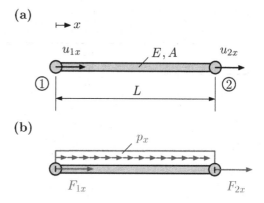

3.2   Rods and Trusses

3.2.1   Rod Elements

Let us consider in the following a rod element which is composed of two nodes as schematically shown in Fig. 3.3. Each node has only one degree of freedom, i.e., a displacement u_x in the direction of the x-axis (i.e., the direction of the principal axis, see Fig. 3.3a) and each node can be only loaded by single forces acting in x-direction (cf. Fig. 3.3b). In the case of distributed loads $p_x(x)$, a transformation to equivalent nodal loads is required.

Different methods can be found in the literature to derive the principal finite element equation (see [2, 5]). All these methods result in the same formulation, which is given in the following for constant material and geometrical properties:

$$\frac{EA}{L}\begin{bmatrix} 1 & -1 \\ -1 & 1 \end{bmatrix}\begin{bmatrix} u_{1x} \\ u_{2x} \end{bmatrix} = \begin{bmatrix} F_{1x} \\ F_{2x} \end{bmatrix} + \int_0^L \begin{bmatrix} N_1 \\ N_2 \end{bmatrix} p_x(x)\,\mathrm{d}x\,, \qquad (3.1)$$

or in abbreviated form

$$\boldsymbol{K}^{\mathrm{e}}\boldsymbol{u}_{\mathrm{p}}^{\mathrm{e}} = \boldsymbol{f}^{\mathrm{e}}\,, \qquad (3.2)$$

where $\boldsymbol{K}^{\mathrm{e}}$ is the elemental stiffness matrix, $\boldsymbol{u}_{\mathrm{p}}^{\mathrm{e}}$ is the elemental column matrix of unknowns and $\boldsymbol{f}^{\mathrm{e}}$ is the elemental column matrix of loads. The interpolation functions in Eq. (3.1) are given by $N_1(x) = 1 - \frac{x}{L}$ and $N_2(x) = \frac{x}{L}$ and Table 3.1 summarizes for some simple shapes of distributed loads the equivalent nodal loads.

Several single finite elements can be combined to form a finite element mesh and the assembly of the elemental equations result in the global system of equations, i.e.

$$\boldsymbol{K}\boldsymbol{u}_{\mathrm{p}} = \boldsymbol{f}\,, \qquad (3.3)$$

Table 3.1 Equivalent nodal loads for a linear rod element (x-axis: right facing)

Loading	Equivalent axial force
	$F_{1x} = \dfrac{pL}{2}$ $F_{2x} = \dfrac{pL}{2}$
	$F_{1x} = -\dfrac{pa^2}{2L} + pa$ $F_{2x} = \dfrac{pa^2}{2L}$
	$F_{1x} = \dfrac{pL}{6}$ $F_{2x} = \dfrac{pL}{3}$
	$F_{1x} = \dfrac{pL}{12}$ $F_{2x} = \dfrac{pL}{4}$
	$F_{1x} = \dfrac{F(L-a)}{L}$ $F_{2x} = \dfrac{Fa}{L}$

where K is the global stiffness matrix, u_{p} is the global column matrix of unknowns and f is the global column matrix of loads. The global system of equations in the form of Eq. (3.3) cannot be solved without the consideration of the support conditions (this results in the reduced system of equations). A few methods to consider different types of boundary conditions are summarized in the following:

- Homogenous DIRICHLET boundary condition $u_x = 0$
  A homogenous DIRICHLET[3] boundary condition at node n ($u_{nX} = 0$) can be considered in the non-reduced system of equations by eliminating the nth row and n^{th} column of the system, see Eq. (3.4).

---

[3] Alternatively known as 1st kind, essential, geometric or kinematic boundary condition.

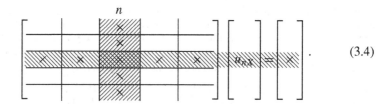

$$\tag{3.4}$$

- Non-homogeneous DIRICHLET boundary condition $u_x \neq 0$

  First possibility: A non-homogeneous DIRICHLET boundary condition ($u_{nX} = u_0 \neq 0$) at node n can be introduced in the system of equations by modifying the nth row in such a way that at the position of the nth column a '1' is obtained while all other entries of the nth row are set to zero. On the right-hand side, the given value u_0 is introduced at the nth position of the column matrix of the external loads as follows.

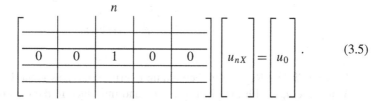

$$\tag{3.5}$$

  Second possibility: If the boundary condition is specified at node n, the nth column of the stiffness matrix is multiplied by the given value u_0. Now we bring the nth column of the stiffness matrix to the right-hand side of the system and delete the nth row of the system of equations. These steps can be identified in the following equations:

$$\tag{3.6}$$

$$\tag{3.7}$$

$$\tag{3.8}$$

Table 3.2 Post-processing of nodal values for a linear rod element (defined by element length L, cross-sectional area A, and YOUNG's modulus E). The distributions are given as being dependent on the nodal values as a function of the physical coordinate $0 \leq x \leq L$ and the natural coordinate $-1 \leq \xi \leq 1$

Axial displacement (Elongation) u_x
$u_x^e(x) = \left[1 - \frac{x}{L}\right] u_{1x} + \left[\frac{x}{L}\right] u_{2x}$
$u_x^e(\xi) = \left[\frac{1}{2}(1-\xi)\right] u_{1x} + \left[\frac{1}{2}(1+\xi)\right] u_{2x}$
Axial strain $\varepsilon_x = \dfrac{du_x}{dx} = \dfrac{d\xi}{dx}\dfrac{du_x}{d\xi}$
$\varepsilon_x^e(x) = \frac{1}{L}(u_{2x} - u_{1x})$
$\varepsilon_x^e(\xi) = \frac{1}{L}(u_{2x} - u_{1x})$
Axial stress $\sigma_x = E\varepsilon_x = E\dfrac{du_x}{dx} = E\dfrac{d\xi}{dx}\dfrac{du_x}{d\xi}$
$\sigma_x^e(x) = \frac{E}{L}(u_{2x} - u_{1x})$
$\sigma_x^e(\xi) = \frac{E}{L}(u_{2x} - u_{1x})$
Normal force $N_x = EA\varepsilon_x = EA\dfrac{du_x}{dx} = EA\dfrac{d\xi}{dx}\dfrac{du_x}{d\xi}$
$N_x^e(x) = \frac{EA}{L}(u_{2x} - u_{1x})$
$N_x^e(\xi) = \frac{EA}{L}(u_{2x} - u_{1x})$

Third possibility: Replace in the column matrix of unknowns the variable of the nodal value u_{nX} with the given value u_0 and introduce in the column matrix of the external loads at the nth position the corresponding reaction force F_{nX}^R. Split the column matrix of the external loads into a component with the given external loads and a component which contains the unknown reaction force F_{nX}^R. Now we bring the nth column of the stiffness matrix to the right-hand side of the system and the component of the load matrix with F_{nX}^R to the left-hand side:

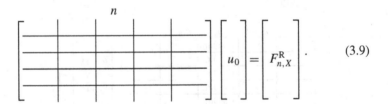

$$\begin{bmatrix} \ \end{bmatrix}\begin{bmatrix} u_0 \end{bmatrix} = \begin{bmatrix} F_{n,X}^R \end{bmatrix} \cdot \qquad (3.9)$$

- NEUMANN boundary condition $F_x = F_0$
  A NEUMANN[4] boundary condition at node n ($F_{nX} = F_0$) can be considered on the right-hand side, i.e., in the column matrix of the external loads.

Once the nodal displacements (u_{1x}, u_{2x}) are known, further quantities and their distributions can be calculated within an element (so-called post-processing), see Table 3.2. As we can see from this table, the distributions and other field variables depend only on the nodal displacement values.

---

[4]Alternatively known as 2nd kind, natural or static boundary condition.

Let us summarize here the recommended steps for a linear finite element solution ('hand calculation') [4]:

① Sketch the free-body diagram of the problem, including a global coordinate system.

② Subdivide the geometry into finite elements. Indicate the node and element numbers (the user may choose any numbering order), local coordinate systems, and equivalent nodal loads.

③ Write separately all elemental stiffness matrices expressed in the global coordinate system. Indicate for each element the nodal unknowns (degrees of freedom) on the right-hand side and over the matrix. In this step, the DOFs must be chosen according to the global coordinate system—conventionally, in the positive direction.

④ Determine the dimensions of the global stiffness matrix and sketch the structure of this matrix with global unknowns on the right-hand side and over the matrix. The dimensions of the matrix are equal to the total number of degrees of freedom which can be determined by multiplying the number of nodes by the number of degrees of freedom per node. After assembling, the validity of the assembled stiffness matrix can be tested by the following check list:

K is symmetrical,
K has only positive components on the main diagonal, and
the coupled DOFs have non-zero values as their corresponding components.

⑤ Insert step-by-step the values of the elemental stiffness matrices into the global stiffness matrix. This process is called assembling the global stiffness matrix.

⑥ Add the column matrix of unknowns and external loads to complete the global system of equations.

⑦ Introduce the boundary conditions to obtain the reduced system of equations.

⑧ Solve the reduced system of equations to obtain the unknown nodal deformations.

⑨ Post-computation or post-processing: determination of reaction forces, stresses and strains.

⑩ Check the global equilibrium between the external loads and the support reactions (optional step for checking the results).

It should be noted that some steps may be combined or omitted depending on the problem and the experience of the finite element user. The above steps can be seen as an initial structured guide to master the solution of finite element problems.

3.1 Example: Rod structure with a point load

Given is a rod structure as shown in Fig. 3.4. The structure has a uniform cross-sectional area A and YOUNG's modulus E. The structure is fixed at its left-hand end and loaded by a single force F_0.

Model the rod structure with two linear finite elements of equal length L and determine:

Fig. 3.4 Rod structure with
a point load

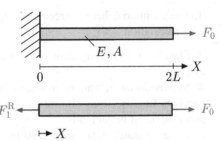

Fig. 3.5 Free-body diagram
of the rod structure with a
point load

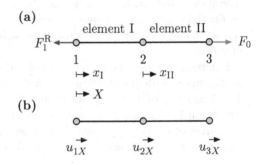

Fig. 3.6 a Free-body
diagram of the discretized
structure with point loads.
b Nodal unknowns

- the displacements at the nodes,
- the reaction force at the left-hand support,
- the strain, stress, and normal force in the elements and
- check the global force equilibrium.

3.1 Solution
The solution will follow the recommended 10 steps outlined on Sect. 3.2.

① Sketch the free-body diagram of the problem, including a global coordinate system.

Remove the support at the left-hand end and introduce the corresponding reaction
force, see Fig. 3.5. Note that the direction of the reaction force can be arbitrarily
chosen. The sign of the result will confirm ($F_1^R > 0$) or not ($F_1^R < 0$) the assumed
direction.
    The rod structure with a total length of $2L$ is divided in the middle into two
elements, see Fig. 3.6. This corresponds to step ②.
③ Write separately all elemental stiffness matrices expressed in the global coordinate
system. Indicate the nodal unknowns on the right-hand sides and over the matrices.

$$K_I^e = \frac{EA}{L}\begin{matrix} u_{1X} & u_{2X} \\ \begin{bmatrix} 1 & -1 \\ -1 & 1 \end{bmatrix} & \begin{matrix} u_{1X} \\ u_{2X} \end{matrix} \end{matrix}, \qquad (3.10)$$

$$\boldsymbol{K}_{\text{II}}^{\text{e}} = \frac{EA}{L}\begin{matrix}u_{2X} & u_{3X} \\ \begin{bmatrix} 1 & -1 \\ -1 & 1 \end{bmatrix} & \begin{matrix}u_{2X} \\ u_{3X}\end{matrix}\end{matrix}. \tag{3.11}$$

④ Determine the dimensions of the global stiffness matrix and sketch the structure of this matrix with global unknowns on the right-hand side and over the matrix.

The finite element structure is composed of 3 nodes, each having one degree of freedom (i.e., the axial displacement). Thus, the dimensions of the global stiffness matrix are $(3 \times 1) \times (3 \times 1) = (3 \times 3)$:

$$\boldsymbol{K} = \begin{matrix} u_{1X} & u_{2X} & u_{3X} \\ \begin{bmatrix} & & \\ \hline & & \\ \hline & & \end{bmatrix} & \begin{matrix} u_{1X} \\ u_{2X} \\ u_{3X} \end{matrix} \end{matrix}. \tag{3.12}$$

⑤ Insert the values of the elemental stiffness matrices step-by-step into the global stiffness matrix.

$$\boldsymbol{K} = \frac{EA}{L}\begin{matrix}u_{1X} & u_{2X} & u_{3X} \\ \begin{bmatrix} 1 & -1 & 0 \\ -1 & 2 & -1 \\ 0 & -1 & 1 \end{bmatrix} & \begin{matrix}u_{1X} \\ u_{2X} \\ u_{3X}\end{matrix}\end{matrix}. \tag{3.13}$$

⑥ Add the column matrix of unknowns and external loads to complete the global system of equations.

$$\frac{EA}{L}\begin{bmatrix} 1 & -1 & 0 \\ -1 & 2 & -1 \\ 0 & -1 & 1 \end{bmatrix}\begin{bmatrix} u_{1X} \\ u_{2X} \\ u_{3X} \end{bmatrix} = \begin{bmatrix} -F_1^{\text{R}} \\ 0 \\ F_0 \end{bmatrix}. \tag{3.14}$$

⑦ Introduce the boundary conditions to obtain the reduced system of equations.

There is no displacement possible at the left-hand end of the structure (i.e., $u_{1X} = 0$ at node 1). Thus, cancel the first row and first column from the linear system to obtain:

$$\frac{EA}{L}\begin{bmatrix} 2 & -1 \\ -1 & 1 \end{bmatrix}\begin{bmatrix} u_{2X} \\ u_{3X} \end{bmatrix} = \begin{bmatrix} 0 \\ F_0 \end{bmatrix}. \tag{3.15}$$

⑧ Solve the reduced system of equations to obtain the unknown nodal deformations.

The solution can be obtained based on the matrix approach $\boldsymbol{u}_{\text{p}} = \boldsymbol{K}^{-1}\boldsymbol{f}$:

$$\begin{bmatrix} u_{2X} \\ u_{3X} \end{bmatrix} = \frac{L}{EA} \times \frac{1}{2-1} \begin{bmatrix} +1 & +1 \\ +1 & 2 \end{bmatrix} \begin{bmatrix} 0 \\ F_0 \end{bmatrix} = \frac{F_0 L}{EA} \begin{bmatrix} 1 \\ 2 \end{bmatrix}. \tag{3.16}$$

⑨ Post-computation: determination of reaction forces, stresses and strains.

Take into account the non-reduced system of equations as given in step ⑥ under the consideration of the known nodal displacements. The first equation of this system reads:

$$\frac{EA}{L}(-u_{2X}) = -F_1^R, \tag{3.17}$$

or finally for the reaction force:

$$F_1^R = F_0. \tag{3.18}$$

The obtained positive value confirms the assumption of the selected initial direction for the reaction force.

The equations for the elemental strains, stresses, and normal forces can be extracted from Table 3.2:

$$\varepsilon_I^e = \frac{1}{L}(u_{2X} - u_{1X}) = \frac{F_0}{EA}, \tag{3.19}$$

$$\varepsilon_{II}^e = \frac{1}{L}(u_{3X} - u_{2X}) = \frac{F_0}{EA}, \tag{3.20}$$

$$\sigma_I^e = \frac{E}{L}(u_{2X} - u_{1X}) = \frac{F_0}{A}, \tag{3.21}$$

$$\sigma_{II}^e = \frac{E}{L}(u_{3X} - u_{2X}) = \frac{F_0}{A}, \tag{3.22}$$

$$N_I^e = \frac{EA}{L}(u_{2X} - u_{1X}) = F_0, \tag{3.23}$$

$$N_{II}^e = \frac{EA}{L}(u_{3X} - u_{2X}) = F_0. \tag{3.24}$$

⑩ Check the global equilibrium between the external loads and the support reactions.

$$\sum_i F_{iX} = 0 \quad \Leftrightarrow \quad \underbrace{-F_1^R}_{\text{reaction force}} + \underbrace{F_0}_{\text{external load}} = 0 \checkmark \tag{3.25}$$

Fig. 3.7 Rod structure with changing distributed load

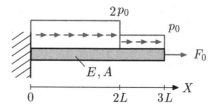

3.2 Example: Rod structure with changing distributed load

Given is a rod structure as shown in Fig. 3.7. The structure has a uniform cross-sectional area A and YOUNG's modulus E. The structure is fixed at its left-hand end and loaded by a single force F_0 at $X = 3L$ as well as

(a) a uniform distributed load $2p_0$ in the range $0 \leq X \leq 2L$, and

(b) a uniform distributed load p_0 in the range $2L \leq X \leq 3L$.

Model the rod structure with two linear finite elements and determine

- the displacements at the nodes,
- the reaction force at the left-hand support,
- the strain, stress, and normal force in each element, and
- check the global force equilibrium.

3.2 Solution

The solution will follow the recommended 10 steps outlined on Sect. 3.2.

① Sketch the free-body diagram of the problem, including a global coordinate system.

Remove the support at the left-hand end and introduce the corresponding reaction force, see Fig. 3.8. Note that the direction of the reaction force can be arbitrarily chosen. The sign of the result will confirm $(F_1^R > 0)$ or not $(F_1^R < 0)$ the assumed direction. Looking from a different angle at the problem, we can say that nodes are introduced at the locations of the single forces $(F_1^R$ and $F_0)$ and the discontinuity of the distributed load.

② Subdivide the geometry into finite elements. Indicate the node and element numbers, local coordinate systems, and equivalent nodal loads.

Fig. 3.8 Free-body diagram of the rod structure with changing distributed load

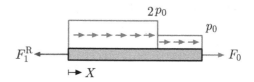

Fig. 3.9 a Free-body diagram of the discretized structure with equivalent nodal loads. **b** Nodal unknowns

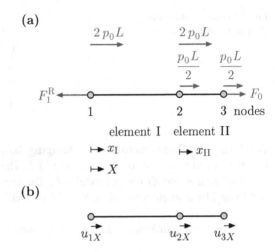

The rod structure with a total length of $3L$ is divided at the discontinuity of the distributed load (i.e., at $X = 2L$) into two elements and the corresponding equivalent nodal loads are calculated from Table 3.1, see Fig. 3.9. The two force contributions of magnitude $2p_0L$ result from the distributed load $2p_0$ and the two force contributions of magnitude $\frac{p_0L}{L}$ result from the distributed load p_0.

③ Write separately all elemental stiffness matrices expressed in the global coordinate system. Indicate the nodal unknowns on the right-hand sides and over the matrices.

$$
K_I^e = \frac{EA}{2L}\begin{matrix} u_{1X} & u_{2X} \\ \begin{bmatrix} 1 & -1 \\ -1 & 1 \end{bmatrix} & \begin{matrix} u_{1X} \\ u_{2X} \end{matrix} \end{matrix} = \frac{EA}{L}\begin{matrix} u_{1X} & u_{2X} \\ \begin{bmatrix} \frac{1}{2} & -\frac{1}{2} \\ -\frac{1}{2} & \frac{1}{2} \end{bmatrix} & \begin{matrix} u_{1X} \\ u_{2X} \end{matrix} \end{matrix},
\tag{3.26}
$$

$$
K_{II}^e = \frac{EA}{L}\begin{matrix} u_{2X} & u_{3X} \\ \begin{bmatrix} 1 & -1 \\ -1 & 1 \end{bmatrix} & \begin{matrix} u_{2X} \\ u_{3X} \end{matrix} \end{matrix}.
\tag{3.27}
$$

④ Determine the dimensions of the global stiffness matrix and sketch the structure of this matrix with global unknowns on the right-hand side and over the matrix.

The finite element structure is composed of 3 nodes, each having one degree of freedom (i.e., the axial displacement). Thus, the dimensions of the global stiffness matrix are $(3 \times 1) \times (3 \times 1) = (3 \times 3)$:

$$K = \begin{array}{c} \begin{array}{ccc} u_{1X} & u_{2X} & u_{3X} \end{array} \\ \left[\begin{array}{c|c|c} & & \\ \hline & & \\ \hline & & \end{array} \right] \begin{array}{c} u_{1X} \\ u_{2X} \\ u_{3X} \end{array} \end{array} . \tag{3.28}$$

⑤ Insert the values of the elemental stiffness matrices step-by-step into the global stiffness matrix.

$$K = \frac{EA}{L} \begin{array}{c} \begin{array}{ccc} u_{1X} & u_{2X} & u_{3X} \end{array} \\ \left[\begin{array}{ccc} \frac{1}{2} & -\frac{1}{2} & 0 \\ -\frac{1}{2} & \frac{1}{2}+1 & -1 \\ 0 & -1 & +1 \end{array} \right] \begin{array}{c} u_{1X} \\ u_{2X} \\ u_{3X} \end{array} \end{array} . \tag{3.29}$$

⑥ Add the column matrix of unknowns and external loads to complete the global system of equations.

$$\frac{EA}{L} \begin{bmatrix} \frac{1}{2} & -\frac{1}{2} & 0 \\ -\frac{1}{2} & \frac{1}{2}+1 & -1 \\ 0 & -1 & +1 \end{bmatrix} \begin{bmatrix} u_{1X} \\ u_{2X} \\ u_{3X} \end{bmatrix} = \begin{bmatrix} -F_1^R + 2p_0L \\ \frac{5}{2}p_0L \\ F_0 + \frac{p_0L}{2} \end{bmatrix} . \tag{3.30}$$

⑦ Introduce the boundary conditions to obtain the reduced system of equations.

There is no displacement possible at the left-hand end of the structure (i.e., $u_{1X} = 0$ at node 1). Thus, cancel the first row and first column from the linear system to obtain:

$$\frac{EA}{L} \begin{bmatrix} \frac{3}{2} & -1 \\ -1 & +1 \end{bmatrix} \begin{bmatrix} u_{2X} \\ u_{3X} \end{bmatrix} = \begin{bmatrix} \frac{5}{2}p_0L \\ F_0 + \frac{p_0L}{2} \end{bmatrix} . \tag{3.31}$$

⑧ Solve the reduced system of equations to obtain the unknown nodal deformations.

The solution can be obtained based on the matrix approach $u_{\mathrm{p}} = K^{-1}f$:

$$\begin{bmatrix} u_{2X} \\ u_{3X} \end{bmatrix} = \frac{L}{EA} \times \frac{1}{\frac{3}{2}-1} \begin{bmatrix} +1 & +1 \\ +1 & \frac{3}{2} \end{bmatrix} \begin{bmatrix} \frac{5}{2}p_0L \\ F_0 + \frac{p_0L}{2} \end{bmatrix} = \frac{2L}{EA} \begin{bmatrix} 3p_0L + F_0 \\ \frac{13}{4}p_0L + \frac{3}{2}F_0 \end{bmatrix} . \tag{3.32}$$

⑨ Post-computation: determination of reaction forces, stresses and strains.

Take into account the non-reduced system of equations as given in step ⑥ under the consideration of the known nodal displacements. The first equation of this system reads:

$$\frac{EA}{L} \left(\frac{1}{2}u_{1X} - \frac{1}{2}u_{2X} + 0 \right) = -F_1^R + 2p_0L , \tag{3.33}$$

or finally for the reaction force:

$$F_1^R = 5p_0L + F_0 \,. \tag{3.34}$$

The equations for the elemental strains, stresses, and normal forces can be extracted from Table 3.2:

$$\varepsilon_I^e = \frac{1}{L_I}(u_{2X} - u_{1X}) = \frac{1}{EA}(3p_0L + F_0) \,, \tag{3.35}$$

$$\varepsilon_{II}^e = \frac{1}{L_{II}}(u_{3X} - u_{2X}) = \frac{1}{EA}\left(\tfrac{1}{2}\,p_0L + F_0\right) \,, \tag{3.36}$$

$$\sigma_I^e = \frac{E}{L_I}(u_{2X} - u_{1X}) = \frac{1}{A}(3p_0L + F_0) \,, \tag{3.37}$$

$$\sigma_{II}^e = \frac{E}{L_{II}}(u_{3X} - u_{2X}) = \frac{1}{A}\left(\tfrac{1}{2}\,p_0L + F_0\right) \,, \tag{3.38}$$

$$N_I^e = \frac{EA}{L_I}(u_{2X} - u_{1X}) = 3p_0L + F_0 \,, \tag{3.39}$$

$$N_{II}^e = \frac{EA}{L_{II}}(u_{3X} - u_{2X}) = \tfrac{1}{2}\,p_0L + F_0 \,. \tag{3.40}$$

⑩ Check the global equilibrium between the external loads and the support reactions.

$$\sum_i F_{iX} = 0 \quad \Leftrightarrow \quad \underbrace{-(5p_0L + F_0)}_{\text{reaction force}} + \underbrace{F_0 + 4p_0L + p_0L}_{\text{external loads}} = 0 \ \checkmark \tag{3.41}$$

3.3 Example: Rod structure with displacement and force boundary conditions

Given is a rod structure as shown in Fig. 3.10. The structure has a uniform cross-sectional area A and YOUNG's modulus E. The structure is fixed at both ends and loaded by a single force F_0 at $X = \tfrac{3}{5}L$ as well as a displacement u_0 at $X = \tfrac{2}{5}L$.

Model the rod structure with five linear finite elements of equal length and determine

- the displacements at the nodes,
- the reaction forces at the supports,
- the strain, stress, and normal force in each element, and
- check the global force equilibrium.
- Assume now that only u_0 is given. Adjust the value of F_0 in such a way that element III is in a stress-free state.

Fig. 3.10 Rod structure with displacement and force boundary conditions

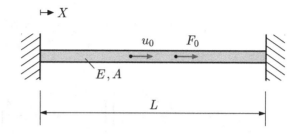

3.3 Solution
The solution will follow the recommended 10 steps outlined on Sect. 3.2.

① Sketch the free-body diagram of the problem, including a global coordinate system.

Remove the supports at both ends and introduce the corresponding reaction forces, see Fig. 3.11.

② Subdivide the geometry into finite elements. Indicate the node and element numbers, local coordinate systems, and equivalent nodal loads, see Fig. 3.12.

③ Write separately all elemental stiffness matrices expressed in the global coordinate system. Indicate the nodal unknowns on the right-hand sides and over the matrices.

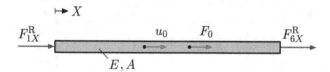

Fig. 3.11 Free-body diagram of the rod structure with displacement and force boundary condition

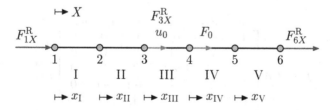

Fig. 3.12 Free-body diagram of the discretized structure

$$\boldsymbol{K}_{\mathrm{I}}^{\mathrm{e}} = \frac{EA}{\frac{L}{5}}\begin{array}{cc}u_{1X} & u_{2X}\\ \begin{bmatrix} 1 & -1\\ -1 & 1\end{bmatrix} & \begin{array}{c}u_{1X}\\ u_{2X}\end{array}\end{array}, \quad \boldsymbol{K}_{\mathrm{II}}^{\mathrm{e}} = \frac{EA}{\frac{L}{5}}\begin{array}{cc}u_{2X} & u_{3X}\\ \begin{bmatrix} 1 & -1\\ -1 & 1\end{bmatrix} & \begin{array}{c}u_{2X}\\ u_{3X}\end{array}\end{array}, \tag{3.42}$$

$$\boldsymbol{K}_{\mathrm{III}}^{\mathrm{e}} = \frac{EA}{\frac{L}{5}}\begin{array}{cc}u_{3X} & u_{4X}\\ \begin{bmatrix} 1 & -1\\ -1 & 1\end{bmatrix} & \begin{array}{c}u_{3X}\\ u_{4X}\end{array}\end{array}, \quad \boldsymbol{K}_{\mathrm{IV}}^{\mathrm{e}} = \frac{EA}{\frac{L}{5}}\begin{array}{cc}u_{4X} & u_{5X}\\ \begin{bmatrix} 1 & -1\\ -1 & 1\end{bmatrix} & \begin{array}{c}u_{4X}\\ u_{5X}\end{array}\end{array}, \tag{3.43}$$

$$\boldsymbol{K}_{\mathrm{V}}^{\mathrm{e}} = \frac{EA}{\frac{L}{5}}\begin{array}{cc}u_{5X} & u_{6X}\\ \begin{bmatrix} 1 & -1\\ -1 & 1\end{bmatrix} & \begin{array}{c}u_{5X}\\ u_{6X}\end{array}\end{array}. \tag{3.44}$$

④ Determine the dimensions of the global stiffness matrix and sketch the structure of this matrix with global unknowns on the right-hand side and over the matrix.

The finite element structure is composed of 6 nodes, each having one degree of freedom (i.e., the axial displacement). Thus, the dimensions of the global stiffness matrix are $(6 \times 1) \times (6 \times 1) = (6 \times 6)$:

$$\boldsymbol{K} = \begin{array}{cccccc}u_{1X} & u_{2X} & u_{3X} & u_{4X} & u_{5X} & u_{6X}\\ \left[\begin{array}{c|c|c|c|c|c} & & & & & \\ \hline & & & & & \\ \hline & & & & & \\ \hline & & & & & \\ \hline & & & & & \\ \hline & & & & & \end{array}\right] & \begin{array}{c}u_{1X}\\ u_{2X}\\ u_{3X}\\ u_{4X}\\ u_{5X}\\ u_{6X}\end{array}\end{array}. \tag{3.45}$$

⑤ Insert the values of the elemental stiffness matrices step-by-step into the global stiffness matrix.

$$\boldsymbol{K} = \frac{EA}{\frac{L}{5}}\begin{array}{cccccc}u_{1X} & u_{2X} & u_{3X} & u_{4X} & u_{5X} & u_{6X}\\ \begin{bmatrix} 1 & -1 & 0 & 0 & 0 & 0\\ -1 & 1+1 & -1 & 0 & 0 & 0\\ 0 & -1 & 1+1 & -1 & 0 & 0\\ 0 & 0 & -1 & 1+1 & -1 & 0\\ 0 & 0 & 0 & -1 & 1+1 & -1\\ 0 & 0 & 0 & 0 & -1 & 1\\ r & & & & & \end{bmatrix} & \begin{array}{c}u_{1X}\\ u_{2X}\\ u_{3X}\\ u_{4X}\\ u_{5X}\\ u_{6X}\end{array}\end{array}. \tag{3.46}$$

⑥ Add the column matrix of unknowns and external loads to complete the global system of equations.

$$
\frac{EA}{\frac{L}{5}}
\begin{bmatrix}
1 & -1 & 0 & 0 & 0 & 0 \\
-1 & 1+1 & -1 & 0 & 0 & 0 \\
0 & -1 & 1+1 & -1 & 0 & 0 \\
0 & 0 & -1 & 1+1 & -1 & 0 \\
0 & 0 & 0 & -1 & 1+1 & -1 \\
0 & 0 & 0 & 0 & -1 & 1
\end{bmatrix}
\begin{bmatrix}
u_{1X} \\ u_{2X} \\ u_{3X} \\ u_{4X} \\ u_{5X} \\ u_{6X}
\end{bmatrix}
=
\begin{bmatrix}
F_{1X}^{R} \\ 0 \\ F_{3X}^{R} \\ F_0 \\ 0 \\ F_{6X}^{R}
\end{bmatrix} .
\tag{3.47}
$$

⑦ and ⑧ Introduce the boundary conditions to obtain the reduced system of equations. Solve the reduced system of equations to obtain the unknown nodal deformations.

There is no displacement possible at either ends of the structure (i.e., $u_{1X} = 0$ at node 1 and $u_{6X} = 0$ at node 6). Thus, cancel the first and last rows and the first and last columns from the linear system to obtain:

$$
\frac{5EA}{L}
\begin{bmatrix}
2 & -1 & 0 & 0 \\
-1 & 2 & -1 & 0 \\
0 & -1 & 2 & -1 \\
0 & 0 & -1 & 2
\end{bmatrix}
\begin{bmatrix}
u_{2X} \\ u_{3X} \\ u_{4X} \\ u_{5X}
\end{bmatrix}
=
\begin{bmatrix}
0 \\ F_{3X}^{R} \\ F_0 \\ 0
\end{bmatrix} .
\tag{3.48}
$$

The first possibility to consider the non-homogeneous DIRICHLET boundary condition $u_{3X} = u_0$ is:

$$
\frac{5EA}{L}
\begin{bmatrix}
2 & -1 & 0 & 0 \\
0 & \frac{L}{5EA} & 0 & 0 \\
0 & -1 & 2 & -1 \\
0 & 0 & -1 & 2
\end{bmatrix}
\begin{bmatrix}
u_{2X} \\ u_{3X} \\ u_{4X} \\ u_{5X}
\end{bmatrix}
=
\begin{bmatrix}
0 \\ u_0 \\ F_0 \\ 0
\end{bmatrix} .
\tag{3.49}
$$

The solution can be obtained based on the matrix approach $u_{\mathrm{p}} = K^{-1} f$:

$$
\begin{bmatrix}
u_{2X} \\ u_{3X} \\ u_{4X} \\ u_{5X}
\end{bmatrix}
=
\begin{bmatrix}
\frac{u_0}{2} \\
u_0 \\
\frac{10 E A u_0 + 2 F_0 L}{15 E A} \\
\frac{5 E A u_0 + F_0 L}{15 E A}
\end{bmatrix} .
\tag{3.50}
$$

The second possibility to consider the non-homogeneous DIRICHLET boundary condition $u_{3X} = u_0$ is to multiply the 2nd column of the stiffness matrix with the given value u_0:

$$
\frac{5EA}{L}
\begin{bmatrix}
2 & -1 \times u_0 & 0 & 0 \\
-1 & 2 \times u_0 & -1 & 0 \\
0 & -1 \times u_0 & 2 & -1 \\
0 & 0 \times u_0 & -1 & 2
\end{bmatrix}
\begin{bmatrix}
u_{2X} \\ u_{3X} \\ u_{4X} \\ u_{5X}
\end{bmatrix}
=
\begin{bmatrix}
0 \\ F_{3X}^{R} \\ F_0 \\ 0
\end{bmatrix} .
\tag{3.51}
$$

Bring the second column to the right-hand side of the system of equations:

$$\frac{5EA}{L} \begin{bmatrix} 2 & 0 & 0 \\ -1 & -1 & 0 \\ 0 & 2 & -1 \\ 0 & -1 & 2 \end{bmatrix} \begin{bmatrix} u_{2X} \\ u_{3X} \\ u_{4X} \\ u_{5X} \end{bmatrix} = \begin{bmatrix} 0 \\ F_{3X}^R \\ F_0 \\ 0 \end{bmatrix} - \frac{5EA}{L} \begin{bmatrix} -u_0 \\ 2u_0 \\ -u_0 \\ 0 \end{bmatrix}. \tag{3.52}$$

Now, let us cancel the second row of the system:

$$\frac{5EA}{L} \begin{bmatrix} 2 & 0 & 0 \\ 0 & 2 & -1 \\ 0 & -1 & 2 \end{bmatrix} \begin{bmatrix} u_{2X} \\ u_{4X} \\ u_{5X} \end{bmatrix} = \begin{bmatrix} 0 \\ F_0 \\ 0 \end{bmatrix} - \frac{5EA}{L} \begin{bmatrix} -u_0 \\ -u_0 \\ 0 \end{bmatrix}. \tag{3.53}$$

The solution can be obtained based on the matrix approach $\boldsymbol{u}_\mathrm{p} = \boldsymbol{K}^{-1}\boldsymbol{f}$:

$$\begin{bmatrix} u_{2X} \\ u_{4X} \\ u_{5X} \end{bmatrix} = \begin{bmatrix} \frac{u_0}{2} \\ \frac{10EAu_0 + 2F_0L}{15EA} \\ \frac{5EAu_0 + F_0L}{15EA} \end{bmatrix}. \tag{3.54}$$

The third possibility to consider the non-homogeneous DIRICHLET boundary condition $u_{3X} = u_0$ is to introduce the prescribed u_0 in the column matrix of unknowns:

$$\frac{5EA}{L} \begin{bmatrix} 2 & -1 & 0 & 0 \\ -1 & 2 & -1 & 0 \\ 0 & -1 & 2 & -1 \\ 0 & 0 & -1 & 2 \end{bmatrix} \begin{bmatrix} u_{2X} \\ u_0 \\ u_{4X} \\ u_{5X} \end{bmatrix} = \begin{bmatrix} 0 \\ R_{3X} \\ F_0 \\ 0 \end{bmatrix}. \tag{3.55}$$

The column matrix of the nodal displacements $\boldsymbol{u}_\mathrm{p}$ contains now unknown quantities (u_{2X}, u_{4X}, u_{5X}) and the given nodal boundary condition (u_0). On the other hand, the right-hand side contains the unknown reaction force F_{3X}^R. Thus, the structure of the linear system of equations is unfavorable for the solution. To rearrange the system to the classical structure where all unknowns are collected on the left and given quantities on the right-hand side, the following steps can be applied:

Let us first split the right-hand side in known and unknowns quantities:

$$\frac{5EA}{L} \begin{bmatrix} 2 & -1 & 0 & 0 \\ -1 & 2 & -1 & 0 \\ 0 & -1 & 2 & -1 \\ 0 & 0 & -1 & 2 \end{bmatrix} \begin{bmatrix} u_{2X} \\ u_0 \\ u_{4X} \\ u_{5X} \end{bmatrix} = \begin{bmatrix} 0 \\ 0 \\ F_0 \\ 0 \end{bmatrix} + \begin{bmatrix} 0 \\ F_{3X}^R \\ 0 \\ 0 \end{bmatrix}. \tag{3.56}$$

Let us now multiply the second column of the stiffness matrix with the given value u_0:

$$\frac{5EA}{L} \begin{bmatrix} 2 & -1 \times u_0 & 0 & 0 \\ -1 & 2 \times u_0 & -1 & 0 \\ 0 & -1 \times u_0 & 2 & -1 \\ 0 & 0 \times u_0 & -1 & 2 \end{bmatrix} \begin{bmatrix} u_{2X} \\ u_0 \\ u_{4X} \\ u_{5X} \end{bmatrix} = \begin{bmatrix} 0 \\ 0 \\ F_0 \\ 0 \end{bmatrix} + \begin{bmatrix} 0 \\ F_{3X}^R \\ 0 \\ 0 \end{bmatrix}. \tag{3.57}$$

The final step is to bring the second column of the stiffness matrix to the right-hand side of the system (known values) and the known column matrix with F_{3X}^R into the stiffness matrix:

$$\frac{5EA}{L}\begin{bmatrix} 2 & 0 & 0 & 0 \\ -1 & -\frac{L}{5EA} & -1 & 0 \\ 0 & 0 & 2 & -1 \\ 0 & 0 & -1 & 2 \end{bmatrix}\begin{bmatrix} u_{2X} \\ F_{3X}^R \\ u_{4X} \\ u_{5X} \end{bmatrix} = \begin{bmatrix} \frac{5EAu_0}{L} \\ -\frac{10EAu_0}{L} \\ F_0 + \frac{5EAu_0}{L} \\ 0 \end{bmatrix}. \tag{3.58}$$

Now we can obtain the solution via the classical matrix approach $u_p = K^{-1}f$:

$$\begin{bmatrix} u_{2X} \\ F_{3X}^R \\ u_{4X} \\ u_{5X} \end{bmatrix} = \begin{bmatrix} \frac{u_0}{2} \\ \frac{25EAu_0 - 4F_0L}{6L} \\ \frac{10EAu_0 + 2F_0L}{15EA} \\ \frac{5EAu_0 + F_0L}{15EA} \end{bmatrix}. \tag{3.59}$$

⑨ Post-computation: determination of reaction forces, stresses and strains.

Take into account the non-reduced system of equations as given in step ⑥ under the consideration of the known nodal displacements. The first equation of this system reads:

$$\frac{5EA}{L}(u_{1X} - u_{2X}) = F_{1X}^R \quad \Rightarrow \quad F_{1X}^R = -\frac{5EAu_0}{2L}. \tag{3.60}$$

In a similar way, we obtain from the other equations:

$$F_{3X}^R = \frac{25EAu_0 - 4F_0L}{6L}, \quad F_{6X}^R = -\frac{5EAu_0 + F_0L}{3L}. \tag{3.61}$$

The equations for the elemental strains, stresses, and normal forces can be extracted from Table 3.2:

$$\varepsilon_I^e = \frac{1}{L_I}(u_{2X} - u_{1X}) = \frac{5u_0}{2L}, \tag{3.62}$$

$$\varepsilon_{II}^e = \frac{1}{L_{II}}(u_0 - u_{2X}) = \frac{5u_0}{2L}, \tag{3.63}$$

$$\varepsilon_{III}^e = \frac{1}{L_{III}}(u_{4X} - u_0) = -\frac{5EAu_0 - 2F_0L}{3EAL}, \tag{3.64}$$

$$\varepsilon_{IV}^e = \frac{1}{L_{IV}}(u_{5X} - u_{4X}) = -\frac{5EAu_0 + F_0L}{3EAL}, \tag{3.65}$$

$$\varepsilon_V^e = \frac{1}{L_V}(u_{6X} - u_{5X}) = -\frac{5EAu_0 + F_0L}{3EAL}, \tag{3.66}$$

$$\sigma_{\mathrm{I}}^{\mathrm{e}} = \frac{E}{L_{\mathrm{I}}}(u_{2X} - u_{1X}) = \frac{5Eu_0}{2L}, \tag{3.67}$$

$$\sigma_{\mathrm{II}}^{\mathrm{e}} = \frac{E}{L_{\mathrm{II}}}(u_0 - u_{2X}) = \frac{5Eu_0}{2L}, \tag{3.68}$$

$$\sigma_{\mathrm{III}}^{\mathrm{e}} = \frac{E}{L_{\mathrm{III}}}(u_{4X} - u_0) = -\frac{5EAu_0 - 2F_0L}{3AL}, \tag{3.69}$$

$$\sigma_{\mathrm{IV}}^{\mathrm{e}} = \frac{E}{L_{\mathrm{IV}}}(u_{5X} - u_{4X}) = -\frac{5EAu_0 + F_0L}{3AL}, \tag{3.70}$$

$$\sigma_{\mathrm{V}}^{\mathrm{e}} = \frac{E}{L_{\mathrm{V}}}(u_{6X} - u_{5X}) = -\frac{5EAu_0 + F_0L}{3AL}, \tag{3.71}$$

$$N_{\mathrm{I}}^{\mathrm{e}} = \frac{EA}{L_{\mathrm{I}}}(u_{2X} - u_{1X}) = \frac{5EAu_0}{2L}, \tag{3.72}$$

$$N_{\mathrm{II}}^{\mathrm{e}} = \frac{EA}{L_{\mathrm{II}}}(u_0 - u_{2X}) = \frac{5EAu_0}{2L}, \tag{3.73}$$

$$N_{\mathrm{III}}^{\mathrm{e}} = \frac{EA}{L_{\mathrm{III}}}(u_{4X} - u_0) = -\frac{5EAu_0 - 2F_0L}{3L}, \tag{3.74}$$

$$N_{\mathrm{IV}}^{\mathrm{e}} = \frac{EA}{L_{\mathrm{IV}}}(u_{5X} - u_{4X}) = -\frac{5EAu_0 + F_0L}{3L}, \tag{3.75}$$

$$N_{\mathrm{V}}^{\mathrm{e}} = \frac{EA}{L_{\mathrm{V}}}(u_{6X} - u_{5X}) = -\frac{5EAu_0 + F_0L}{3L}. \tag{3.76}$$

⑩ Check the global equilibrium between the external loads and the support reactions.

$$\sum_i F_{iX} = 0 \quad \Leftrightarrow \quad \underbrace{(F_{1X}^{\mathrm{R}} + F_{3X}^{\mathrm{R}} + F_{6X}^{\mathrm{R}})}_{\text{reaction forces}} + \underbrace{F_0}_{\text{external load}} = 0. \checkmark \tag{3.77}$$

Additional question: Assume now that only u_0 is given. Adjust the value of F_0 in such a way that element III is in a stress-free state. From ⑨ we can get the stress in element 3:

$$\sigma_{\mathrm{III}}^{\mathrm{e}} = -\frac{5AEu_0 - 2F_0L}{AL} \overset{!}{=} 0. \tag{3.78}$$

From the last equation we can conclude: $F_0 = \dfrac{5AEu_0}{2L}$.

3.2.2 Truss Structures

Let us consider a rod element which can deform in the global X-Z plane. The local x-coordinate is rotated by an angle α against the global coordinate system (X, Z), see Fig. 3.13. If the rotation of the global coordinate system to the local coordinate system is clockwise, a positive rotational angle is obtained.

Each node has now in the global coordinate system two degrees of freedom, i.e., a displacement in the X- and a displacement in the Z-direction. These two global displacements at each node can be used to calculate the displacement in the direction of the rod axis, i.e., in the direction of the local x-axis. The transformation of components of the principal finite element equation between the elemental and global coordinate system in summarized in Table 3.3 whereas the transformation matrix T is given by

$$T = \begin{bmatrix} \cos\alpha & -\sin\alpha & 0 & 0 \\ 0 & 0 & \cos\alpha & -\sin\alpha \end{bmatrix}. \tag{3.79}$$

The triple matrix product for the stiffness matrix results in the following formulation for a rotated rod element:

$$\frac{EA}{L} \begin{bmatrix} \cos^2\alpha & -\cos\alpha\sin\alpha & -\cos^2\alpha & \cos\alpha\sin\alpha \\ -\cos\alpha\sin\alpha & \sin^2\alpha & \cos\alpha\sin\alpha & -\sin^2\alpha \\ -\cos^2\alpha & \cos\alpha\sin\alpha & \cos^2\alpha & -\cos\alpha\sin\alpha \\ \cos\alpha\sin\alpha & -\sin^2\alpha & -\cos\alpha\sin\alpha & \sin^2\alpha \end{bmatrix} \begin{bmatrix} u_{1X} \\ u_{1Z} \\ u_{2X} \\ u_{2Z} \end{bmatrix} = \begin{bmatrix} F_{1X} \\ F_{1Z} \\ F_{2X} \\ F_{2Z} \end{bmatrix}. \tag{3.80}$$

To simplify the solution of simple truss structures, Table 3.4 collects expressions for the global stiffness matrix for some common angles α.

Fig. 3.13 Rotational transformation of a rod element in the X-Z plane

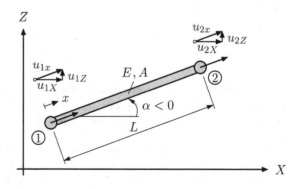

Table 3.3 Transformation of matrices between the elemental (x, z) and global coordinate (X, Z) system

Stiffness matrix	
$K^e_{xz} = T K^e_{XZ} T^T,$	$K^e_{XZ} = T^T K^e_{xz} T$
Column matrix of nodal unknowns	
$u^e_{xz} = T u^e_{XZ},$	$u^e_{XZ} = T^T u^e_{xz}$
Column matrix of external loads	
$f^e_{xz} = T f^e_{XZ},$	$f^e_{XZ} = T^T f^e_{xz}$

Table 3.4 Elemental stiffness matrices for truss elements given for different rotation angles α, cf. Eq. (3.80)

$0°$	$180°$
$\dfrac{EA}{L}\begin{bmatrix} 1 & 0 & -1 & 0 \\ 0 & 0 & 0 & 0 \\ -1 & 0 & 1 & 0 \\ 0 & 0 & 0 & 0 \end{bmatrix}$	$\dfrac{EA}{L}\begin{bmatrix} 1 & 0 & -1 & 0 \\ 0 & 0 & 0 & 0 \\ -1 & 0 & 1 & 0 \\ 0 & 0 & 0 & 0 \end{bmatrix}$
$-30°$	$30°$
$\dfrac{EA}{L}\begin{bmatrix} \frac{3}{4} & \frac{1}{4}\sqrt{3} & -\frac{3}{4} & -\frac{1}{4}\sqrt{3} \\ \frac{1}{4}\sqrt{3} & \frac{1}{4} & -\frac{1}{4}\sqrt{3} & -\frac{1}{4} \\ -\frac{3}{4} & -\frac{1}{4}\sqrt{3} & \frac{3}{4} & \frac{1}{4}\sqrt{3} \\ -\frac{1}{4}\sqrt{3} & -\frac{1}{4} & \frac{1}{4}\sqrt{3} & \frac{1}{4} \end{bmatrix}$	$\dfrac{EA}{L}\begin{bmatrix} \frac{3}{4} & -\frac{1}{4}\sqrt{3} & -\frac{3}{4} & \frac{1}{4}\sqrt{3} \\ -\frac{1}{4}\sqrt{3} & \frac{1}{4} & \frac{1}{4}\sqrt{3} & -\frac{1}{4} \\ -\frac{3}{4} & \frac{1}{4}\sqrt{3} & \frac{3}{4} & -\frac{1}{4}\sqrt{3} \\ \frac{1}{4}\sqrt{3} & -\frac{1}{4} & -\frac{1}{4}\sqrt{3} & \frac{1}{4} \end{bmatrix}$
$-45°$	$45°$
$\dfrac{EA}{L}\begin{bmatrix} \frac{1}{2} & \frac{1}{2} & -\frac{1}{2} & -\frac{1}{2} \\ \frac{1}{2} & \frac{1}{2} & -\frac{1}{2} & -\frac{1}{2} \\ -\frac{1}{2} & -\frac{1}{2} & \frac{1}{2} & \frac{1}{2} \\ -\frac{1}{2} & -\frac{1}{2} & \frac{1}{2} & \frac{1}{2} \end{bmatrix}$	$\dfrac{EA}{L}\begin{bmatrix} \frac{1}{2} & -\frac{1}{2} & -\frac{1}{2} & \frac{1}{2} \\ -\frac{1}{2} & \frac{1}{2} & \frac{1}{2} & -\frac{1}{2} \\ -\frac{1}{2} & \frac{1}{2} & \frac{1}{2} & -\frac{1}{2} \\ \frac{1}{2} & -\frac{1}{2} & -\frac{1}{2} & \frac{1}{2} \end{bmatrix}$
$-90°$	$90°$
$\dfrac{EA}{L}\begin{bmatrix} 0 & 0 & 0 & 0 \\ 0 & 1 & 0 & -1 \\ 0 & 0 & 0 & 0 \\ 0 & -1 & 0 & 1 \end{bmatrix}$	$\dfrac{EA}{L}\begin{bmatrix} 0 & 0 & 0 & 0 \\ 0 & 1 & 0 & -1 \\ 0 & 0 & 0 & 0 \\ 0 & -1 & 0 & 1 \end{bmatrix}$

The results for the transformation of matrices given in Table 3.3 can be combined with the relationships for post-processing of nodal values in Table 3.2 to express the distributions in global coordinates, see Table 3.5.

3.4 Example: Simple truss structure with two members

Given is a plane truss structure as shown in Fig. 3.14. Both members have a uniform cross-sectional area A and YOUNG's modulus E. The length of the members can be calculated from the given values (horizontal and vertical length a) in the figure. The structure is supported at its lower end and loaded by the single force F_0 at the top of

Table 3.5 Post-processing of nodal values in *global coordinates* for a linear rod element (defined by element length L, cross-sectional area A, and YOUNG's modulus E)

Axial displacement (Elongation) u_x
$u_x^e(x) = \left[1 - \frac{x}{L}\right](\cos(\alpha)u_{1X} - \sin(\alpha)u_{1Z}) + \left[\frac{x}{L}\right](\cos(\alpha)u_{2X} - \sin(\alpha)u_{2Z})$
$u_x^e(\xi) = \left[\frac{1}{2}(1 - \xi)\right](\cos(\alpha)u_{1X} - \sin(\alpha)u_{1Z}) + \left[\frac{1}{2}(1 + \xi)\right](\cos(\alpha)u_{2X} - \sin(\alpha)u_{2Z})$

Axial strain ε_x
$\varepsilon_x^e(x) = \frac{1}{L}\left((\cos(\alpha)u_{2X} - \sin(\alpha)u_{2Z}) - (\cos(\alpha)u_{1X} - \sin(\alpha)u_{1Z})\right)$
$\varepsilon_x^e(\xi) = \frac{1}{L}\left((\cos(\alpha)u_{2X} - \sin(\alpha)u_{2Z}) - (\cos(\alpha)u_{1X} - \sin(\alpha)u_{1Z})\right)$

Axial stress σ_x
$\sigma_x^e(x) = \frac{E}{L}\left((\cos(\alpha)u_{2X} - \sin(\alpha)u_{2Z}) - (\cos(\alpha)u_{1X} - \sin(\alpha)u_{1Z})\right)$
$\sigma_x^e(\xi) = \frac{E}{L}\left((\cos(\alpha)u_{2X} - \sin(\alpha)u_{2Z}) - (\cos(\alpha)u_{1X} - \sin(\alpha)u_{1Z})\right)$

Normal force N_x
$N_x^e(x) = \frac{EA}{L}\left((\cos(\alpha)u_{2X} - \sin(\alpha)u_{2Z}) - (\cos(\alpha)u_{1X} - \sin(\alpha)u_{1Z})\right)$
$N_x^e(\xi) = \frac{EA}{L}\left((\cos(\alpha)u_{2X} - \sin(\alpha)u_{2Z}) - (\cos(\alpha)u_{1X} - \sin(\alpha)u_{1Z})\right)$

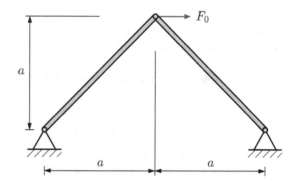

Fig. 3.14 Simple truss structure composed of two straight inclined members

the structure.

Model the truss structure with two linear finite elements and determine

- the displacement of the free node,
- the reaction forces at the supports,
- the strain, stress, and normal force in each element, and
- check the global force equilibrium.

3.4 Solution

① and ② Sketch the free-body diagram of the problem, including a global coordinate system. Subdivide the geometry into finite elements. Indicate the node and element numbers, local coordinate systems, and equivalent nodal loads, see Fig. 3.15.

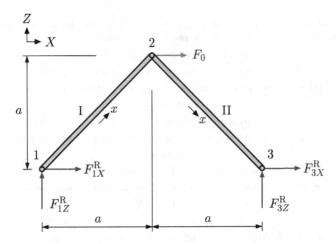

Fig. 3.15 Free-body diagram of the truss structure composed of two straight inclined members

③ Write separately all elemental stiffness matrices expressed in the global coordinate system. Indicate the nodal unknowns on the right-hand sides and over the matrices.

Element I is rotated by an angle of $\alpha = -45°$:

$$
\boldsymbol{K}_{\mathrm{I}}^{\mathrm{e}} = \frac{EA}{\sqrt{2}a}
\begin{matrix}
& \overset{u_{1X}}{} & \overset{u_{1Z}}{} & \overset{u_{2X}}{} & \overset{u_{2Z}}{} & \\
\left[\begin{matrix}
\frac{1}{2} & \frac{1}{2} & -\frac{1}{2} & -\frac{1}{2} \\
\frac{1}{2} & \frac{1}{2} & -\frac{1}{2} & -\frac{1}{2} \\
-\frac{1}{2} & -\frac{1}{2} & \frac{1}{2} & \frac{1}{2} \\
-\frac{1}{2} & -\frac{1}{2} & \frac{1}{2} & \frac{1}{2}
\end{matrix}\right] &
\begin{matrix}
u_{1X} \\ u_{1Z} \\ u_{2X} \\ u_{2Z}
\end{matrix}
\end{matrix} ,
\tag{3.81}
$$

Element II is rotated by an angle of $\alpha = +45°$:

$$
\boldsymbol{K}_{\mathrm{II}}^{\mathrm{e}} = \frac{EA}{\sqrt{2}a}
\begin{matrix}
& \overset{u_{2X}}{} & \overset{u_{2Z}}{} & \overset{u_{3X}}{} & \overset{u_{3Z}}{} & \\
\left[\begin{matrix}
\frac{1}{2} & -\frac{1}{2} & -\frac{1}{2} & \frac{1}{2} \\
-\frac{1}{2} & \frac{1}{2} & \frac{1}{2} & -\frac{1}{2} \\
-\frac{1}{2} & \frac{1}{2} & \frac{1}{2} & -\frac{1}{2} \\
\frac{1}{2} & -\frac{1}{2} & -\frac{1}{2} & \frac{1}{2}
\end{matrix}\right] &
\begin{matrix}
u_{2X} \\ u_{2Z} \\ u_{3X} \\ u_{3Z}
\end{matrix}
\end{matrix} .
\tag{3.82}
$$

④ Determine the dimensions of the global stiffness matrix and sketch the structure of this matrix with global unknowns on the right-hand side and over the matrix.

The finite element structure is composed of 3 nodes, each having two degrees of freedom (i.e., the horizontal and vertical displacements). Thus, the dimensions of the global stiffness matrix are $(3 \times 2) \times (3 \times 2) = (6 \times 6)$:

$$
K = \begin{array}{c}
\begin{array}{cccccc} u_{1X} & u_{1Z} & u_{2X} & u_{2Z} & u_{3X} & u_{3Z} \end{array} \\
\left[\begin{array}{cccccc}
 & & & & & \\
 & & & & & \\
 & & & & & \\
 & & & & & \\
 & & & & & \\
 & & & & &
\end{array} \right]
\begin{array}{c} u_{1X} \\ u_{1Z} \\ u_{2X} \\ u_{2Z} \\ u_{3X} \\ u_{3Z} \end{array}
\end{array} . \tag{3.83}
$$

⑤ Insert the values of the elemental stiffness matrices step-by-step into the global stiffness matrix.

$$
K = \frac{EA}{\sqrt{2}a} \begin{array}{c}
\begin{array}{cccccc} u_{1X} & u_{1Z} & u_{2X} & u_{2Z} & u_{3X} & u_{3Z} \end{array} \\
\left[\begin{array}{cccccc}
\frac{1}{2} & \frac{1}{2} & -\frac{1}{2} & -\frac{1}{2} & 0 & 0 \\
\frac{1}{2} & \frac{1}{2} & -\frac{1}{2} & -\frac{1}{2} & 0 & 0 \\
-\frac{1}{2} & -\frac{1}{2} & \frac{1}{2}+\frac{1}{2} & \frac{1}{2}-\frac{1}{2} & -\frac{1}{2} & \frac{1}{2} \\
-\frac{1}{2} & -\frac{1}{2} & \frac{1}{2}-\frac{1}{2} & \frac{1}{2}+\frac{1}{2} & \frac{1}{2} & -\frac{1}{2} \\
0 & 0 & -\frac{1}{2} & \frac{1}{2} & \frac{1}{2} & -\frac{1}{2} \\
0 & 0 & \frac{1}{2} & -\frac{1}{2} & -\frac{1}{2} & \frac{1}{2}
\end{array} \right]
\begin{array}{c} u_{1X} \\ u_{1Z} \\ u_{2X} \\ u_{2Z} \\ u_{3X} \\ u_{3Z} \end{array}
\end{array} . \tag{3.84}
$$

⑥ Add the column matrix of unknowns and external loads to complete the global system of equations.

$$
\frac{EA}{\sqrt{2}a} \begin{bmatrix}
\frac{1}{2} & \frac{1}{2} & -\frac{1}{2} & -\frac{1}{2} & 0 & 0 \\
\frac{1}{2} & \frac{1}{2} & -\frac{1}{2} & -\frac{1}{2} & 0 & 0 \\
-\frac{1}{2} & -\frac{1}{2} & \frac{2}{2} & 0 & -\frac{1}{2} & \frac{1}{2} \\
-\frac{1}{2} & -\frac{1}{2} & 0 & \frac{2}{2} & \frac{1}{2} & -\frac{1}{2} \\
0 & 0 & -\frac{1}{2} & \frac{1}{2} & \frac{1}{2} & -\frac{1}{2} \\
0 & 0 & \frac{1}{2} & -\frac{1}{2} & -\frac{1}{2} & \frac{1}{2}
\end{bmatrix}
\begin{bmatrix} u_{1X} \\ u_{1Z} \\ u_{2X} \\ u_{2Z} \\ u_{3X} \\ u_{3Z} \end{bmatrix}
=
\begin{bmatrix} F_{1X}^{R} \\ F_{1Z}^{R} \\ F_0 \\ 0 \\ F_{3X}^{R} \\ F_{3Z}^{R} \end{bmatrix} . \tag{3.85}
$$

⑦ Introduce the boundary conditions to obtain the reduced system of equations.

There is no displacement possible at the lower left-hand and lower right-hand of the structure (i.e., $u_{1X} = u_{1Z} = 0$ at node 1 and $u_{3X} = u_{3Z} = 0$ at node 3). Thus, cancel the first two and last two columns and rows from the linear system to obtain:

$$
\frac{EA}{\sqrt{2}a} \begin{bmatrix} 1 & 0 \\ 0 & 1 \end{bmatrix} \begin{bmatrix} u_{2X} \\ u_{2Z} \end{bmatrix} = \begin{bmatrix} F_0 \\ 0 \end{bmatrix} . \tag{3.86}
$$

⑧ Solve the reduced system of equations to obtain the unknown nodal deformations.

The solution can be obtained based on the matrix approach $u_p = K^{-1} f$:

$$
\begin{bmatrix} u_{2X} \\ u_{2Z} \end{bmatrix} = \frac{\sqrt{2}a}{EA} \frac{1}{1-0} \begin{bmatrix} 1 & 0 \\ 0 & 1 \end{bmatrix} \begin{bmatrix} F_0 \\ 0 \end{bmatrix} = \frac{\sqrt{2}a F_0}{EA} \begin{bmatrix} 1 \\ 0 \end{bmatrix} . \tag{3.87}
$$

⑨ Post-computation: determination of reaction forces, stresses and strains.

Take into account the non-reduced system of equations as given in step ⑥ under the consideration of the known nodal displacements. The first equation of this system reads:

$$\frac{EA}{\sqrt{2}a}\left(-\frac{1}{2}u_{2X}\right) = F_{1X}^{R} \quad \Rightarrow \quad F_{1X}^{R} = -\frac{F}{2}. \tag{3.88}$$

In a similar way, we obtain from the other equations:

$$\frac{EA}{\sqrt{2}a}\left(-\frac{1}{2}u_{2X}\right) = F_{1Z}^{R} \quad \Rightarrow \quad F_{1Z}^{R} = -\frac{F}{2}, \tag{3.89}$$

$$\frac{EA}{\sqrt{2}a}\left(-\frac{1}{2}u_{2X}\right) = F_{3X}^{R} \quad \Rightarrow \quad F_{3X}^{R} = -\frac{F}{2}, \tag{3.90}$$

$$\frac{EA}{\sqrt{2}a}\left(\frac{1}{2}u_{2X}\right) = F_{3Z}^{R} \quad \Rightarrow \quad F_{3Z}^{R} = \frac{F}{2}. \tag{3.91}$$

The elemental stresses can be obtained from the displacements of the start ('s') and end ('e') node as:

$$\sigma = \frac{E}{\sqrt{2}a}(-\cos(\alpha)u_{sX} + \sin(\alpha)u_{sZ} + \cos(\alpha)u_{eX} - \sin(\alpha)u_{eZ}) . \tag{3.92}$$

In the case of element I ($\alpha_{I} = -45°$), we should consider $u_{1X} = u_{1Z} = u_{2Z} = 0$ to obtain:

$$\sigma_{I} = \frac{E}{\sqrt{2}a}\cos(\alpha_{I})u_{2X} = \frac{F_0}{\sqrt{2}A}. \tag{3.93}$$

Similarly, $u_{1X} = u_{1Z} = u_{2Z} = 0$ for element II ($\alpha_{II} = +45°$):

$$\sigma_{II} = -\frac{E}{\sqrt{2}a}\cos(\alpha_{II})u_{2X} = -\frac{F_0}{\sqrt{2}A}. \tag{3.94}$$

Application of HOOKE's law, i.e., $\sigma = E\varepsilon$, allows the calculation of the elemental strains:

$$\varepsilon_I = \frac{\sigma_I}{E} = +\frac{F_0}{\sqrt{2} E A}, \tag{3.95}$$

$$\varepsilon_{II} = \frac{\sigma_{II}}{E} = -\frac{F_0}{\sqrt{2} E A}. \tag{3.96}$$

The normal forces can be obtained from the normal stresses in each element:

$$N_I = \sigma_I A = \frac{F_0}{\sqrt{2}}, \tag{3.97}$$

$$N_{II} = \sigma_{II} A = -\frac{F_0}{\sqrt{2}}. \tag{3.98}$$

⑩ Check the global equilibrium between the external loads and the support reactions.

$$\sum_i F_{iX} = 0 \quad \Leftrightarrow \quad \underbrace{(F_{1X}^R + F_{3X}^R)}_{\text{reaction force}} + \underbrace{F_0}_{\text{external loads}} = 0, \ \checkmark \tag{3.99}$$

$$\sum_i F_{iZ} = 0 \quad \Leftrightarrow \quad \underbrace{(F_{1Z}^R + F_{3Z}^R)}_{\text{reaction force}} + \underbrace{0}_{\text{external load}} = 0. \ \checkmark \tag{3.100}$$

3.5 Example: Approximation of a solid using a truss structure

Given is an isotropic and homogeneous solid as shown in Fig. 3.16a. This solid should be modeled with the plane truss structure shown Fig. 3.16b. The six truss members have a uniform cross-sectional area A and YOUNG's modulus E. The length of each member can be taken from the figure. The structure is supported at its left-hand side and the bottom. A uniform displacement u_0 is applied at the top nodes in the vertical direction.

Determine:

- the displacements of the nodes,
- the reaction forces at the supports and nodes where displacements are prescribed,
- the 'macroscopic' POISSON's ratio of the truss structure, and
- check the global force equilibrium.

3.5 Solution

The solution will follow the recommended 10 steps outlined on Sect. 3.2.

① and ② Sketch the free-body diagram of the problem, including a global coordinate system. Subdivide the geometry into finite elements. Indicate the node and element numbers, local coordinate systems, and equivalent nodal loads, see Fig. 3.17.

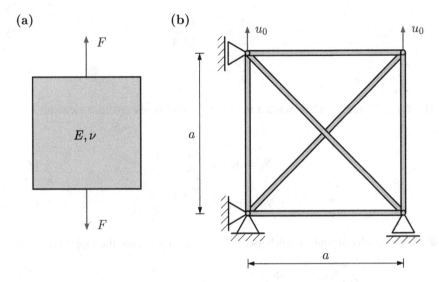

Fig. 3.16 Approximation of a solid using a truss: **a** solid, and **b** truss structure

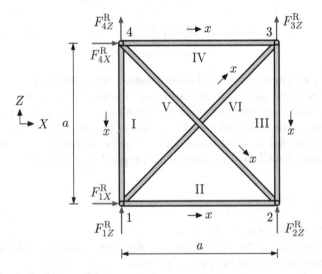

Fig. 3.17 Free-body diagram of the truss structure

③ Write separately all elemental stiffness matrices expressed in the global coordinate system. Indicate the nodal unknowns on the right-hand sides and over the matrices.

Elements II and IV do not require any rotation ($\alpha = 0°$) and thus, the simple elemental stiffness matrix given in Eq. (3.1) can be used:

$$\boldsymbol{K}_{\mathrm{II}}^{\mathrm{e}} = \frac{EA}{a} \begin{array}{cc} u_{1X} & u_{2X} \\ \begin{bmatrix} 1 & -1 \\ -1 & 1 \end{bmatrix} & \begin{array}{c} u_{1X} \\ u_{2X} \end{array} \end{array}, \tag{3.101}$$

$$\boldsymbol{K}_{\mathrm{IV}}^{\mathrm{e}} = \frac{EA}{a} \begin{array}{cc} u_{4X} & u_{3X} \\ \begin{bmatrix} 1 & -1 \\ -1 & 1 \end{bmatrix} & \begin{array}{c} u_{4X} \\ u_{3X} \end{array} \end{array}. \tag{3.102}$$

Elements I and III are rotated by an angle of $\alpha = +90°$ and Eq. (3.80) allows us to express the elemental stiffness matrices as:

$$\boldsymbol{K}_{\mathrm{I}}^{\mathrm{e}} = \frac{EA}{a} \begin{array}{cccc} u_{4X} & u_{4Z} & u_{1X} & u_{1Z} \\ \begin{bmatrix} 0 & 0 & 0 & 0 \\ 0 & 1 & 0 & -1 \\ 0 & 0 & 0 & 0 \\ 0 & -1 & 0 & 1 \end{bmatrix} & \begin{array}{c} u_{4X} \\ u_{4Z} \\ u_{1X} \\ u_{1Z} \end{array} \end{array}, \tag{3.103}$$

$$\boldsymbol{K}_{\mathrm{III}}^{\mathrm{e}} = \frac{EA}{a} \begin{array}{cccc} u_{3X} & u_{3Z} & u_{2X} & u_{2Z} \\ \begin{bmatrix} 0 & 0 & 0 & 0 \\ 0 & 1 & 0 & -1 \\ 0 & 0 & 0 & 0 \\ 0 & -1 & 0 & 1 \end{bmatrix} & \begin{array}{c} u_{3X} \\ u_{3Z} \\ u_{2X} \\ u_{2Z} \end{array} \end{array}. \tag{3.104}$$

Element V is rotated by an angle of $\alpha = +45°$:

$$\boldsymbol{K}_{\mathrm{V}}^{\mathrm{e}} = \frac{EA}{\sqrt{2}a} \begin{array}{cccc} u_{4X} & u_{4Z} & u_{2X} & u_{2Z} \\ \begin{bmatrix} \frac{1}{2} & -\frac{1}{2} & -\frac{1}{2} & \frac{1}{2} \\ -\frac{1}{2} & \frac{1}{2} & \frac{1}{2} & -\frac{1}{2} \\ -\frac{1}{2} & \frac{1}{2} & \frac{1}{2} & -\frac{1}{2} \\ \frac{1}{2} & -\frac{1}{2} & -\frac{1}{2} & \frac{1}{2} \end{bmatrix} & \begin{array}{c} u_{4X} \\ u_{4Z} \\ u_{2X} \\ u_{2Z} \end{array} \end{array}. \tag{3.105}$$

Element VI is rotated by an angle of $\alpha = -45°$:

$$
K_{VI}^{e} = \frac{EA}{\sqrt{2}a}
\begin{array}{cccc}
u_{1X} & u_{1Z} & u_{3X} & u_{3Z}
\end{array}
\begin{bmatrix}
\frac{1}{2} & \frac{1}{2} & -\frac{1}{2} & -\frac{1}{2} \\
\frac{1}{2} & \frac{1}{2} & -\frac{1}{2} & -\frac{1}{2} \\
-\frac{1}{2} & -\frac{1}{2} & \frac{1}{2} & \frac{1}{2} \\
-\frac{1}{2} & -\frac{1}{2} & \frac{1}{2} & \frac{1}{2}
\end{bmatrix}
\begin{array}{l}
u_{1X} \\
u_{1Z} \\
u_{3X} \\
u_{3Z}
\end{array}
\tag{3.106}
$$

④ Determine the dimensions of the global stiffness matrix and sketch the structure of this matrix with the global unknowns on the right-hand side and over the matrix.

The finite element structure is composed of 4 nodes, each having two degrees of freedom (i.e., the horizontal and vertical displacements). Thus, the dimensions of the global stiffness matrix are $(4 \times 2) \times (4 \times 2) = (8 \times 8)$:

$$
\boldsymbol{K} =
\begin{array}{cccccccc}
u_{1X} & u_{1Z} & u_{2X} & u_{2Z} & u_{3X} & u_{3Z} & u_{4X} & u_{4Z}
\end{array}
\begin{bmatrix}
 & & & & & & & \\
 & & & & & & & \\
 & & & & & & & \\
 & & & & & & & \\
 & & & & & & & \\
 & & & & & & & \\
 & & & & & & & \\
 & & & & & & &
\end{bmatrix}
\begin{array}{l}
u_{1X} \\
u_{1Z} \\
u_{2X} \\
u_{2Z} \\
u_{3X} \\
u_{3Z} \\
u_{4X} \\
u_{4Z}
\end{array}
\tag{3.107}
$$

⑤ Insert the values of the elemental stiffness matrices step-by-step into the global stiffness matrix.

$$
\frac{\boldsymbol{K}}{\frac{EA}{a}} =
\begin{array}{cccccccc}
u_{1X} & u_{1Z} & u_{2X} & u_{2Z} & u_{3X} & u_{3Z} & u_{4X} & u_{4Z}
\end{array}
\begin{bmatrix}
1+\frac{1}{2\sqrt{2}} & \frac{1}{2\sqrt{2}} & -1 & 0 & -\frac{1}{2\sqrt{2}} & -\frac{1}{2\sqrt{2}} & 0 & 0 \\
\frac{1}{2\sqrt{2}} & 1+\frac{1}{2\sqrt{2}} & 0 & 0 & -\frac{1}{2\sqrt{2}} & -\frac{1}{2\sqrt{2}} & 0 & -1 \\
-1 & 0 & 1+\frac{1}{2\sqrt{2}} & -\frac{1}{2\sqrt{2}} & 0 & 0 & -\frac{1}{2\sqrt{2}} & \frac{1}{2\sqrt{2}} \\
0 & 0 & -\frac{1}{2\sqrt{2}} & 1+\frac{1}{2\sqrt{2}} & 0 & -1 & \frac{1}{2\sqrt{2}} & -\frac{1}{2\sqrt{2}} \\
-\frac{1}{2\sqrt{2}} & -\frac{1}{2\sqrt{2}} & 0 & 0 & 1+\frac{1}{2\sqrt{2}} & \frac{1}{2\sqrt{2}} & -1 & 0 \\
-\frac{1}{2\sqrt{2}} & -\frac{1}{2\sqrt{2}} & 0 & -1 & \frac{1}{2\sqrt{2}} & 1+\frac{1}{2\sqrt{2}} & 0 & 0 \\
0 & 0 & -\frac{1}{2\sqrt{2}} & \frac{1}{2\sqrt{2}} & -1 & 0 & 1+\frac{1}{2\sqrt{2}} & -\frac{1}{2\sqrt{2}} \\
0 & -1 & \frac{1}{2\sqrt{2}} & -\frac{1}{2\sqrt{2}} & 0 & 0 & -\frac{1}{2\sqrt{2}} & 1+\frac{1}{2\sqrt{2}}
\end{bmatrix}
\begin{array}{l}
u_{1X} \\
u_{1Z} \\
u_{2X} \\
u_{2Z} \\
u_{3X} \\
u_{3Z} \\
u_{4X} \\
u_{4Z}
\end{array}
$$
$$
\tag{3.108}
$$

⑥ Add the column matrix of unknowns and external loads to complete the global system of equations.

The global system of equations can be expressed in matrix from as

$$\boldsymbol{K}\boldsymbol{u}_{\mathrm{p}} = \boldsymbol{f}\,, \tag{3.109}$$

where the column matrix of the external loads reads:

$$\boldsymbol{f} = \begin{bmatrix} F_{1X}^{\mathrm{R}} & F_{1Z}^{\mathrm{R}} & 0 & F_{2Z}^{\mathrm{R}} & 0 & F_{3Z}^{\mathrm{R}} & F_{4X}^{\mathrm{R}} & F_{4Z}^{\mathrm{R}} \end{bmatrix}^{\mathrm{T}}. \tag{3.110}$$

⑦ Introduce the boundary conditions to obtain the reduced system of equations.

The consideration of the support conditions, i.e., $u_{1X} = u_{1Z} = u_{2Z} = u_{4X} = 0$, results in the following 4×4 system:

$$\frac{EA}{a}\begin{bmatrix} 1+\frac{1}{2\sqrt{2}} & 0 & 0 & \frac{1}{2\sqrt{2}} \\ 0 & 1+\frac{1}{2\sqrt{2}} & \frac{1}{2\sqrt{2}} & 0 \\ 0 & \frac{1}{2\sqrt{2}} & 1+\frac{1}{2\sqrt{2}} & 0 \\ \frac{1}{2\sqrt{2}} & 0 & 0 & 1+\frac{1}{2\sqrt{2}} \end{bmatrix}\begin{bmatrix} u_{2X} \\ u_{3X} \\ u_{3Z} \\ u_{4Z} \end{bmatrix} = \begin{bmatrix} 0 \\ 0 \\ F_{3Z}^{\mathrm{R}} \\ F_{4Z}^{\mathrm{R}} \end{bmatrix}. \tag{3.111}$$

The consideration of the displacement boundary condition $u_{3Z} = u_0$ allows a further reduction of the dimensions of the system of equations. Multiplication of the third column of the coefficient matrix by the given displacement u_0 and bringing this column to the right-hand side of the system gives after the deletion of the third row the following equation:

$$\frac{EA}{a}\begin{bmatrix} 1+\frac{1}{2\sqrt{2}} & 0 & \frac{1}{2\sqrt{2}} \\ 0 & 1+\frac{1}{2\sqrt{2}} & 0 \\ \frac{1}{2\sqrt{2}} & 0 & 1+\frac{1}{2\sqrt{2}} \end{bmatrix}\begin{bmatrix} u_{2X} \\ u_{3X} \\ u_{4Z} \end{bmatrix} = \begin{bmatrix} 0 \\ 0 \\ F_{4Z}^{\mathrm{R}} \end{bmatrix} - \frac{EAu_0}{a}\begin{bmatrix} 0 \\ \frac{1}{2\sqrt{2}} \\ 0 \end{bmatrix}. \tag{3.112}$$

A further reduction can be achieved under the consideration of the displacement boundary conditions $u_{4Z} = u_0$. Multiplication of the third column of the coefficient matrix by the given displacement u_0 and bringing this column to the right-hand side of the system gives after the deletion of the third row:

$$\frac{EA}{a}\begin{bmatrix} 1+\frac{1}{2\sqrt{2}} & 0 \\ 0 & 1+\frac{1}{2\sqrt{2}} \end{bmatrix}\begin{bmatrix} u_{2X} \\ u_{3X} \end{bmatrix} = -\frac{EAu_0}{a}\begin{bmatrix} \frac{1}{2\sqrt{2}} \\ \frac{1}{2\sqrt{2}} \end{bmatrix}. \tag{3.113}$$

⑧ Solve the reduced system of equations to obtain the unknown nodal deformations.

The solution can be obtained based on the matrix approach $\boldsymbol{u}_{\mathrm{p}} = \boldsymbol{K}^{-1}\boldsymbol{f}$:

$$
\begin{bmatrix} u_{2X} \\ u_{3X} \end{bmatrix} = \frac{a}{EA} \times \frac{1}{\left(1+\frac{1}{2\sqrt{2}}\right)^2 - 0} \begin{bmatrix} 1+\frac{1}{2\sqrt{2}} & 0 \\ 0 & 1+\frac{1}{2\sqrt{2}} \end{bmatrix} \left(-\frac{EAu_0}{a}\right) \begin{bmatrix} \frac{1}{2\sqrt{2}} \\ \frac{1}{2\sqrt{2}} \end{bmatrix}
$$

$$
= -\frac{u_0}{1+2\sqrt{2}} \begin{bmatrix} 1 \\ 1 \end{bmatrix}. \tag{3.114}
$$

⑨ Post-computation: determination of reaction forces, stresses and strains.

Take into account the non-reduced system of equations as given in step ⑥ under the consideration of the known nodal displacements. The first equation of this system reads:

$$
\frac{EA}{a}\left(-u_{2X} - \frac{u_{3X}}{2\sqrt{2}} - \frac{u_{3Z}}{2\sqrt{2}} + 0\right) = F_{1X}^{\mathrm{R}} \quad \Rightarrow \quad F_{1X}^{\mathrm{R}} = 0. \tag{3.115}
$$

The second equation of this system reads:

$$
\frac{EA}{a}\left(0 - \frac{u_{3X}}{2\sqrt{2}} - \frac{u_{3Z}}{2\sqrt{2}} - u_{4Z}\right) = F_{1Z}^{\mathrm{R}} \quad \Rightarrow \quad F_{1Z}^{\mathrm{R}} = -\frac{4+2\sqrt{2}}{4+\sqrt{2}} \times \frac{EAu_0}{a}. \tag{3.116}
$$

In a similar way,[5] the other reactions are obtained as:

$$
F_{2Z}^{\mathrm{R}} = F_{1Z}^{\mathrm{R}}, \; F_{3Z}^{\mathrm{R}} = F_{4Z}^{\mathrm{R}} = -F_{1Z}^{\mathrm{R}}, \; F_{4X}^{\mathrm{R}} = 0. \tag{3.117}
$$

'Macroscopic' POISSON's ratio of the truss structure:

$$
\nu = -\frac{\varepsilon_X}{\varepsilon_Z} = -\frac{-\frac{u_0/(1+2\sqrt{2})}{a}}{\frac{u_0}{a}} = \frac{1}{1+2\sqrt{2}} \approx 0.261. \tag{3.118}
$$

⑩ Check the global equilibrium between the external loads and the support reactions.

$$
\sum_i F_{iX} = 0 \quad \Leftrightarrow \quad \underbrace{(F_{1X}^{\mathrm{R}} + F_{4X}^{\mathrm{R}})}_{\text{reaction force}} + \underbrace{0}_{\text{external loads}} = 0, \; \checkmark \tag{3.119}
$$

$$
\sum_i F_{iZ} = 0 \quad \Leftrightarrow \quad \underbrace{(F_{1Z}^{\mathrm{R}} + F_{2Z}^{\mathrm{R}} + F_{3Z}^{\mathrm{R}} + F_{4Z}^{\mathrm{R}})}_{\text{reaction force}} + \underbrace{0}_{\text{external load}} = 0. \; \checkmark \tag{3.120}
$$

3.6 Example: Truss structure with six members (computational problem)
Given is a plane truss structure as shown in Fig. 3.18. The members have a uniform cross-sectional area A and YOUNG's modulus E. The length of each member can be

---

[5]The relation $\frac{4+2\sqrt{2}}{4+\sqrt{2}} = \frac{2(1+\sqrt{2})}{1+2\sqrt{2}}$ might be useful to show the identities.

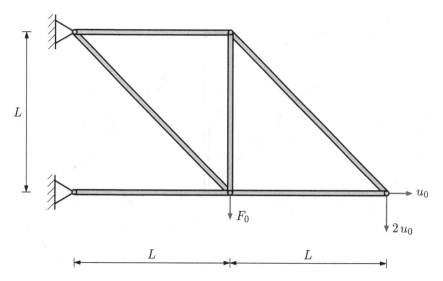

Fig. 3.18 Truss structure composed of six straight members

taken from the figure. The structure is fixed at its left-hand side and loaded by

- two prescribed displacements u_0 and $2u_0$ at the very right-hand corner, and
- a vertical point load F_0.

Model the truss structure with six linear finite elements and determine

- the displacements of the nodes,
- the reaction forces at the supports and nodes where displacements are prescribed,
- the strain, stress, and normal force in each element, and
- check the global force equilibrium.

Simplify all your results for the following special cases:

(a) $u_0 = 0$,
(b) $F_0 = 0$.

3.6 Solution

The solution will follow the recommended 10 steps outlined on Sect. 3.2.

① and ② Sketch the free-body diagram of the problem, including a global coordinate system. Subdivide the geometry into finite elements. Indicate the node and element numbers, local coordinate systems, and equivalent nodal loads, see Fig. 3.19.

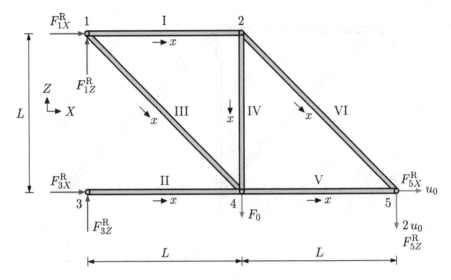

Fig. 3.19 Free-body diagram of the truss structure composed of six axial members

③ Write separately all elemental stiffness matrices expressed in the global coordinate system. Indicate the nodal unknowns on the right-hand sides and over the matrices.

Elements I, II and V do not require any rotation ($\alpha = 0°$) and the simple elemental stiffness matrix given in Eq. (3.1) can be used:

$$
\boldsymbol{K}_{\mathrm{I}}^{\mathrm{e}} = \frac{EA}{L}
\begin{matrix}
u_{1X} \;\; u_{2X} \\
\begin{bmatrix} 1 & -1 \\ -1 & 1 \end{bmatrix}
\begin{matrix} u_{1X} \\ u_{2X} \end{matrix}
\end{matrix} ,
\tag{3.121}
$$

$$
\boldsymbol{K}_{\mathrm{II}}^{\mathrm{e}} = \frac{EA}{L}
\begin{matrix}
u_{3X} \;\; u_{4X} \\
\begin{bmatrix} 1 & -1 \\ -1 & 1 \end{bmatrix}
\begin{matrix} u_{3X} \\ u_{4X} \end{matrix}
\end{matrix} ,
\tag{3.122}
$$

$$
\boldsymbol{K}_{\mathrm{V}}^{\mathrm{e}} = \frac{EA}{L}
\begin{matrix}
u_{4X} \;\; u_{5X} \\
\begin{bmatrix} 1 & -1 \\ -1 & 1 \end{bmatrix}
\begin{matrix} u_{4X} \\ u_{5X} \end{matrix}
\end{matrix} .
\tag{3.123}
$$

Element IV is rotated by an angle of $\alpha = +90°$ and Eq. (3.80) allows us to express the elemental stiffness matrix as:

$$\boldsymbol{K}_{\mathrm{IV}}^{\mathrm{e}} = \frac{EA}{L} \begin{bmatrix} \overset{u_{2X}}{0} & \overset{u_{2Z}}{0} & \overset{u_{4X}}{0} & \overset{u_{4Z}}{0} \\ 0 & 1 & 0 & -1 \\ 0 & 0 & 0 & 0 \\ 0 & -1 & 0 & 1 \end{bmatrix} \begin{matrix} u_{2X} \\ u_{2Z} \\ u_{4X} \\ u_{4Z} \end{matrix} \tag{3.124}$$

Elements III and VI are both rotated by an angle of $\alpha = +45°$:

$$\boldsymbol{K}_{\mathrm{III}}^{\mathrm{e}} = \frac{EA}{\sqrt{2}L} \begin{bmatrix} \overset{u_{1X}}{\tfrac{1}{2}} & \overset{u_{1Z}}{-\tfrac{1}{2}} & \overset{u_{4X}}{-\tfrac{1}{2}} & \overset{u_{4Z}}{\tfrac{1}{2}} \\ -\tfrac{1}{2} & \tfrac{1}{2} & \tfrac{1}{2} & -\tfrac{1}{2} \\ -\tfrac{1}{2} & \tfrac{1}{2} & \tfrac{1}{2} & -\tfrac{1}{2} \\ \tfrac{1}{2} & -\tfrac{1}{2} & -\tfrac{1}{2} & \tfrac{1}{2} \end{bmatrix} \begin{matrix} u_{1X} \\ u_{1Z} \\ u_{4X} \\ u_{4Z} \end{matrix} \tag{3.125}$$

$$= \frac{EA}{L} \begin{bmatrix} \overset{u_{1X}}{\tfrac{\sqrt{2}}{4}} & \overset{u_{1Z}}{-\tfrac{\sqrt{2}}{4}} & \overset{u_{4X}}{-\tfrac{\sqrt{2}}{4}} & \overset{u_{4Z}}{\tfrac{\sqrt{2}}{4}} \\ -\tfrac{\sqrt{2}}{4} & \tfrac{\sqrt{2}}{4} & \tfrac{\sqrt{2}}{4} & -\tfrac{\sqrt{2}}{4} \\ -\tfrac{\sqrt{2}}{4} & \tfrac{\sqrt{2}}{4} & \tfrac{\sqrt{2}}{4} & -\tfrac{\sqrt{2}}{4} \\ \tfrac{\sqrt{2}}{4} & -\tfrac{\sqrt{2}}{4} & -\tfrac{\sqrt{2}}{4} & \tfrac{\sqrt{2}}{4} \end{bmatrix} \begin{matrix} u_{1X} \\ u_{1Z} \\ u_{4X} \\ u_{4Z} \end{matrix} \ , \tag{3.126}$$

$$\boldsymbol{K}_{\mathrm{VI}}^{\mathrm{e}} = \frac{EA}{\sqrt{2}L} \begin{bmatrix} \overset{u_{2X}}{\tfrac{1}{2}} & \overset{u_{2Z}}{-\tfrac{1}{2}} & \overset{u_{5X}}{-\tfrac{1}{2}} & \overset{u_{5Z}}{\tfrac{1}{2}} \\ -\tfrac{1}{2} & \tfrac{1}{2} & \tfrac{1}{2} & -\tfrac{1}{2} \\ -\tfrac{1}{2} & \tfrac{1}{2} & \tfrac{1}{2} & -\tfrac{1}{2} \\ \tfrac{1}{2} & -\tfrac{1}{2} & -\tfrac{1}{2} & \tfrac{1}{2} \end{bmatrix} \begin{matrix} u_{2X} \\ u_{2Z} \\ u_{5X} \\ u_{5Z} \end{matrix} \tag{3.127}$$

$$= \frac{EA}{L} \begin{bmatrix} \overset{u_{2X}}{\tfrac{\sqrt{2}}{4}} & \overset{u_{2Z}}{-\tfrac{\sqrt{2}}{4}} & \overset{u_{5X}}{-\tfrac{\sqrt{2}}{4}} & \overset{u_{5Z}}{\tfrac{\sqrt{2}}{4}} \\ -\tfrac{\sqrt{2}}{4} & \tfrac{\sqrt{2}}{4} & \tfrac{\sqrt{2}}{4} & -\tfrac{\sqrt{2}}{4} \\ -\tfrac{\sqrt{2}}{4} & \tfrac{\sqrt{2}}{4} & \tfrac{\sqrt{2}}{4} & -\tfrac{\sqrt{2}}{4} \\ \tfrac{\sqrt{2}}{4} & -\tfrac{\sqrt{2}}{4} & -\tfrac{\sqrt{2}}{4} & \tfrac{\sqrt{2}}{4} \end{bmatrix} \begin{matrix} u_{2X} \\ u_{2Z} \\ u_{5X} \\ u_{5Z} \end{matrix} \ . \tag{3.128}$$

④ Determine the dimensions of the global stiffness matrix and sketch the structure of this matrix with global unknowns on the right-hand side and over the matrix.

The finite element structure is composed of 5 nodes, each having two degree of freedom (i.e., the horizontal and vertical displacements). Thus, the dimensions of the global stiffness matrix are $(5 \times 2) \times (5 \times 2) = (10 \times 10)$:

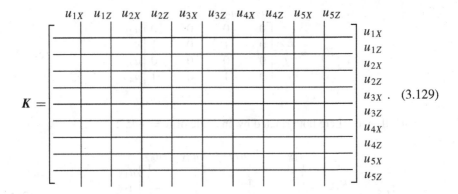

$$K = \quad (3.129)$$

⑤ Insert the values of the elemental stiffness matrices step-by-step into the global stiffness matrix.

$$
\frac{K}{\frac{EA}{L}} =
\begin{bmatrix}
1+\frac{\sqrt{2}}{4} & -\frac{\sqrt{2}}{4} & -1 & 0 & 0 & 0 & -\frac{\sqrt{2}}{4} & \frac{\sqrt{2}}{4} & 0 & 0 \\
-\frac{\sqrt{2}}{4} & \frac{\sqrt{2}}{4} & 0 & 0 & 0 & 0 & \frac{\sqrt{2}}{4} & -\frac{\sqrt{2}}{4} & 0 & 0 \\
-1 & 0 & 1+\frac{\sqrt{2}}{4} & -\frac{\sqrt{2}}{4} & 0 & 0 & 0 & 0 & -\frac{\sqrt{2}}{4} & \frac{\sqrt{2}}{4} \\
0 & 0 & -\frac{\sqrt{2}}{4} & 1+\frac{\sqrt{2}}{4} & 0 & 0 & 0 & -1 & \frac{\sqrt{2}}{4} & -\frac{\sqrt{2}}{4} \\
0 & 0 & 0 & 0 & 1 & 0 & -1 & 0 & 0 & 0 \\
0 & 0 & 0 & 0 & 0 & 0 & 0 & 0 & 0 & 0 \\
-\frac{\sqrt{2}}{4} & \frac{\sqrt{2}}{4} & 0 & 0 & -1 & 0 & 1+1+\frac{\sqrt{2}}{4} & -\frac{\sqrt{2}}{4} & -1 & 0 \\
\frac{\sqrt{2}}{4} & -\frac{\sqrt{2}}{4} & 0 & -1 & 0 & 0 & -\frac{\sqrt{2}}{4} & 1+\frac{\sqrt{2}}{4} & 0 & 0 \\
0 & 0 & -\frac{\sqrt{2}}{4} & \frac{\sqrt{2}}{4} & 0 & 0 & -1 & 0 & 1+\frac{\sqrt{2}}{4} & -\frac{\sqrt{2}}{4} \\
0 & 0 & \frac{\sqrt{2}}{4} & -\frac{\sqrt{2}}{4} & 0 & 0 & 0 & 0 & -\frac{\sqrt{2}}{4} & \frac{\sqrt{2}}{4}
\end{bmatrix}
\begin{matrix}
u_{1X} \\ u_{1Z} \\ u_{2X} \\ u_{2Z} \\ u_{3X} \\ u_{3Z} \\ u_{4X} \\ u_{4Z} \\ u_{5X} \\ u_{5Z}
\end{matrix}
$$

$$(3.130)$$

⑥ and ⑦ Add the column matrix of unknowns and external loads to complete the global system of equations. Introduce the boundary conditions to obtain the reduced system of equations.

The global system of equations can be expressed in matrix from as

$$K u_{\mathrm{p}} = f , \qquad (3.131)$$

where the column matrix of the external loads reads:

$$f = \begin{bmatrix} F_{1X}^{R} & F_{1Z}^{R} & 0 & 0 & F_{3X}^{R} & F_{3Z}^{R} & 0 & -F_0 & F_{5X}^{R} & -F_{5Z}^{R} \end{bmatrix}^{T} . \qquad (3.132)$$

The boundary conditions $u_{1X} = u_{1Z} = u_{3X} = u_{3Z} = 0$ allow to delete four rows and columns from the system of equations:

$$
\frac{EA}{L}
\begin{bmatrix}
1+\frac{\sqrt{2}}{4} & -\frac{\sqrt{2}}{4} & 0 & 0 & -\frac{\sqrt{2}}{4} & \frac{\sqrt{2}}{4} \\
-\frac{\sqrt{2}}{4} & 1+\frac{\sqrt{2}}{4} & 0 & -1 & \frac{\sqrt{2}}{4} & -\frac{\sqrt{2}}{4} \\
0 & 0 & 1+1+\frac{\sqrt{2}}{4} & -\frac{\sqrt{2}}{4} & -1 & 0 \\
0 & -1 & -\frac{\sqrt{2}}{4} & 1+\frac{\sqrt{2}}{4} & 0 & 0 \\
-\frac{\sqrt{2}}{4} & \frac{\sqrt{2}}{4} & -1 & 0 & 1+\frac{\sqrt{2}}{4} & -\frac{\sqrt{2}}{4} \\
\frac{\sqrt{2}}{4} & -\frac{\sqrt{2}}{4} & 0 & 0 & -\frac{\sqrt{2}}{4} & \frac{\sqrt{2}}{4}
\end{bmatrix}
\begin{bmatrix}
u_{2X} \\ u_{2Z} \\ u_{4X} \\ u_{4Z} \\ u_{5X} \\ u_{5Z}
\end{bmatrix}
=
\begin{bmatrix}
0 \\ 0 \\ 0 \\ -F_0 \\ F_{5X}^R \\ -F_{5Z}^R
\end{bmatrix}.
$$
(3.133)

Let us first consider the displacement boundary condition $u_{5Z} = -2u_0$. Multiplication of the column corresponding to u_{5Z} with the prescribed value $-2u_0$ gives:

$$
\frac{EA}{L}
\begin{bmatrix}
1+\frac{\sqrt{2}}{4} & -\frac{\sqrt{2}}{4} & 0 & 0 & -\frac{\sqrt{2}}{4} & \frac{\sqrt{2}}{4}(-2u_0) \\
-\frac{\sqrt{2}}{4} & 1+\frac{\sqrt{2}}{4} & 0 & -1 & \frac{\sqrt{2}}{4} & -\frac{\sqrt{2}}{4}(-2u_0) \\
0 & 0 & 2+\frac{\sqrt{2}}{4} & -\frac{\sqrt{2}}{4} & -1 & 0(-2u_0) \\
0 & -1 & -\frac{\sqrt{2}}{4} & 1+\frac{\sqrt{2}}{4} & 0 & 0(-2u_0) \\
-\frac{\sqrt{2}}{4} & \frac{\sqrt{2}}{4} & -1 & 0 & 1+\frac{\sqrt{2}}{4} & -\frac{\sqrt{2}}{4}(-2u_0) \\
\frac{\sqrt{2}}{4} & -\frac{\sqrt{2}}{4} & 0 & 0 & -\frac{\sqrt{2}}{4} & \frac{\sqrt{2}}{4}(-2u_0)
\end{bmatrix}
\begin{bmatrix}
u_{2X} \\ u_{2Z} \\ u_{4X} \\ u_{4Z} \\ u_{5X} \\ u_{5Z}
\end{bmatrix}
=
\begin{bmatrix}
0 \\ 0 \\ 0 \\ -F_0 \\ F_{5X}^R \\ -F_{5Z}^R
\end{bmatrix}.
$$
(3.134)

Let us bring the last column of the stiffness matrix to the right-hand side and cancel the last row of the system of equations:

$$
\frac{EA}{L}
\begin{bmatrix}
1+\frac{\sqrt{2}}{4} & -\frac{\sqrt{2}}{4} & 0 & 0 & -\frac{\sqrt{2}}{4} \\
-\frac{\sqrt{2}}{4} & 1+\frac{\sqrt{2}}{4} & 0 & -1 & \frac{\sqrt{2}}{4} \\
0 & 0 & 2+\frac{\sqrt{2}}{4} & -\frac{\sqrt{2}}{4} & -1 \\
0 & -1 & -\frac{\sqrt{2}}{4} & 1+\frac{\sqrt{2}}{4} & 0 \\
-\frac{\sqrt{2}}{4} & \frac{\sqrt{2}}{4} & -1 & 0 & 1+\frac{\sqrt{2}}{4}
\end{bmatrix}
\begin{bmatrix}
u_{2X} \\ u_{2Z} \\ u_{4X} \\ u_{4Z} \\ u_{5X}
\end{bmatrix}
=
\begin{bmatrix}
0 \\ 0 \\ 0 \\ -F_0 \\ F_{5X}^R
\end{bmatrix}
-\frac{EA}{L}
\begin{bmatrix}
-\frac{\sqrt{2}}{2}u_0 \\ \frac{\sqrt{2}}{2}u_0 \\ 0 \\ 0 \\ \frac{\sqrt{2}}{2}u_0
\end{bmatrix}.
$$
(3.135)

Now, let us consider the second prescribed displacement boundary condition $u_{5X} = u_0$. Multiplication of the column corresponding to u_{5X} with the prescribed value u_0 gives:

$$
\frac{EA}{L}
\begin{bmatrix}
1+\frac{\sqrt{2}}{4} & -\frac{\sqrt{2}}{4} & 0 & 0 & -\frac{\sqrt{2}}{4}u_0 \\
-\frac{\sqrt{2}}{4} & 1+\frac{\sqrt{2}}{4} & 0 & -1 & \frac{\sqrt{2}}{4}u_0 \\
0 & 0 & 2+\frac{\sqrt{2}}{4} & -\frac{\sqrt{2}}{4} & -1u_0 \\
0 & -1 & -\frac{\sqrt{2}}{4} & 1+\frac{\sqrt{2}}{4} & 0u_0 \\
-\frac{\sqrt{2}}{4} & \frac{\sqrt{2}}{4} & -1 & 0 & (1+\frac{\sqrt{2}}{4})u_0
\end{bmatrix}
\begin{bmatrix}
u_{2X} \\ u_{2Z} \\ u_{4X} \\ u_{4Z} \\ u_{5X}
\end{bmatrix}
=
\begin{bmatrix}
0 \\ 0 \\ 0 \\ -F_0 \\ F_{5X}^R
\end{bmatrix}
-\frac{EA}{L}
\begin{bmatrix}
-\frac{\sqrt{2}}{2}u_0 \\ \frac{\sqrt{2}}{2}u_0 \\ 0 \\ 0 \\ \frac{\sqrt{2}}{2}u_0
\end{bmatrix}.
$$
(3.136)

Let us bring the last column of the stiffness matrix to the right-hand side and cancel the last row of the system of equations:

$$
\frac{EA}{L}
\begin{bmatrix}
1+\frac{\sqrt{2}}{4} & -\frac{\sqrt{2}}{4} & 0 & 0 \\
-\frac{\sqrt{2}}{4} & 1+\frac{\sqrt{2}}{4} & 0 & -1 \\
0 & 0 & 2+\frac{\sqrt{2}}{4} & -\frac{\sqrt{2}}{4} \\
0 & -1 & -\frac{\sqrt{2}}{4} & 1+\frac{\sqrt{2}}{4}
\end{bmatrix}
\begin{bmatrix} u_{2X} \\ u_{2Z} \\ u_{4X} \\ u_{4Z} \end{bmatrix}
=
\begin{bmatrix} 0 \\ 0 \\ 0 \\ -F_0 \end{bmatrix}
-
\frac{EA}{L}
\left(
\begin{bmatrix} -\frac{\sqrt{2}}{2}u_0 \\ \frac{\sqrt{2}}{2}u_0 \\ 0 \\ 0 \end{bmatrix}
+
\begin{bmatrix} -\frac{\sqrt{2}}{4}u_0 \\ \frac{\sqrt{2}}{4}u_0 \\ -u_0 \\ 0 \end{bmatrix}
\right),
$$

$$(3.137)$$

$$
\frac{EA}{L}
\begin{bmatrix}
1+\frac{\sqrt{2}}{4} & -\frac{\sqrt{2}}{4} & 0 & 0 \\
-\frac{\sqrt{2}}{4} & 1+\frac{\sqrt{2}}{4} & 0 & -1 \\
0 & 0 & 2+\frac{\sqrt{2}}{4} & -\frac{\sqrt{2}}{4} \\
0 & -1 & -\frac{\sqrt{2}}{4} & 1+\frac{\sqrt{2}}{4}
\end{bmatrix}
\begin{bmatrix} u_{2X} \\ u_{2Z} \\ u_{4X} \\ u_{4Z} \end{bmatrix}
=
\begin{bmatrix} 0 \\ 0 \\ 0 \\ -F_0 \end{bmatrix}
+
\frac{EAu_0}{L}
\begin{bmatrix} +\frac{3\sqrt{2}}{4} \\ -\frac{3\sqrt{2}}{4} \\ 1 \\ 0 \end{bmatrix}.
$$

$$(3.138)$$

⑧ Solve the reduced system of equations to obtain the unknown nodal deformations.

The solution can be obtained based on the matrix approach $u_\text{p} = K^{-1}f$:

$$
u_{2X} = 0.429\,u_0 - 0.408\,\frac{FL}{EA},
$$

$$(3.139)$$

$$
u_{2Z} = -1.357\,u_0 - 1.562\,\frac{FL}{EA},
$$

$$(3.140)$$

$$
u_{4X} = 0.285\,u_0 - 0.296\,\frac{FL}{EA},
$$

$$(3.141)$$

$$
u_{4Z} = -0.928\,u_0 - 1.970\,\frac{FL}{EA}.
$$

$$(3.142)$$

⑨ Post-computation: determination of reaction forces, stresses and strains.

Take into account the non-reduced system of equations as given in step ⑥ under the consideration of the known nodal displacements. The evaluation of the first, second, fifth, sixth, ninth and tenth equation of this system gives the following results, respectively:

$$
F_{1X}^{R} = -0.184\,F_0 - 0.858\,\frac{EAu_0}{L},
$$

$$(3.143)$$

$$
F_{1Z}^{R} = 0.592\,F_0 + 0.429\,\frac{EAu_0}{L},
$$

$$(3.144)$$

$$
F_{3X}^{R} = 0.296\,F_0 - 0.285\,\frac{EAu_0}{L},
$$

$$(3.145)$$

$$
F_{3Z}^{R} = 0,
$$

$$(3.146)$$

$$
F_{5X}^{R} = -0.112\,F_0 + 1.144\,\frac{EAu_0}{L},
$$

$$(3.147)$$

$$
F_{5Z}^{R} = -0.408\,F_0 + 0.429\,\frac{EAu_0}{L}.
$$

$$(3.148)$$

The elemental stresses can be obtained from the displacements of the start ('s') and end ('e') node as:

$$\sigma = \frac{E}{L}(-\cos(\alpha)u_{sX} + \sin(\alpha)u_{sZ} + \cos(\alpha)u_{eX} - \sin(\alpha)u_{eZ}) \ . \tag{3.149}$$

Application of this general equation (pay attention to the length of element III and VI which is equal to $\sqrt{2}L$) to the six elements under the consideration of the given nodal displacements gives:

$$\sigma_{\mathrm{I}} = -0.408 \frac{F_0}{A} + 0.429 \frac{Eu_0}{L}, \tag{3.150}$$

$$\sigma_{\mathrm{II}} = -0.296 \frac{F_0}{A} + 0.285 \frac{Eu_0}{L}, \tag{3.151}$$

$$\sigma_{\mathrm{III}} = 0.837 \frac{F_0}{A} + 0.607 \frac{Eu_0}{L}, \tag{3.152}$$

$$\sigma_{\mathrm{IV}} = 0.408 \frac{F_0}{A} - 0.429 \frac{Eu_0}{L}, \tag{3.153}$$

$$\sigma_{\mathrm{V}} = 0.296 \frac{F_0}{A} + 0.715 \frac{Eu_0}{L}, \tag{3.154}$$

$$\sigma_{\mathrm{VI}} = -0.577 \frac{F_0}{A} + 0.607 \frac{Eu_0}{L}. \tag{3.155}$$

Application of HOOKE's law, i.e., $\sigma = E\varepsilon$, allows the calculation of the elemental strains:

$$\varepsilon_{\mathrm{I}} = -0.408 \frac{F_0}{EA} + 0.429 \frac{u_0}{L}, \tag{3.156}$$

$$\varepsilon_{\mathrm{II}} = -0.296 \frac{F_0}{EA} + 0.285 \frac{u_0}{L}, \tag{3.157}$$

$$\varepsilon_{\mathrm{III}} = 0.837 \frac{F_0}{EA} + 0.607 \frac{u_0}{L}, \tag{3.158}$$

$$\varepsilon_{\mathrm{IV}} = 0.408 \frac{F_0}{EA} - 0.429 \frac{u_0}{L}, \tag{3.159}$$

$$\varepsilon_{\mathrm{V}} = 0.296 \frac{F_0}{EA} + 0.715 \frac{u_0}{L}, \tag{3.160}$$

$$\varepsilon_{\mathrm{VI}} = -0.577 \frac{F_0}{EA} + 0.607 \frac{u_0}{L}. \tag{3.161}$$

⑩ Check the global equilibrium between the external loads and the support reactions.

$$\sum_i F_{iX} = 0 \quad \Leftrightarrow \quad \underbrace{(F_{1X}^{\mathrm{R}} + F_{3X}^{\mathrm{R}} + F_{5X}^{\mathrm{R}})}_{\text{reaction force}} + \underbrace{0}_{\text{external loads}} = 0, \ \checkmark \qquad (3.162)$$

$$\sum_i F_{iZ} = 0 \quad \Leftrightarrow \quad \underbrace{(F_{1Z}^{\mathrm{R}} + F_{3Z}^{\mathrm{R}} - F_{5Z}^{\mathrm{R}})}_{\text{reaction force}} + \underbrace{-F_0}_{\text{external load}} = 0. \ \checkmark \qquad (3.163)$$

3.3 Beams and Frames

3.3.1 Euler–Bernoulli Beam Elements

Let us consider an EULER–BERNOULLI beam element which is composed of two nodes as schematically shown in Fig. 3.20. Each node has two degrees of freedom, i.e., a displacement u_z in the direction of the z-axis (i.e., perpendicular to the principal beam axis) and a rotation φ_y around the y-axis, see Fig. 3.20a. Each node can be loaded by single forces acting in the z-direction or single moments around the y-axis, see Fig. 3.20b. In the case of distributed loads $q_z(x)$, a transformation must be made to calculate the equivalent nodal loads.

Different methods can be found in the literature to derive the principal finite element equation (see [2, 5]). All these methods result in the same elemental formulation, which is given in the following for constant material (E) and geometrical properties (I_y):

Fig. 3.20 Definition of the EULER–BERNOULLI beam element for deformation in the x-z plane: **a** deformations; **b** external loads. The nodes are symbolized by two circles at the ends (○)

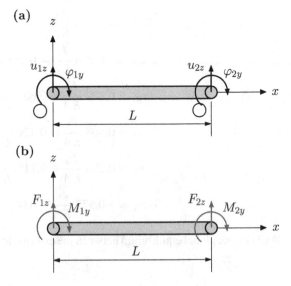

$$\frac{EI_y}{L^3} \begin{bmatrix} 12 & -6L & -12 & -6L \\ -6L & 4L^2 & 6L & 2L^2 \\ -12 & 6L & 12 & 6L \\ -6L & 2L^2 & 6L & 4L^2 \end{bmatrix} \begin{bmatrix} u_{1z} \\ \varphi_{1y} \\ u_{2z} \\ \varphi_{2y} \end{bmatrix} = \begin{bmatrix} F_{1z} \\ M_{1y} \\ F_{2z} \\ M_{2y} \end{bmatrix} + \int_0^L q_z(x) \begin{bmatrix} N_{1u} \\ N_{1\varphi} \\ N_{2u} \\ N_{2\varphi} \end{bmatrix} dx, \quad (3.164)$$

or in the abbreviated form

$$\boldsymbol{K}^e \boldsymbol{u}_p^e = \boldsymbol{f}^e, \tag{3.165}$$

where $\boldsymbol{K}^e$ is the elemental stiffness matrix, $\boldsymbol{u}_p^e$ is the elemental column matrix of unknowns and $\boldsymbol{f}^e$ is the elemental column matrix of loads. The interpolation functions in Eq. (3.164) are given by $N_{1u}(x) = 1 - 3\left(\frac{x}{L}\right)^2 + 2\left(\frac{x}{L}\right)^3$, $N_{1\varphi}(x) = -x + 2\frac{x^2}{L} - \frac{x^3}{L^2}$, $N_{2u}(x) = 3\left(\frac{x}{L}\right)^2 - 2\left(\frac{x}{L}\right)^3$ and $N_{2\varphi}(x) = \frac{x^2}{L} - \frac{x^3}{L^2}$. Table 3.6 summarizes for some simple shapes of distributed loads the equivalent nodal loads.

Once the nodal displacements $(u_{1z}, \varphi_{1y}, u_{2z}, \varphi_{2y})$ are known, further quantities and their distributions can be calculated within an element (so-called postprocessing), see Tables 3.7 and 3.8.

3.7 Cantilever beam with point loads

The cantilever beam shown in Fig. 3.21 is loaded by a force F_0 and a moment M_0 at the free right-hand end. The bending stiffness EI is constant and the total length of the beam is equal to L. Model the beam with one single finite element to determine:

- the unknowns at the nodes,
- the equation of the bending line,
- the reactions at the support,
- the internal reactions (shear force and bending moment distribution) in the element,
- the strain and stress distributions in the element, and
- the global force and moment equilibrium.

Repeat the solution procedure for an approach based on two equal elements of length $\frac{L}{2}$.

3.7 Solution

The solution will follow the recommended 10 steps outlined on Sect. 3.2.

① Sketch the free-body diagram of the problem, including a global coordinate system, see Fig. 3.22.
② Subdivide the geometry into finite elements. Indicate the node and element numbers, local coordinate systems, and equivalent nodal loads, see Fig. 3.23.
Steps ③ to ⑥ can be combined since we have only a single element problem. The global system of equations reads:

$$\frac{EI_Y}{L^3} \begin{bmatrix} 12 & -6L & -12 & -6L \\ -6L & 4L^2 & 6L & 2L^2 \\ -12 & 6L & 12 & 6L \\ -6L & 2L^2 & 6L & 4L^2 \end{bmatrix} \begin{bmatrix} u_{1Z} \\ \varphi_{1Y} \\ u_{2Z} \\ \varphi_{2Y} \end{bmatrix} = \begin{bmatrix} F_{1Z}^R \\ M_{1Y}^R \\ -F_0 \\ -M_0 \end{bmatrix}. \tag{3.166}$$

Table 3.6 Equivalent nodal loads for an EULER–BERNOULLI beam element (x-axis: right facing; z-axis: upward facing), partially adapted from [1]

Loading	Shear force	Bending moment
	$F_{1z} = -\dfrac{qL}{2}$	$M_{1y} = +\dfrac{qL^2}{12}$
	$F_{2z} = -\dfrac{qL}{2}$	$M_{2y} = -\dfrac{qL^2}{12}$
	$F_{1z} = -\dfrac{qa}{2L^3}(a^3 - 2a^2L + 2L^3)$	$M_{1y} = +\dfrac{qa^2}{12L^2}(3a^2 - 8aL + 6L^2)$
	$F_{2z} = -\dfrac{qa^3}{2L^3}(2L - a)$	$M_{2y} = -\dfrac{qa^3}{12L^2}(4L - 3a)$
	$F_{1z} = -\dfrac{3}{20}qL$	$M_{1y} = +\dfrac{qL^2}{30}$
	$F_{2z} = -\dfrac{7}{20}qL$	$M_{2y} = -\dfrac{qL^2}{20}$
	$F_{1z} = -\dfrac{1}{4}qL$	$M_{1y} = +\dfrac{5qL^2}{96}$
	$F_{2z} = -\dfrac{1}{4}qL$	$M_{2y} = -\dfrac{5qL^2}{96}$
	$F_{1z} = -\dfrac{F}{2}$	$M_{1y} = +\dfrac{FL}{8}$
	$F_{2z} = -\dfrac{F}{2}$	$M_{2y} = -\dfrac{FL}{8}$
	$F_{1z} = -\dfrac{Fb^2(3a + b)}{L^3}$	$M_{1y} = +\dfrac{Fb^2a}{L^2}$
	$F_{2z} = -\dfrac{Fa^2(a + 3b)}{L^3}$	$M_{2y} = -\dfrac{Fa^2b}{L^2}$
	$F_{1z} = -\dfrac{3M}{2L}$	$M_{1y} = +\dfrac{M}{4}$
	$F_{2z} = +\dfrac{3M}{2L}$	$M_{2y} = +\dfrac{M}{4}$
	$F_{1z} = -6M\dfrac{ab}{L^3}$	$M_{1y} = +M\dfrac{b(2a - b)}{L^2}$
	$F_{2z} = +6M\dfrac{ab}{L^3}$	$M_{2y} = +M\dfrac{a(2b - a)}{L^2}$

Table 3.7 Post-processing quantities (part 1) for a BERNOULLI beam element given as being dependent on the nodal values as a function of the physical coordinate $0 \le x \le L$ and natural coordinate $-1 \le \xi \le 1$. Bending in the x-z plane

Vertical displacement (Deflection) u_z

$$u_z^e(x) = \left[1 - 3\left(\frac{x}{L}\right)^2 + 2\left(\frac{x}{L}\right)^3\right] u_{1z} + \left[-x + \frac{2x^2}{L} - \frac{x^3}{L^2}\right] \varphi_{1y}$$

$$+ \left[3\left(\frac{x}{L}\right)^2 - 2\left(\frac{x}{L}\right)^3\right] u_{2z} + \left[+\frac{x^2}{L} - \frac{x^3}{L^2}\right] \varphi_{2y}$$

$$u_z^e(\xi) = \frac{1}{4}\left[2 - 3\xi + \xi^3\right] u_{1z} - \frac{1}{4}\left[1 - \xi - \xi^2 + \xi^3\right] \frac{L}{2} \varphi_{1y}$$

$$+ \frac{1}{4}\left[2 + 3\xi - \xi^3\right] u_{2z} - \frac{1}{4}\left[-1 - \xi + \xi^2 + \xi^3\right] \frac{L}{2} \varphi_{2y}$$

Rotation (Slope) $\varphi_y = -\dfrac{du_z}{dx} = -\dfrac{2}{L}\dfrac{du_z}{d\xi}$

$$\varphi_y^e(x) = \left[+\frac{6x}{L^2} - \frac{6x^2}{L^3}\right] u_{1z} + \left[1 - \frac{4x}{L} + \frac{3x^2}{L^2}\right] \varphi_{1y}$$

$$+ \left[-\frac{6x}{L^2} + \frac{6x^2}{L^3}\right] u_{2z} + \left[-\frac{2x}{L} + \frac{3x^2}{L^2}\right] \varphi_{2y}$$

$$\varphi_y^e(\xi) = \frac{1}{2L}\left[+3 - 3\xi^2\right] u_{1z} + \frac{1}{4}\left[-1 - 2\xi + 3\xi^2\right] \varphi_{1y}$$

$$+ \frac{1}{2L}\left[-3 + 3\xi^2\right] u_{2z} + \frac{1}{4}\left[-1 + 2\xi + 3\xi^2\right] \varphi_{2y}$$

Curvature $\kappa_y = -\dfrac{d^2 u_z}{dx^2} = -\dfrac{4}{L^2}\dfrac{d^2 u_z}{d\xi^2}$

$$\kappa_y^e(x) = \left[+\frac{6}{L^2} - \frac{12x}{L^3}\right] u_{1z} + \left[-\frac{4}{L} + \frac{6x}{L^2}\right] \varphi_{1y}$$

$$+ \left[-\frac{6}{L^2} + \frac{12x}{L^3}\right] u_{2z} + \left[-\frac{2}{L} + \frac{6x}{L^2}\right] \varphi_{2y}$$

$$\kappa_y^e(\xi) = \frac{6}{L^2}[-\xi] u_{1z} + \frac{1}{L}[-1 + 3\xi] \varphi_{1y}$$

$$+ \frac{6}{L^2}[\xi] u_{2z} + \frac{1}{L}[1 + 3\xi] \varphi_{2y}$$

⑦ Introduce the boundary conditions to obtain the reduced system of equations.

$$\frac{EI_Y}{L^3}\begin{bmatrix} 12 & 6L \\ 6L & 4L^2 \end{bmatrix}\begin{bmatrix} u_{2Z} \\ \varphi_{2Y} \end{bmatrix} = \begin{bmatrix} -F_0 \\ -M_0 \end{bmatrix}. \tag{3.167}$$

⑧ Solve the reduced system of equations to obtain the unknown nodal deformations.

Table 3.8 Post-processing quantities (part 2) for an EULER–BERNOULLI beam element given as being dependent on the nodal values as function of the physical coordinate $0 \le x \le L$ and natural coordinate $-1 \le \xi \le 1$. Bending in the x-z plane

$$\text{Bending moment } M_y = -EI_y \frac{d^2 u_z}{dx^2} = -\frac{4}{L^2} EI_y \frac{d^2 u_z}{d\xi^2}$$

$$
M_y^e(x) = EI_y \left(\left[+\frac{6}{L^2} - \frac{12x}{L^3} \right] u_{1z} + \left[-\frac{4}{L} + \frac{6x}{L^2} \right] \varphi_{1y} \right.
$$
$$
\left. + \left[-\frac{6}{L^2} + \frac{12x}{L^3} \right] u_{2z} + \left[-\frac{2}{L} + \frac{6x}{L^2} \right] \varphi_{2y} \right)
$$
$$
M_y^e(\xi) = EI_y \left(\frac{6}{L^2}[-\xi] u_{1z} + \frac{1}{L}[-1+3\xi] \varphi_{1y} + \frac{6}{L^2}[\xi] u_{2z} + \frac{1}{L}[1+3\xi] \varphi_{2y} \right)
$$

$$\text{Shear force } Q_z = -EI_y \frac{d^3 u_z}{dx^3} = -\frac{8}{L^3} EI_y \frac{d^3 u_z}{d\xi^3}$$

$$
Q_z^e(x) = EI_y \left(\left[-\frac{12}{L^3} \right] u_{1z} + \left[+\frac{6}{L^2} \right] \varphi_{1y} + \left[+\frac{12}{L^3} \right] u_{2z} + \left[+\frac{6}{L^2} \right] \varphi_{2y} \right)
$$
$$
Q_z^e(\xi) = EI_y \left(\frac{12}{L^3}[-1] u_{1z} + \frac{2}{L^2}[+3] \varphi_{1y} + \frac{12}{L^3}[1] u_{2z} + \frac{2}{L^2}[+3] \varphi_{2y} \right)
$$

$$\text{Normal strain } \varepsilon_x^e(x, z) = -z \frac{d^2 u_z^e(x)}{dx^2} = -z \frac{4}{L^2} \frac{d^2 u_z}{d\xi^2}$$

$$
\varepsilon_x^e(x, z) = \left(\left[+\frac{6}{L^2} - \frac{12x}{L^3} \right] u_{1z} + \left[-\frac{4}{L} + \frac{6x}{L^2} \right] \varphi_{1y} \right.
$$
$$
\left. + \left[-\frac{6}{L^2} + \frac{12x}{L^3} \right] u_{2z} + \left[-\frac{2}{L} + \frac{6x}{L^2} \right] \varphi_{2y} \right) z
$$
$$
\varepsilon_x^e(\xi, z) = \left(\frac{6}{L^2}[-\xi] u_{1z} + \frac{1}{L}[-1+3\xi] \varphi_{1y} + \frac{6}{L^2}[\xi] u_{2z} + \frac{1}{L}[1+3\xi] \varphi_{2y} \right) z
$$

$$\text{Normal stress } \sigma_x^e(x, z) = E\varepsilon_x^e(x, z) = E\varepsilon_x^e(\xi, z) \quad \left(= \frac{M_y}{I_y} z \right)$$

$$
\sigma_x^e(x, z) = E \left(\left[+\frac{6}{L^2} - \frac{12x}{L^3} \right] u_{1z} + \left[-\frac{4}{L} + \frac{6x}{L^2} \right] \varphi_{1y} \right.
$$
$$
\left. + \left[-\frac{6}{L^2} + \frac{12x}{L^3} \right] u_{2z} + \left[-\frac{2}{L} + \frac{6x}{L^2} \right] \varphi_{2y} \right) z
$$
$$
\sigma_x^e(\xi, z) = E \left(\frac{6}{L^2}[-\xi] u_{1z} + \frac{1}{L}[-1+3\xi] \varphi_{1y} + \frac{6}{L^2}[\xi] u_{2z} + \frac{1}{L}[1+3\xi] \varphi_{2y} \right) z
$$

Fig. 3.21 Cantilever beam with two point loads at the free end

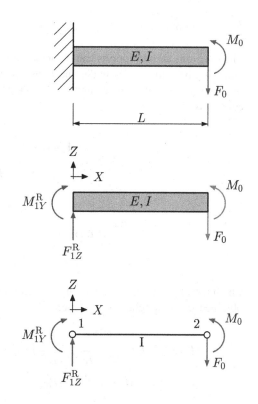

Fig. 3.22 Free-body diagram of the cantilever beam with two point loads at the free end

Fig. 3.23 Free-body diagram of the discretized structure

$$
\begin{bmatrix} u_{2Z} \\ \varphi_{2Y} \end{bmatrix} = \frac{L^3}{EI_Y} \times \frac{1}{12 \times 4L^2 - 6L \times 6L} \begin{bmatrix} 4L^2 & -6L \\ -6L & 12 \end{bmatrix} \begin{bmatrix} -F_0 \\ -M_0 \end{bmatrix}
$$
$$
= \frac{L}{12EI_Y} \begin{bmatrix} -4F_0L^2 + 6LM_0 \\ 6LF_0 - 12M_0 \end{bmatrix}. \tag{3.168}
$$

Based on these nodal unknowns, the bending line (deflection) can be obtained from the general equation provided in Table 3.7:

$$
u_Z(X) = 0 + 0 + \left[3\left(\frac{X}{L}\right)^2 - 2\left(\frac{X}{L}\right)^3 \right] u_{2Z} + \left[\frac{X^2}{L} - \frac{X^3}{L^2} \right] \varphi_{2Y}
$$
$$
= \frac{L}{12EI_Y} \left\{ \left[3\left(\frac{X}{L}\right)^2 - 2\left(\frac{X}{L}\right)^3 \right] (-4F_0L^2 + 6LM_0) + \right.
$$
$$
\left. + \left[\frac{X^2}{L} - \frac{X^3}{L^2} \right] (6LF_0 - 12M_0) \right\}. \tag{3.169}
$$

⑨ Post-computation: determination of reactions, stresses and strains.

The reactions at the supports can be obtained from the non-reduced system of equations as given in step ⑥ under the consideration of the known nodal degrees of freedom (i.e., displacements and rotations). The evaluation of the first and second equations gives:

$$\frac{EI_Y}{L^3}(-12u_{2Z} - 6L\varphi_{2Y}) = F_{1Z}^R \tag{3.170}$$

$$\Rightarrow \quad F_{1Z}^R = F_0 , \tag{3.171}$$

$$\frac{EI_Y}{L^3}(6Lu_{2Z} + 2L^2\varphi_{2Y}) = M_{1Y}^R \tag{3.172}$$

$$\Rightarrow \quad M_{1Y}^R = M_0 - F_0L . \tag{3.173}$$

The internal reactions (i.e., bending moment and shear force) in the element can be obtained from the relations provided in Table 3.8:

$$M_Y^e(X) = EI_Y\left(\left[-\frac{6}{L^2} + \frac{12X}{L^3}\right]u_{2Z} + \left[-\frac{2}{L} + \frac{6X}{L^2}\right]\varphi_{2Y}\right) \tag{3.174}$$

$$= -M_0 + (L - X)F_0 , \tag{3.175}$$

$$Q_Z^e(X) = EI_Y\left(\left[\frac{12}{L^3}\right]u_{2Z} + \left[\frac{6}{L^2}\right]\varphi_{2Y}\right) \tag{3.176}$$

$$= -F_0 . \tag{3.177}$$

The normal stress distribution can be obtained from the bending moment given in Eq. (3.175):

$$\sigma_X^e(X, Z) = \frac{M_Y^e(X)}{I_Y}Z = \frac{1}{I_Y}(-M_0 + (L - X)F_0)\,Z . \tag{3.178}$$

The strain distribution results from HOOKE's law:

$$\varepsilon_X^e(X, Z) = \frac{\sigma_x^e(X, Z)}{E} = \frac{1}{EI_Y}(-M_0 + (L - X)F_0)\,Z . \tag{3.179}$$

⑩ Check the global equilibrium between the external loads and the support reactions.

Fig. 3.24 Cantilever beam with simple supports and distributed load

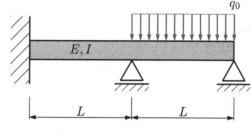

Fig. 3.25 Free-body diagram of the cantilevered beam with simple supports and distributed load

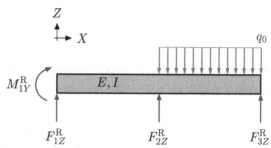

$$\sum_i F_{iZ} = 0 \quad \Leftrightarrow \quad \underbrace{(F_{1Z}^{\mathrm{R}})}_{\text{reaction force}} + \underbrace{(-F_0)}_{\text{external load}} = 0, \ \checkmark \qquad (3.180)$$

$$\sum_i M_{iY} = 0 \quad \Leftrightarrow \quad \underbrace{(M_{1Y}^{\mathrm{R}})}_{\text{reaction}} + \underbrace{(F_0 L - M_0)}_{\text{external load}} = 0. \ \checkmark \qquad (3.181)$$

3.8 Cantilever beam with simple supports and distributed load

The beam shown in Fig. 3.24 is loaded by a constant distributed load q_0. The bending stiffness EI is constant and the total length of the beam is equal to $2L$. Model the beam with two finite elements of length L to determine:

- the unknowns at the nodes,
- the reactions at the supports,
- the internal reactions (shear force and bending moment) in each element,
- the strain and stress distributions in each element, and
- the global force and moment equilibrium.

3.8 Solution

The solution will follow the recommended 10 steps outlined on Sect. 3.2.

① Sketch the free-body diagram of the problem, including a global coordinate system, see Fig. 3.25.

② Subdivide the geometry into finite elements. Indicate the node and element numbers, local coordinate systems, and equivalent nodal loads, see Fig. 3.26.

Fig. 3.26 Free-body diagram of the discretized structure with equivalent nodal loads

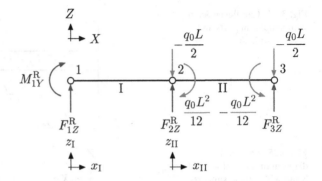

③ Write separately all elemental stiffness matrices expressed in the global coordinate system. Indicate the nodal unknowns on the right-hand sides and over the matrices.

$$K_I^e = \frac{EI_Y}{L^3} \begin{array}{c} \begin{array}{cccc} u_{1Z} & \varphi_{1Y} & u_{2Z} & \varphi_{2Y} \end{array} \\ \begin{bmatrix} 12 & -6L & -12 & -6L \\ -6L & 4L^2 & 6L & 2L^2 \\ -12 & 6L & 12 & 6L \\ -6L & 2L^2 & 6L & 4L^2 \end{bmatrix} \begin{array}{c} u_{1Z} \\ \varphi_{1Y} \\ u_{2Z} \\ \varphi_{2Y} \end{array} \end{array}, \tag{3.182}$$

$$K_{II}^e = \frac{EI_Y}{L^3} \begin{array}{c} \begin{array}{cccc} u_{2Z} & \varphi_{2Y} & u_{3Z} & \varphi_{3Y} \end{array} \\ \begin{bmatrix} 12 & -6L & -12 & -6L \\ -6L & 4L^2 & 6L & 2L^2 \\ -12 & 6L & 12 & 6L \\ -6L & 2L^2 & 6L & 4L^2 \end{bmatrix} \begin{array}{c} u_{2Z} \\ \varphi_{2Y} \\ u_{3Z} \\ \varphi_{3Y} \end{array} \end{array}. \tag{3.183}$$

④ Determine the dimensions of the global stiffness matrix and sketch the structure of this matrix with global unknowns on the right-hand side and over the matrix.

The finite element structure is composed of 3 nodes, each having two degrees of freedom (i.e., the vertical displacement and rotation). Thus, the dimensions of the global stiffness matrix are $(3 \times 2) \times (3 \times 2) = (6 \times 6)$:

$$K = \begin{array}{c} \begin{array}{cccccc} u_{1Z} & \varphi_{1Y} & u_{2Z} & \varphi_{2Y} & u_{3Z} & \varphi_{3Y} \end{array} \\ \begin{bmatrix} & & & & & \\ & & & & & \\ & & & & & \\ & & & & & \\ & & & & & \\ & & & & & \end{bmatrix} \begin{array}{c} u_{1Z} \\ \varphi_{1Y} \\ u_{2Z} \\ \varphi_{2Y} \\ u_{3Z} \\ \varphi_{3Y} \end{array} \end{array}. \tag{3.184}$$

⑤ Insert the values of the elemental stiffness matrices step-by-step into the global stiffness matrix.

$$K = \frac{EI_Y}{L^3} \begin{bmatrix} \begin{smallmatrix} u_{1Z} & \varphi_{1Y} & u_{2Z} & \varphi_{2Y} & u_{3Z} & \varphi_{3Y} \end{smallmatrix} \\ \begin{array}{cccccc} 12 & -6L & -12 & -6L & 0 & 0 \\ -6L & 4L^2 & 6L & 2L^2 & 0 & 0 \\ -12 & 6L & 24 & 0 & -12 & -6L \\ -6L & 2L^2 & 0 & 8L^2 & 6L & 2L^2 \\ 0 & 0 & -12 & 6L & 12 & 6L \\ 0 & 0 & -6L & 2L^2 & 6L & 4L^2 \end{array} \end{bmatrix} \begin{matrix} u_{1Z} \\ \varphi_{1Y} \\ u_{2Z} \\ \varphi_{2Y} \\ u_{3Z} \\ \varphi_{3Y} \end{matrix} \quad . \tag{3.185}$$

⑥ Add the column matrix of unknowns and external loads to complete the global system of equations.

$$\frac{EI_Y}{L^3} \begin{bmatrix} 12 & -6L & -12 & -6L & 0 & 0 \\ -6L & 4L^2 & 6L & 2L^2 & 0 & 0 \\ -12 & 6L & 24 & 0 & -12 & -6L \\ -6L & 2L^2 & 0 & 8L^2 & 6L & 2L^2 \\ 0 & 0 & -12 & 6L & 12 & 6L \\ 0 & 0 & -6L & 2L^2 & 6L & 4L^2 \end{bmatrix} \begin{bmatrix} u_{1Z} \\ \varphi_{1Y} \\ u_{2Z} \\ \varphi_{2Y} \\ u_{3Z} \\ \varphi_{3Y} \end{bmatrix} = \begin{bmatrix} F_{1Z}^R \\ M_{1Y}^R \\ F_{2Z}^R - \frac{q_0 L}{2} \\ \frac{q_0 L^2}{12} \\ F_{3Z}^R - \frac{q_0 L}{2} \\ -\frac{q_0 L^2}{12} \end{bmatrix} . \tag{3.186}$$

⑦ Introduce the boundary conditions to obtain the reduced system of equations.

$$\frac{EI_Y}{L^3} \begin{bmatrix} 8L^2 & 2L^2 \\ 2L^2 & 4L^2 \end{bmatrix} \begin{bmatrix} \varphi_{2Y} \\ \varphi_{3Y} \end{bmatrix} = \begin{bmatrix} \frac{q_0 L^2}{12} \\ -\frac{q_0 L^2}{12} \end{bmatrix} . \tag{3.187}$$

⑧ Solve the reduced system of equations to obtain the unknown nodal deformations.

The solution can be obtained based on the matrix approach $u_p = K^{-1}f$:

$$\begin{bmatrix} \varphi_{2Y} \\ \varphi_{3Y} \end{bmatrix} = \frac{L^3}{EI_Y} \frac{1}{32L^4 - 4L^4} \begin{bmatrix} 4L^2 & -2L^2 \\ -2L^2 & 8L^2 \end{bmatrix} \begin{bmatrix} \frac{q_0 L^2}{12} \\ -\frac{q_0 L^2}{12} \end{bmatrix}$$

$$= \frac{L}{28EI_Y} \begin{bmatrix} 4 & -2 \\ -2 & 8 \end{bmatrix} \begin{bmatrix} \frac{q_0 L^2}{12} \\ -\frac{q_0 L^2}{12} \end{bmatrix} = \frac{q_0 L^3}{28EI_Y} \begin{bmatrix} \frac{1}{2} \\ -\frac{5}{6} \end{bmatrix} . \tag{3.188}$$

The obtained nodal unknowns allow to calculate, for example, the bending curve based on the nodal approach provided in Table 3.7, see Fig. 3.27. It can be seen that all the support conditions, i.e., $u_{1Z} = u_{2Z} = u_{3Z} = 0$ and $\varphi_{1Y} = 0$, are fulfilled.
⑨ Post-computation: determination of reactions, stresses and strains.

The reactions at the supports can be obtained from the non-reduced system of equations as given in step ⑥ under the consideration of the known nodal degrees of freedom (i.e., displacements and rotations). The evaluation of the first, second, third and fifth equation gives:

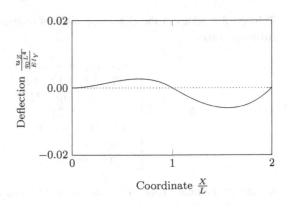

Fig. 3.27 Beam deflection along the major axis

$$\frac{EI_Y}{L^3}(12u_{1Z}-6L\varphi_{1Y} - 12u_{2Z} - 6L\varphi_{2Y}) = F_{1Z}^{R} \tag{3.189}$$

$$\Rightarrow \ F_{1Z}^{R} = -\frac{3}{28}q_0L\,, \tag{3.190}$$

$$\frac{EI_Y}{L^3}(-6Lu_{1Z}+4L^2\varphi_{1Y} + 6Lu_{2Z} + 2L^2\varphi_{2Y}) = M_{1Y}^{R} \tag{3.191}$$

$$\Rightarrow \ M_{1Y}^{R} = \frac{1}{28}q_0L^2\,, \tag{3.192}$$

$$\frac{EI_Y}{L^3}(-12u_{1Z}+6L\varphi_{1Y} + 24u_{2Z} - 12u_{3Z} - 6L\varphi_{3Y}) = F_{2Z}^{R} - \frac{q_0L}{2} \tag{3.193}$$

$$\Rightarrow \ F_{2Z}^{R} = \frac{19}{28}q_0L\,, \tag{3.194}$$

$$\frac{EI_Y}{L^3}(-12u_{2Z}+6L\varphi_{2Y} + 12u_{3Z} + 6L\varphi_{3Y}) = F_{3Z}^{R} - \frac{q_0L}{2} \tag{3.195}$$

$$\Rightarrow \ F_{3Z}^{R} = \frac{3}{7}q_0L\,. \tag{3.196}$$

The internal reactions (i.e., bending moment and shear force) in each element can be obtained from the relations provided in Table 3.8.

$$M_y^e(x_I) = EI_y\left(-\frac{2}{L}+\frac{6x_I}{L^2}\right)\varphi_{2y} = \frac{q_0L^3}{56}\left(-\frac{2}{L}+\frac{6x_I}{L^2}\right), \tag{3.197}$$

$$M_y^e(x_{\text{II}}) = EI_y \left(\left[-\frac{4}{L} + \frac{6x_{\text{II}}}{L^2} \right] \varphi_{2y} + \left[-\frac{2}{L} + \frac{6x_{\text{II}}}{L^2} \right] \varphi_{3y} \right)$$

$$= \frac{q_0 L^3}{28} \left(\frac{1}{2} \left[-\frac{4}{L} + \frac{6x_{\text{II}}}{L^2} \right] - \frac{5}{6} \left[-\frac{2}{L} + \frac{6x_{\text{II}}}{L^2} \right] \right). \tag{3.198}$$

It is easy to check that the values at the very left- and right-hand boundary correspond in magnitude to the external loads: $M_y^e(x_{\text{I}} = 0) = -\frac{q_0 L^2}{28}$ and $M_y^e(x_{\text{II}} = L) = -\frac{q_0 L^2}{12}$. The graphical representation of the bending moment is shown in Fig. 3.28. It can be seen that the magnitude of the bending moment equals the external single moments at the left and right-hand end. Furthermore, the jump in the middle equals the sum of the single moments at $X = L$.

$$Q_z^e(x_{\text{I}}) = EI_y \frac{6}{L^2} \varphi_{2y} = \frac{3}{28} q_0 L, \tag{3.199}$$

$$Q_z^e(x_{\text{II}}) = EI_y \left(\frac{6}{L^2} \varphi_{2y} + \frac{6}{L^2} \varphi_{3y} \right) = -\frac{1}{14} q_0 L. \tag{3.200}$$

The graphical representation of the shear force is shown in Fig. 3.29. It can be seen that the shear force corresponds in magnitude to the external forces at the very left- and right-hand boundary, as well as in the middle of the beam structure.

The normal stress distribution can be obtained from the bending moments given in Eqs. (3.197) and (3.198):

$$\sigma_x^e(x_{\text{I}}, z) = \frac{M_y^e(x_{\text{I}})}{I_y} z = \frac{q_0 L^3}{56 I_y} \left(-\frac{2}{L} + \frac{6x_{\text{I}}}{L^2} \right) z, \tag{3.201}$$

Fig. 3.28 Bending moment distribution

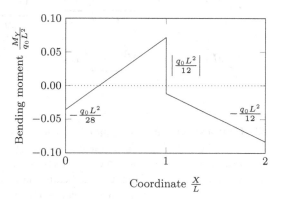

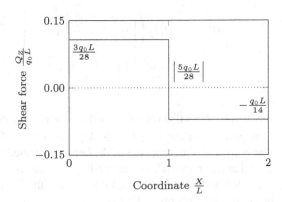

Fig. 3.29 Shear force distribution

$$\sigma_x^e(x_{II}, z) = \frac{M_y^e(x_{II})}{I_y} z = \frac{q_0 L^3}{28 I_y}\left(\frac{1}{2}\left[-\frac{4}{L}+\frac{6x_{II}}{L^2}\right] - \frac{5}{6}\left[-\frac{2}{L}+\frac{6x_{II}}{L^2}\right]\right) z . \quad (3.202)$$

The strains result from HOOKE's law:

$$\varepsilon_x^e(x_I, z) = \frac{\sigma_x^e(x_I, z)}{E} = \frac{q_0 L^3}{56 E I_y}\left(-\frac{2}{L}+\frac{6x_I}{L^2}\right) z , \quad (3.203)$$

$$\varepsilon_x^e(x_{II}, z) = \frac{\sigma_x^e(x_{II}, z)}{E} = \frac{q_0 L^3}{28 E I_y}\left(\frac{1}{2}\left[-\frac{4}{L}+\frac{6x_{II}}{L^2}\right] - \frac{5}{6}\left[-\frac{2}{L}+\frac{6x_{II}}{L^2}\right]\right) z . \quad (3.204)$$

⑩ Check the global equilibrium between the external loads and the support reactions.

$$\sum_i F_{iZ} = 0 \quad \Leftrightarrow \quad \underbrace{(F_{1Z}^R + F_{2Z}^R + F_{3Z}^R)}_{\text{reaction force}} + \underbrace{\left(-\frac{q_0 L}{2} - \frac{q_0 L}{2}\right)}_{\text{external load}} = 0, \checkmark \quad (3.205)$$

$$\sum_i M_{iY} = 0 \quad \Leftrightarrow \quad \underbrace{(M_{1Y}^R - F_{2Z}^R L - F_{3Z}^R 2L)}_{\text{reaction}} + \underbrace{\left(\frac{q_0 L}{2}L + \frac{q_0 L}{2}2L\right)}_{\text{external load}} = 0 . \checkmark \quad (3.206)$$

3.9 Cantilever beam with supporting rod

The beam shown in Fig. 3.30 is loaded by a triangular shaped distributed load (maximum value q_0) and a single force F_0 at its right-hand end. The bending stiffness EI is constant and the total length of the beam is equal to L. The beam is supported at its right-hand end by a rod, which is inclined by 45°. The rod is characterized by its

Fig. 3.30 Cantilever beam with supporting rod

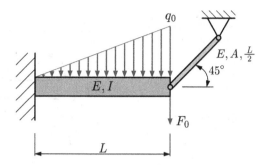

constant tensile stiffness EA and length $\frac{1}{2}L$.

Model the beam/rod structure with two finite elements to determine:

- the free-body diagram,
- the unknowns at the nodes,
- the reactions at the supports,
- the internal reactions in the beam (shear force and bending moment) and in the rod (normal force), and
- the global force and moment equilibrium.

3.9 Solution

The solution will follow the recommended 10 steps outlined on Sect. 3.2.

① Sketch the free-body diagram of the problem, including a global coordinate system, see Fig. 3.31.
② Subdivide the geometry into finite elements. Indicate the node and element numbers, local coordinate systems, and equivalent nodal loads, see Fig. 3.32.
③ Write separately all elemental stiffness matrices expressed in the global coordinate system. Indicate the nodal unknowns on the right-hand sides and over the matrices.

Since the beam element is horizontal, it does not require any transformation to the global coordinate system and thus, the standard stiffness matrix as given in Eq. (3.164) can be used:

$$
\boldsymbol{K}_I^e = \frac{EI_Y}{L^3}
\begin{array}{c}
\begin{array}{cccc} u_{1Z} & \varphi_{1Y} & u_{2Z} & \varphi_{2Y} \end{array} \\
\begin{bmatrix}
12 & -6L & -12 & -6L \\
-6L & 4L^2 & 6L & 2L^2 \\
-12 & 6L & 12 & 6L \\
-6L & 2L^2 & 6L & 4L^2
\end{bmatrix}
\end{array}
\begin{array}{c}
u_{1Z} \\ \varphi_{1Y} \\ u_{2Z} \\ \varphi_{2Y}
\end{array} ,
\tag{3.207}
$$

Fig. 3.31 Free-body
diagram of the cantilever
beam with supporting rod

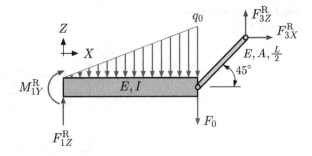

Fig. 3.32 Free-body
diagram of the discretized
structure with equivalent
nodal loads

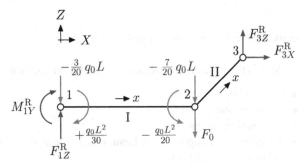

The rod element is rotated by an angle of $\alpha = -45°$:

$$
\mathbf{K}^{e}_{\mathrm{VI}} = \frac{EA}{\frac{1}{2}L}
\begin{array}{c c c c}
u_{2X} & u_{2Z} & u_{3X} & u_{3Z}
\end{array}
\begin{bmatrix}
\frac{1}{2} & \frac{1}{2} & -\frac{1}{2} & -\frac{1}{2} \\
\frac{1}{2} & \frac{1}{2} & -\frac{1}{2} & -\frac{1}{2} \\
-\frac{1}{2} & -\frac{1}{2} & \frac{1}{2} & \frac{1}{2} \\
-\frac{1}{2} & -\frac{1}{2} & \frac{1}{2} & \frac{1}{2}
\end{bmatrix}
\begin{array}{l}
u_{2X} \\ u_{2Z} \\ u_{3X} \\ u_{3Z}
\end{array}
\tag{3.208}
$$

$$
= \frac{EA}{L}
\begin{array}{c c c c}
u_{2X} & u_{2Z} & u_{3X} & u_{3Z}
\end{array}
\begin{bmatrix}
1 & 1 & -1 & -1 \\
1 & 1 & -1 & -1 \\
-1 & -1 & 1 & 1 \\
-1 & -1 & 1 & 1
\end{bmatrix}
\begin{array}{l}
u_{2X} \\ u_{2Z} \\ u_{3X} \\ u_{3Z}
\end{array}
\tag{3.209}
$$

④ Determine the dimensions of the global stiffness matrix and sketch the structure
of this matrix with global unknowns on the right-hand side and over the matrix.

The finite element structure is composed of 3 nodes. The first node has two degrees
of freedom (deflection and rotation), the second node has three degrees of freedom (a
vertical and horizontal displacement as well as a rotation) while the third node has two

degrees of freedom (a vertical and horizontal displacement). Thus, the dimensions of the global stiffness matrix are (7×7):

$$
K =
\begin{bmatrix}
& & & & & & \\
& & & & & & \\
& & & & & & \\
& & & & & & \\
& & & & & & \\
& & & & & & \\
& & & & & &
\end{bmatrix}
\begin{matrix}
u_{1Z} \\ \varphi_{1Y} \\ u_{2Z} \\ \varphi_{2Y} \\ u_{2X} \\ u_{3X} \\ u_{3Z}
\end{matrix}
\qquad (3.210)
$$

with column headers $u_{1Z}\ \varphi_{1Y}\ u_{2Z}\ \varphi_{2Y}\ u_{2X}\ u_{3X}\ u_{3Z}$.

⑤ Insert the values of the elemental stiffness matrices step-by-step into the global stiffness matrix.

$$
K =
\begin{bmatrix}
\frac{12EI}{L^3} & -\frac{6EI}{L^2} & -\frac{12EI}{L^3} & -\frac{6EI}{L^2} & 0 & 0 & 0 \\
-\frac{6EI}{L^2} & \frac{4EI}{L} & \frac{6EI}{L^2} & \frac{2EI}{L} & 0 & 0 & 0 \\
-\frac{12EI}{L^3} & \frac{6EI}{L^2} & \frac{12EI}{L^3}+\frac{EA}{L} & \frac{6EI}{L^2} & \frac{EA}{L} & -\frac{EA}{L} & -\frac{EA}{L} \\
-\frac{6EI}{L^2} & \frac{2EI}{L} & \frac{6EI}{L^2} & \frac{4EI}{L} & 0 & 0 & 0 \\
0 & 0 & \frac{EA}{L} & 0 & \frac{EA}{L} & -\frac{EA}{L} & -\frac{EA}{L} \\
0 & 0 & -\frac{EA}{L} & 0 & -\frac{EA}{L} & \frac{EA}{L} & \frac{EA}{L} \\
0 & 0 & -\frac{EA}{L} & 0 & -\frac{EA}{L} & \frac{EA}{L} & \frac{EA}{L}
\end{bmatrix}
\begin{matrix}
u_{1Z} \\ \varphi_{1Y} \\ u_{2Z} \\ \varphi_{2Y} \\ u_{2X} \\ u_{3X} \\ u_{3Z}
\end{matrix}
\qquad (3.211)
$$

with column headers $u_{1Z}\ \varphi_{1Y}\ u_{2Z}\ \varphi_{2Y}\ u_{2X}\ u_{3X}\ u_{3Z}$.

⑥ Add the column matrix of unknowns and external loads to complete the global system of equations.

$$
\begin{bmatrix}
\frac{12EI}{L^3} & -\frac{6EI}{L^2} & -\frac{12EI}{L^3} & -\frac{6EI}{L^2} & 0 & 0 & 0 \\
-\frac{6EI}{L^2} & \frac{4EI}{L} & \frac{6EI}{L^2} & \frac{2EI}{L} & 0 & 0 & 0 \\
-\frac{12EI}{L^3} & \frac{6EI}{L^2} & \frac{12EI}{L^3}+\frac{EA}{L} & \frac{6EI}{L^2} & \frac{EA}{L} & -\frac{EA}{L} & -\frac{EA}{L} \\
-\frac{6EI}{L^2} & \frac{2EI}{L} & \frac{6EI}{L^2} & \frac{4EI}{L} & 0 & 0 & 0 \\
0 & 0 & \frac{EA}{L} & 0 & \frac{EA}{L} & -\frac{EA}{L} & -\frac{EA}{L} \\
0 & 0 & -\frac{EA}{L} & 0 & -\frac{EA}{L} & \frac{EA}{L} & \frac{EA}{L} \\
0 & 0 & -\frac{EA}{L} & 0 & -\frac{EA}{L} & \frac{EA}{L} & \frac{EA}{L}
\end{bmatrix}
\begin{bmatrix}
u_{1Z} \\ \varphi_{1Y} \\ u_{2Z} \\ \varphi_{2Y} \\ u_{2X} \\ u_{3X} \\ u_{3Z}
\end{bmatrix}
=
\begin{bmatrix}
F_{1Z}^{R} - \frac{3}{20}q_0 L \\
M_{1Y}^{R} + \frac{q_0 L^2}{30} \\
-F_0 - \frac{7}{20}q_0 L \\
-\frac{q_0 L^2}{20} \\
F_{2X}^{R} \\
F_{3X}^{R} \\
F_{3Z}^{R}
\end{bmatrix}
.
$$

$$(3.212)$$

⑦ Introduce the boundary conditions to obtain the reduced system of equations.

The obvious support conditions are at node 1 and 3, i.e., $u_{1Z} = 0$, $\varphi_{1Y} = 0$ and $u_{3X} = u_{3Z} = 0$. However, it is also important to consider that the beam cannot have any elongation in the X-direction at node 2: $u_{2X} = 0$ (see Fig. 3.33 for details). Thus, a reduced 2×2 system is obtained:

Fig. 3.33 Support
conditions for the truss
element ('as seen' from the
truss element)

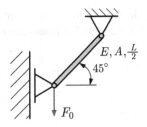

$$\begin{bmatrix} \dfrac{12EI_Y}{L^3} + \dfrac{EA}{L} & \dfrac{6EI_Y}{L^2} \\[3mm] \dfrac{6EI_Y}{L^2} & \dfrac{4EI_Y}{L} \end{bmatrix} \begin{bmatrix} u_{2Z} \\[3mm] \varphi_{2Y} \end{bmatrix} = \begin{bmatrix} -F_0 - \dfrac{7}{20}q_0 L \\[3mm] -\dfrac{q_0 L^2}{20} \end{bmatrix} . \tag{3.213}$$

⑧ Solve the reduced system of equations to obtain the unknown nodal deformations.

The solution can be obtained based on the matrix approach $u_{\mathrm{p}} = K^{-1} f$:

$$\begin{bmatrix} u_{2Z} \\[3mm] \varphi_{2Y} \end{bmatrix} = \dfrac{1}{\frac{4EI_Y}{L}\left(\frac{12EI_Y}{L^3} + \frac{EA}{L}\right) - \left(\frac{6EI_Y}{L^2}\right)^2} \begin{bmatrix} \dfrac{4EI_Y}{L} & -\dfrac{6EI_Y}{L^2} \\[3mm] -\dfrac{6EI_Y}{L^2} & \dfrac{12EI_Y}{L^3} + \dfrac{EA}{L} \end{bmatrix} \begin{bmatrix} -F_0 - \dfrac{7q_0 L}{20} \\[3mm] -\dfrac{q_0 L^2}{20} \end{bmatrix}$$

$$= \dfrac{L^4}{12(EI_Y)^2 + 4(EI_Y)(EA)L^2} \begin{bmatrix} -\dfrac{4EI_Y F_0}{L} - \dfrac{11EI_Y q_0}{10} \\[3mm] \dfrac{6EI_Y F_0}{L^2} + \dfrac{3EI_Y q_0}{2L} - \dfrac{EA q_0 L}{20} \end{bmatrix}$$

$$= \begin{bmatrix} -\dfrac{1}{40} \times \dfrac{L^3(11q_0 L + 40F_0)}{E(AL^2 + 3I_Y)} \\[3mm] -\dfrac{1}{80} \times \dfrac{L^2(Aq_0 L^3 - 30I_Y q_0 L - 120I_Y F_0)}{EI_Y(AL^2 + 3I_Y)} \end{bmatrix} . \tag{3.214}$$

⑨ Post-computation: determination of reactions, stresses and strains.

The reactions at the supports can be obtained from the non-reduced system of equations as given in step ⑥ under the consideration of the known nodal degrees of freedom (i.e., displacements and rotations). The evaluation of the first equation gives:

$$-12\dfrac{EI_Y}{L^3} u_{2Z} - 6\dfrac{EI_Y}{L^2}\varphi_{2Y} = F_{1Z}^{\mathrm{R}} - \dfrac{3}{20}q_0 L \tag{3.215}$$

$$\Rightarrow \quad F_{1Z}^{\mathrm{R}} = \dfrac{3}{40} \times \dfrac{3AL^3 q_0 + 20I_Y q_0 L + 40I_Y F_0}{AL^2 + 3I_Y} . \tag{3.216}$$

The other reactions can be obtained in a similar way as:

$$M_{1Y}^{R} = -\frac{L(7AL^3 q_0 + 120 I_Y q_0 L + 360 I_Y F_0)}{120(AL^2 + 3I_Y)}, \tag{3.217}$$

$$F_{2X}^{R} = -\frac{AL^2(11 q_0 L + 40 F_0)}{40(AL^2 + 3I_Y)}, \tag{3.218}$$

$$F_{3X}^{R} = \frac{AL^2(11 q_0 L + 40 F_0)}{40(AL^2 + 3I_Y)}, \tag{3.219}$$

$$F_{3Z}^{R} = \frac{AL^2(11 q_0 L + 40 F_0)}{40(AL^2 + 3I_Y)}. \tag{3.220}$$

It should be noted that the evaluation of the third and fourth equation can be used to check—to a certain extent—the system of equations since the result must be equal to the given value on the right-hand side.

The internal reactions (i.e., bending moment and shear force) in the beam element can be obtained from the relations provided in Table 3.8.

$$M_Y^e(x_\mathrm{I}) = E I_Y \left(0 + 0 + \left[-\frac{6}{L^2} + \frac{12 x_\mathrm{I}}{L^3} \right] u_{2Z} + \left[\frac{2}{L} + \frac{6 x_\mathrm{I}}{L^2} \right] \varphi_{2Y} \right), \tag{3.221}$$

$$Q_Z^e(x_\mathrm{I}) = E I_Y \left(0 + 0 + \left[\frac{12}{L^3} \right] u_{2Z} + \left[\frac{6}{L^2} \right] \varphi_{2Y} \right) \tag{3.222}$$

$$= -\frac{3}{40} \times \frac{AL^3 q_0 + 14 I_Y q_0 L + 40 I_Y F_0}{AL^2 + 3I_Y}. \tag{3.223}$$

The internal reaction (i.e., normal force) in the rod element can be obtained from the relation provided in Table 3.5:

$$N_x^e(x_\mathrm{II}) = \frac{E A}{L/2} (\sin(\alpha) u_{2Z}) = \frac{AL^2 \sqrt{2}(11 q_0 L + 4 F_0)}{40(AL^2 + 3I_Y)}. \tag{3.224}$$

⑩ Check the global equilibrium between the external loads and the support reactions.

$$\sum_i F_{iX} = 0 \Leftrightarrow \underbrace{(F_{2X}^{R} + F_{3X}^{R})}_{\text{reaction force}} + \underbrace{0}_{\text{external load}} = 0, \ \checkmark \tag{3.225}$$

$$\sum_i F_{iZ} = 0 \Leftrightarrow \underbrace{(F_{1Z}^{R} + F_{3Z}^{R})}_{\text{reaction force}} + \underbrace{\left(-F_0 - \frac{3q_0 L}{20} - \frac{7q_0 L}{20}\right)}_{\text{external load}} = 0, \ \checkmark \tag{3.226}$$

$$\sum_i M_{iY} = 0 \quad \Leftrightarrow \quad \underbrace{(M_{1Y}^{R} - F_{3Z}^{R}(L + L/4\sqrt{2}) + F_{3X}^{R}L/4\sqrt{2})}_{\text{reaction}}$$

$$+ \underbrace{\left(q_0 L^2/30 + F_0 L + 7/20 q_0 L L - q_0 L^2/20\right)}_{\text{external load}} = 0 . \checkmark \qquad (3.227)$$

3.3.2   Timoshenko Beam Elements

There are many different formulations for TIMOSHENKO beams available in the scientific literature [8, 10]. A very early and simple derivation is based on linear interpolation functions for the displacement and rotational fields. For this purpose, let us consider in the following a TIMOSHENKO beam element which is composed of two nodes as schematically shown in Fig. 3.34. Each node has two degrees of freedom, i.e., a displacement u_z in the direction of the z-axis (i.e., perpendicular to the principal beam axis) and a rotation ϕ_y around the y-axis, see Fig. 3.34a. Each node can be loaded by single forces acting in the z-direction or single moments around the y-axis, see Fig. 3.34b. In the case of distributed loads $q_z(x)$, a transformation to equivalent nodal loads is required.

Different methods can be found in the literature to derive the principal finite element equation (see [2, 5]). All these methods result in the same elemental formulation, which is given in the following for constant material (E, G), geometrical (I_y, A, k_s) properties and linear interpolation functions:

Fig. 3.34 Definition of the TIMOSHENKO beam element for deformation in the x-z plane: **a** deformations, and **b** external loads. The nodes are symbolized by two circles at the end ($\bigcirc$)

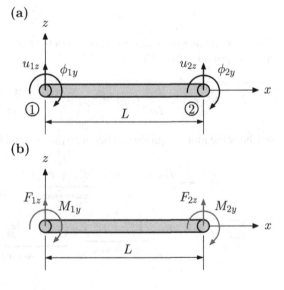

$$\frac{k_s AG}{4L} \begin{bmatrix} 4 & -2L & -4 & -2L \\ -2L & \frac{4}{3}L^2 + \alpha & 2L & \frac{4}{6}L^2 - \alpha \\ -4 & 2L & 4 & 2L \\ -2L & \frac{4}{6}L^2 - \alpha & 2L & \frac{4}{3}L^2 + \alpha \end{bmatrix} \begin{bmatrix} u_{1z} \\ \phi_{1y} \\ u_{2z} \\ \phi_{2y} \end{bmatrix} = \begin{bmatrix} F_{1z} \\ M_{1y} \\ F_{2z} \\ M_{2y} \end{bmatrix} + \int\limits_0^L q_z(x) \begin{bmatrix} N_{1u} \\ 0 \\ N_{2u} \\ 0 \end{bmatrix} dx,$$

$$(3.228)$$

where the abbreviation $\alpha = \frac{4EI_y}{k_s AG}$ was used. In abbreviated form, we can write

$$K^e u_p^e = f^e,$$
                                                                                         $$(3.229)$$

where K^e is the elemental stiffness matrix, u_p^e is the elemental column matrix of unknowns and f^e is the elemental column matrix of loads. The shape functions in Eq. (3.228) are given by $N_{1u}(x) = 1 - \frac{x}{L}$ and $N_{2u}(x) = \frac{x}{L}$. Table 3.9 summarizes for some simple shapes of distributed loads the equivalent nodal loads. It is obvious from Eq. (3.228) that this simple element formulation, based on linear interpolation functions for the displacement and rotational field and exact integration, yields an equivalent load vector only with force contributions whereas moment contributions are not considered, see Table 3.9.

Once the nodal displacements $(u_{1z}, \phi_{1y}, u_{2z}, \phi_{2y})$ are known, further quantities and their distributions can be calculated within an element (so-called postprocessing), see Table 3.10.

3.10 Beam under pure bending load

The cantilever TIMOSHENKO beam shown in Fig. 3.35 is loaded by a moment M_0 at the free right-hand end. The bending stiffness EI and the shear stiffness $k_s AG$ are constant and the total length of the beam is equal to L. Model the beam with one single linear TIMOSHENKO finite element to determine:

- the unknowns at the nodes,
- the equation of the bending line and the distribution of the rotation,
- the reactions at the support,
- the internal reactions (shear force and bending moment distribution) in the element,
- the strain and stress distributions in the element,
- the global force and moment equilibrium, and
- sketch the deflection of the load application point as a function of the slenderness ratio $\frac{h}{L}$ for $\nu = 0.0, 0.3$ and 0.5.

3.10 Solution

The solution will follow the recommended 10 steps outlined on Sect. 3.2.

① Sketch the free-body diagram of the problem, including a global coordinate system (see Fig. 3.36).
② Subdivide the geometry into finite elements. Indicate the node and element numbers, local coordinate systems, and equivalent nodal loads (see Fig. 3.37).

Table 3.9 Equivalent nodal loads for a linear TIMOSHENKO beam element (x-axis: right facing; z-axis: upward facing)

Loading	Shear force	Bending moment
	$F_{1z} = -\dfrac{qL}{2}$	$M_{1y} = 0$
	$F_{2z} = -\dfrac{qL}{2}$	$M_{2y} = 0$
	$F_{1z} = \dfrac{qa^2}{2L} - qa$	$M_{1y} = 0$
	$F_{2z} = -\dfrac{qa^2}{2L}$	$M_{2y} = 0$
	$F_{1z} = -\dfrac{1}{6}qL$	$M_{1y} = 0$
	$F_{2z} = -\dfrac{1}{3}qL$	$M_{2y} = 0$
	$F_{1z} = -\dfrac{1}{4}qL$	$M_{1y} = 0$
	$F_{2z} = -\dfrac{1}{4}qL$	$M_{2y} = 0$
	$F_{1z} = -\dfrac{F}{2}$	$M_{1y} = 0$
	$F_{2z} = -\dfrac{F}{2}$	$M_{2y} = 0$

Steps ③ to ⑥ can be combined since we have only a single element problem. The global system of equations reads:

$$\frac{k_s AG}{4L} \begin{bmatrix} 4 & -2L & -4 & -2L \\ -2L & \frac{4}{3}L^2 + \alpha & 2L & \frac{4}{6}L^2 - \alpha \\ -4 & 2L & 4 & 2L \\ -2L & \frac{4}{6}L^2 - \alpha & 2L & \frac{4}{3}L^2 + \alpha \end{bmatrix} \begin{bmatrix} u_{1Z} \\ \phi_{1Y} \\ u_{2Z} \\ \phi_{2Y} \end{bmatrix} = \begin{bmatrix} F_{1Z}^{\mathrm{R}} \\ M_{1Y}^{\mathrm{R}} \\ 0 \\ -M_0 \end{bmatrix}. \tag{3.230}$$

⑦ Introduce the boundary conditions to obtain the reduced system of equations.

$$\frac{k_s AG}{4L} \begin{bmatrix} 4 & 2L \\ 2L & \frac{4}{3}L^2 + \alpha \end{bmatrix} \begin{bmatrix} u_{2Z} \\ \phi_{2Y} \end{bmatrix} = \begin{bmatrix} 0 \\ -M_0 \end{bmatrix}. \tag{3.231}$$

Table 3.10 Displacement, rotation, curvature, shear strain, shear force and bending moment distribution for a linear TIMOSHENKO beam element given as a function of the nodal values in Cartesian and natural coordinates (bending occurs in the x-z plane)

Vertical displacement (Deflection) u_z
$u_z^e(x) = \left[1 - \frac{x}{L}\right] u_{1z} + \left[\frac{x}{L}\right] u_{2z}$ $u_z^e(\xi) = \left[\frac{1}{2}(1 - \xi)\right] u_{1z} + \left[\frac{1}{2}(1 + \xi)\right] u_{2z}$
Rotation ϕ_y
$\phi_y^e(x) = \left[1 - \frac{x}{L}\right] \phi_{1y} + \left[\frac{x}{L}\right] \phi_{2y}$ $\phi_y^e(\xi) = \left[\frac{1}{2}(1 - \xi)\right] \phi_{1y} + \left[\frac{1}{2}(1 + \xi)\right] \phi_{2y}$
Curvature $\kappa_y = \dfrac{d\phi_y}{dx} = \dfrac{d\phi_y}{d\xi}\dfrac{d\xi}{dx} = \dfrac{2}{L}\dfrac{d\phi_y}{d\xi}$
$\kappa_y^e(x) = \left[-\frac{1}{L}\right] \phi_{1y} + \left[\frac{1}{L}\right] \phi_{2y}$ $\kappa_y^e(\xi) = \left[-\frac{1}{L}\right] \phi_{1y} + \left[\frac{1}{L}\right] \phi_{2y}$
Shear strain $\gamma_{xz} = \dfrac{du_z}{dx} + \phi_y = \dfrac{d\xi}{dx}\dfrac{du_z}{d\xi} + \phi_y$
$\gamma_{xz}^e(x) = \left[-\frac{1}{L}\right] u_{1z} + \left[\frac{1}{L}\right] u_{2z} + \left[1 - \frac{x}{L}\right] \phi_{1y} + \left[\frac{x}{L}\right] \phi_{2y}$ $\gamma_{xz}^e(\xi) = \left[-\frac{1}{L}\right] u_{1z} + \left[\frac{1}{L}\right] u_{2z} + \left[\frac{1}{2}(1 - \xi)\right] \phi_{1y} + \left[\frac{1}{2}(1 + \xi)\right] \phi_{2y}$
Shear force $Q_z = k_s A G \gamma_{xz} = k_s A G \left(\dfrac{du_z}{dx} + \phi_y \right)$
$Q_z^e(x) = k_s A G \left(\left[-\frac{1}{L}\right] u_{1z} + \left[\frac{1}{L}\right] u_{2z} + \left[1 - \frac{x}{L}\right] \phi_{1y} + \left[\frac{x}{L}\right] \phi_{2y} \right)$ $Q_z^e(\xi) = k_s A G \left(\left[-\frac{1}{L}\right] u_{1z} + \left[\frac{1}{L}\right] u_{2z} + \left[\frac{1}{2}(1 - \xi)\right] \phi_{1y} + \left[\frac{1}{2}(1 + \xi)\right] \phi_{2y} \right)$
Bending moment $M_y = +EI_y \kappa_y = EI_y \dfrac{d\phi_y}{dx} = EI_y \dfrac{d\phi_y}{d\xi}\dfrac{d\xi}{dx}$
$M_y^e(x) = EI_y \left(\left[-\frac{1}{L}\right] \phi_{1y} + \left[\frac{1}{L}\right] \phi_{2y} \right)$ $M_y^e(\xi) = EI_y \left(\left[-\frac{1}{L}\right] \phi_{1y} + \left[\frac{1}{L}\right] \phi_{2y} \right)$

Fig. 3.35 Beam loaded under pure bending moment

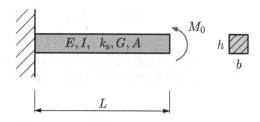

⑧ Solve the reduced system of equations to obtain the unknown nodal deformations.

The solution can be obtained based on the matrix approach $u_p = K^{-1} f$:

Fig. 3.36 Free-body diagram of the beam loaded under pure bending moment

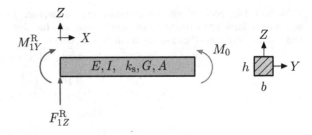

Fig. 3.37 Free-body diagram of the discretized structure

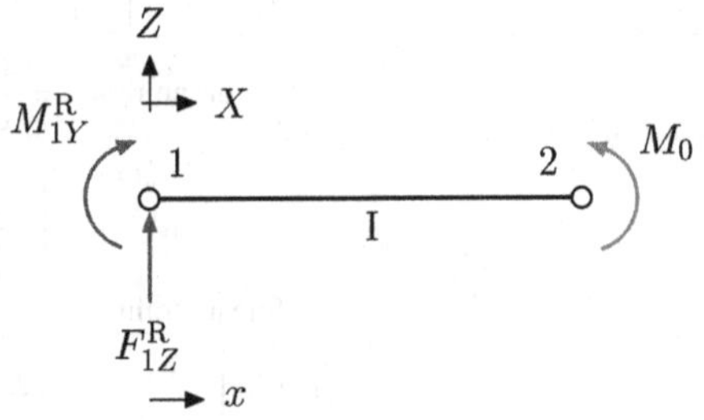

$$
\begin{bmatrix} u_{2Z} \\ \phi_{2Y} \end{bmatrix} = \frac{4L}{k_s AG} \times \frac{1}{4\left(\frac{4}{3}L^2 + \alpha\right) - 4L^2} \begin{bmatrix} \frac{4}{3}L^2 + \alpha & -2L \\ -2L & 4 \end{bmatrix} \begin{bmatrix} 0 \\ -M_0 \end{bmatrix}
$$

$$
= \frac{L}{k_s AG} \times \frac{1}{\frac{L^2}{3} + \alpha} \begin{bmatrix} 2M_0 L \\ -4M_0 \end{bmatrix}
$$

$$
= \frac{3L}{k_s AGL^2 + 12EI} \begin{bmatrix} 2M_0 L \\ -4M_0 \end{bmatrix}. \tag{3.232}
$$

Based on these nodal unknowns, the bending line (deflection) and the rotational distribution can be obtained from the general equations provided in Table 3.10:

$$
u_z^e(x) = \begin{bmatrix} \frac{x}{L} \end{bmatrix} u_{2Z} = \frac{6M_0 L^2}{k_s AGL^2 + 12EI} \times \begin{bmatrix} \frac{x}{L} \end{bmatrix}, \tag{3.233}
$$

$$
\phi_y^e(x) = \begin{bmatrix} \frac{x}{L} \end{bmatrix} \phi_{2Y} = -\frac{12M_0 L}{k_s AGL^2 + 12EI} \times \begin{bmatrix} \frac{x}{L} \end{bmatrix}. \tag{3.234}
$$

⑨ Post-computation: determination of reactions, stresses and strains.

The reactions at the supports can be obtained from the non-reduced system of equations as given in step ⑥ under the consideration of the known nodal degrees of freedom (i.e., displacement and rotation). The evaluation of the first equation gives:

$$\frac{k_s AG}{4L}(-4u_{2Z} - 2L\varphi_{2Y}) = F_{1Z}^R \tag{3.235}$$

$$\Rightarrow \quad F_{1Z}^R = 0. \tag{3.236}$$

The evaluation of the second equation gives in a similar way:

$$M_{1Y}^R = M_0. \tag{3.237}$$

The internal reactions (i.e., bending moment and shear force) in the element can be obtained from the relations provided in Table 3.10.

$$M_y^e(x) = EI \begin{bmatrix} \frac{1}{L} \end{bmatrix} \phi_{2Y} = -\frac{12EI M_0}{k_s AGL^2 + 12EI}, \tag{3.238}$$

$$Q_z^e(x) = k_s AG \left(\begin{bmatrix} \frac{1}{L} \end{bmatrix} u_{2Z} + \begin{bmatrix} \frac{x}{L} \end{bmatrix} \phi_{2Y} \right)$$

$$= \frac{6k_s AG M_0(L - 2x)}{k_s AGL^2 + 12EI}. \tag{3.239}$$

The normal and shear stress distributions can be obtained from Eqs. (2.125) and (2.126):

$$\sigma_x^e(x, z) = \frac{M_y^e(x)}{I} z(x) = -\frac{12E M_0}{k_s AGL^2 + 12EI} z(x), \tag{3.240}$$

$$\tau_{xz}^e(x) = \frac{Q_z^e(x)}{k_s A} = \frac{6G M_0(L - 2x)}{k_s AGL^2 + 12EI}. \tag{3.241}$$

⑩ Check the global equilibrium between the external loads and the support reactions.

$$\sum_i F_{iz} = 0 \quad \Leftrightarrow \quad \underbrace{(F_{1Z}^R)}_{\text{reaction force}} + \underbrace{(0)}_{\text{external load}} = 0, \checkmark \tag{3.242}$$

$$\sum_i M_{iY} = 0 \quad \Leftrightarrow \quad \underbrace{(M_{1Y}^R)}_{\text{reaction}} + \underbrace{(-M_0)}_{\text{external load}} = 0. \checkmark \tag{3.243}$$

Deflection of the load application point as a function of the slenderness ratio $\frac{h}{L}$ for $\nu = 0.0, 0.3$ and 0.5.

Fig. 3.38 Comparison of the finite element solution for a linear TIMOSHENKO element based on analytical integration with the analytical solutions for beam problems

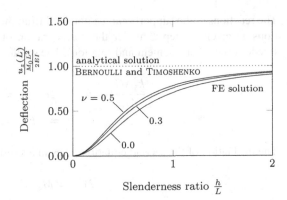

Fig. 3.39 Beam loaded by a single force

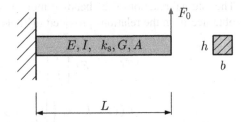

The nodal displacement at node 2 is given in Eq. (3.232):

$$u_{2Z} = \frac{6M_0L^2}{k_s AGL^2 + 12EI}. \tag{3.244}$$

Considering that $G = \frac{E}{2(1+\nu)}$, $A = bh$, $I = \frac{bh^3}{12}$, and $k_s = \frac{5}{6}$ allows us to express the last equation as:

$$u_{2Z} = \frac{1+\nu}{\frac{5}{12}\left(\frac{L}{h}\right)^2 + (1+\nu)} \times \frac{M_0L^2}{2EI}. \tag{3.245}$$

The graphical representation of this equation for different values of POISSON's ratio is given in Fig. 3.38.

3.11 Beam loaded by a single force

The cantilever TIMOSHENKO beam shown in Fig. 3.39 is loaded by a single force F_0 at the free right-hand end. The bending stiffness EI and the shear stiffness $k_s AG$ are constant and the total length of the beam is equal to L. Model the beam with one single linear TIMOSHENKO finite element to determine:

- the unknowns at the nodes,
- the equation of the bending line and the distribution of the rotation,
- the reactions at the support,

- the internal reactions (shear force and bending moment distribution) in the element,
- the strain and stress distributions in the element,
- the global force and moment equilibrium, and
- sketch the deflection of the load application point as a function of the slenderness ratio $\frac{h}{L}$ for $\nu = 0.0, 0.3$ and 0.5.

3.11 Solution

The solution will follow the recommended 10 steps outlined on Sect. 3.2.

① Sketch the free-body diagram of the problem, including a global coordinate system (see Fig. 3.40).

② Subdivide the geometry into finite elements. Indicate the node and element numbers, local coordinate systems, and equivalent nodal loads (see Fig. 3.41).

Steps ③ to ⑥ can be combined since we have only a single element problem. The global system of equations reads:

$$\frac{k_s AG}{4L} \begin{bmatrix} 4 & -2L & -4 & -2L \\ -2L & \frac{4}{3}L^2 + \alpha & 2L & \frac{4}{6}L^2 - \alpha \\ -4 & 2L & 4 & 2L \\ -2L & \frac{4}{6}L^2 - \alpha & 2L & \frac{4}{3}L^2 + \alpha \end{bmatrix} \begin{bmatrix} u_{1Z} \\ \phi_{1Y} \\ u_{2Z} \\ \phi_{2Y} \end{bmatrix} = \begin{bmatrix} F_{1Z}^R \\ M_{1Y}^R \\ F_0 \\ 0 \end{bmatrix} . \tag{3.246}$$

⑦ Introduce the boundary conditions to obtain the reduced system of equations.

$$\frac{k_s AG}{4L} \begin{bmatrix} 4 & 2L \\ 2L & \frac{4}{3}L^2 + \alpha \end{bmatrix} \begin{bmatrix} u_{2Z} \\ \phi_{2Y} \end{bmatrix} = \begin{bmatrix} F_0 \\ 0 \end{bmatrix} . \tag{3.247}$$

⑧ Solve the reduced system of equations to obtain the unknown nodal deformations.

The solution can be obtained based on the matrix approach $\boldsymbol{u}_\mathrm{p} = \boldsymbol{K}^{-1} \boldsymbol{f}$:

Fig. 3.40 Free-body diagram of the beam loaded by a single force

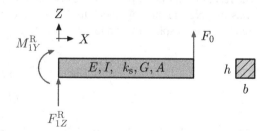

Fig. 3.41 Free-body
diagram of the discretized
structure

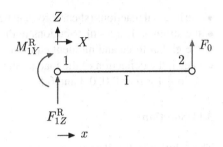

$$
\begin{bmatrix} u_{2Z} \\ \phi_{2Y} \end{bmatrix} = \frac{4L}{k_s AG} \times \frac{1}{4\left(\frac{4}{3}L^2 + \alpha\right) - 4L^2} \begin{bmatrix} \frac{4}{3}L^2 + \alpha & -2L \\ -2L & 4 \end{bmatrix} \begin{bmatrix} F_0 \\ 0 \end{bmatrix}
$$

$$
= \frac{F_0 L}{k_s AGL^2 + 12EI} \begin{bmatrix} \dfrac{4(k_s AGL^2 + 3EI)}{k_s AG} \\ -6L^2 \end{bmatrix}. \tag{3.248}
$$

Based on these nodal unknowns, the bending line (deflection) and the rotational
distribution can be obtained from the general equations provided in Table 3.10:

$$
u_z^e(x) = \begin{bmatrix} \dfrac{x}{L} \end{bmatrix} u_{2Z} = \frac{4F_0 L(k_s AGL^2 + 3EI)}{k_s AG(k_s AGL^2 + 12EI)} \times \begin{bmatrix} \dfrac{x}{L} \end{bmatrix}, \tag{3.249}
$$

$$
\phi_y^e(x) = \begin{bmatrix} \dfrac{x}{L} \end{bmatrix} \phi_{2Y} = -\frac{6F_0 L^2}{k_s AGL^2 + 12EI} \times \begin{bmatrix} \dfrac{x}{L} \end{bmatrix}. \tag{3.250}
$$

⑨ Post-computation: determination of reactions, stresses and strains.

The reactions at the supports can be obtained from the non-reduced system of equations as given in step ⑥ under the consideration of the known nodal degrees of freedom (i.e., displacement and rotation). The evaluation of the first equation gives:

$$
\frac{k_s AG}{4L}(-4u_{2Z} - 2L\varphi_{2Y}) = F_{1Z}^{R} \tag{3.251}
$$

$$
\Rightarrow\ F_{1Z}^{R} = -F_0. \tag{3.252}
$$

The evaluation of the second equation gives in a similar way:

$$
M_{1Y}^{R} = F_0 L. \tag{3.253}
$$

The internal reactions (i.e., bending moment and shear force) in the element can be obtained from the relations provided in Table 3.10.

$$M_y^e(x) = EI \begin{bmatrix} \dfrac{1}{L} \end{bmatrix} \phi_{2Y} = -\frac{6EIF_0L}{k_sAGL^2 + 12EI}, \tag{3.254}$$

$$Q_z^e(x) = k_sAG \left(\begin{bmatrix} \dfrac{1}{L} \end{bmatrix} u_{2Z} + \begin{bmatrix} \dfrac{x}{L} \end{bmatrix} \phi_{2Y} \right)$$

$$= \frac{2F_0(2k_sAGL^2 + 6EI - 3k_sAGLx)}{k_sAGL^2 + 12EI}. \tag{3.255}$$

The normal and shear stress distributions can be obtained from Eqs. (2.125) and (2.126):

$$\sigma_x^e(x, z) = \frac{M_y^e(x)}{I} z(x) = -\frac{6EF_0L}{k_sAGL^2 + 12EI} z(x), \tag{3.256}$$

$$\tau_{xz}^e(x) = \frac{Q_z^e(x)}{k_sA} = \frac{2F_0(2k_sAGL^2 + 6EI - 3k_sAGLx)}{k_sA(k_sAGL^2 + 12EI)}. \tag{3.257}$$

⑩ Check the global equilibrium between the external loads and the support reactions.

$$\sum_i F_{iz} = 0 \quad \Leftrightarrow \quad \underbrace{(F_{12}^R)}_{\text{reaction force}} + \underbrace{(F_0)}_{\text{external load}} = 0, \ \checkmark \tag{3.258}$$

$$\sum_i M_{iY} = 0 \quad \Leftrightarrow \quad \underbrace{(M_{1Y}^R)}_{\text{reaction}} + \underbrace{(-F_0L)}_{\text{external load}} = 0. \ \checkmark \tag{3.259}$$

Deflection of the load application point as a function of the slenderness ratio $\frac{h}{L}$ for $\nu = 0.0, 0.3$ and 0.5.

The nodal displacement at node 2 is given in Eq. (3.248):

$$u_{2Z} = \frac{4F_0L(k_sAGL^2 + 3EI)}{k_sAG(k_sAGL^2 + 12EI)}. \tag{3.260}$$

Considering that $G = \frac{E}{2(1+\nu)}$, $A = bh$, $I = \frac{bh^3}{12}$, and $k_s = \frac{5}{6}$ allows us to express the last equation as:

$$u_{2Z} = \frac{36(1+\nu)\left(\frac{h}{L}\right)^2 + 60}{25\left(\frac{L}{h}\right)^2 \frac{1}{1+\nu} + 60} \times \frac{F_0L^3}{3EI}. \tag{3.261}$$

The graphical representation of this equation for different values of POISSON's ratio is given in Fig. 3.42.

Fig. 3.42 Comparison
between the finite element
solution for a single linear
TIMOSHENKO beam element
with analytical integration
and the corresponding
analytical solutions **a** general
view and **b** magnification for
small slenderness ratios

(a)

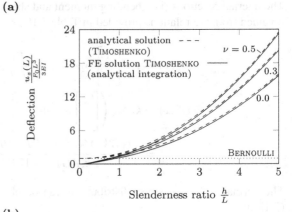

(b)

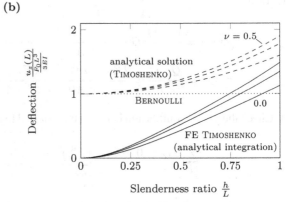

Fig. 3.43 Beam loaded by a
distributed load

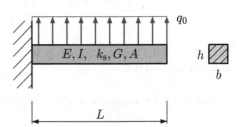

3.12 Beam loaded by a distributed load

The cantilever TIMOSHENKO beam shown in Fig. 3.43 is loaded by a constant distributed load q_0. The bending stiffness EI and the shear stiffness $k_s AG$ are constant and the total length of the beam is equal to L. Model the beam with one single linear TIMOSHENKO finite element to determine:

- the unknowns at the nodes,
- the equation of the bending line and the distribution of the rotation,
- the reactions at the support,

Fig. 3.44 Free-body diagram of the beam loaded by a distributed load

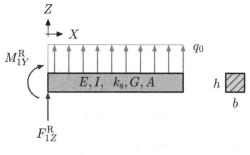

Fig. 3.45 Free-body diagram of the discretized structure

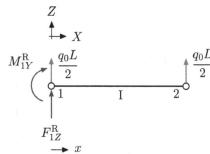

- the internal reactions (shear force and bending moment distribution) in the element,
- the strain and stress distributions in the element,
- the global force and moment equilibrium, and
- sketch the deflection of the right-hand end ($x = L$) as a function of the slenderness ratio $\frac{h}{L}$ for $\nu = 0.0, 0.3$ and 0.5.

3.12 Solution

The solution will follow the recommended 10 steps outlined on Sect. 3.2.

① Sketch the free-body diagram of the problem, including a global coordinate system (see Fig. 3.44).

② Subdivide the geometry into finite elements. Indicate the node and element numbers, local coordinate systems, and equivalent nodal loads (see Fig. 3.45).

Steps ③ to ⑥ can be combined since we have only a single element problem. The global system of equations reads:

$$\frac{k_s A G}{4L} \begin{bmatrix} 4 & -2L & -4 & -2L \\ -2L & \frac{4}{3}L^2 + \alpha & 2L & \frac{4}{6}L^2 - \alpha \\ -4 & 2L & 4 & 2L \\ -2L & \frac{4}{6}L^2 - \alpha & 2L & \frac{4}{3}L^2 + \alpha \end{bmatrix} \begin{bmatrix} u_{1Z} \\ \phi_{1Y} \\ u_{2Z} \\ \phi_{2Y} \end{bmatrix} = \begin{bmatrix} F_{1Z}^R + \frac{q_0 L}{2} \\ M_{1Y}^R \\ \frac{q_0 L}{2} \\ 0 \end{bmatrix}. \tag{3.262}$$

⑦ Introduce the boundary conditions to obtain the reduced system of equations.

$$
\frac{k_s A G}{4L}
\begin{bmatrix}
4 & 2L \\
2L & \frac{4}{3}L^2 + \alpha
\end{bmatrix}
\begin{bmatrix}
u_{2Z} \\
\phi_{2Y}
\end{bmatrix}
=
\begin{bmatrix}
\frac{q_0 L}{2} \\
0
\end{bmatrix} .
\tag{3.263}
$$

⑧ Solve the reduced system of equations to obtain the unknown nodal deformations.

The solution can be obtained based on the matrix approach $u_p = K^{-1} f$:

$$
\begin{bmatrix}
u_{2Z} \\
\phi_{2Y}
\end{bmatrix}
=
\frac{4L}{k_s A G} \times \frac{1}{4\left(\frac{4}{3}L^2 + \alpha\right) - 4L^2}
\begin{bmatrix}
\frac{4}{3}L^2 + \alpha & -2L \\
-2L & 4
\end{bmatrix}
\begin{bmatrix}
\frac{q_0 L}{2} \\
0
\end{bmatrix}
$$

$$
= \frac{q_0 L^2}{k_s A G L^2 + 12 E I}
\begin{bmatrix}
\dfrac{2(k_s A G L^2 + 3 E I)}{k_s A G} \\
-3L
\end{bmatrix} .
\tag{3.264}
$$

Based on these nodal unknowns, the bending line (deflection) and the rotational distribution can be obtained from the general equations provided in Table 3.10:

$$
u_z^e(x) = \begin{bmatrix} \dfrac{x}{L} \end{bmatrix} u_{2Z} = \frac{2 q_0 L^2 (k_s A G L^2 + 3 E I)}{k_s A G (k_s A G L^2 + 12 E I)} \times \begin{bmatrix} \dfrac{x}{L} \end{bmatrix} ,
\tag{3.265}
$$

$$
\phi_y^e(x) = \begin{bmatrix} \dfrac{x}{L} \end{bmatrix} \phi_{2Y} = -\frac{3 q_0 L^3}{k_s A G L^2 + 12 E I} \times \begin{bmatrix} \dfrac{x}{L} \end{bmatrix} .
\tag{3.266}
$$

⑨ Post-computation: determination of reactions, stresses and strains.

The reactions at the supports can be obtained from the non-reduced system of equations as given in step ⑥ under the consideration of the known nodal degrees of freedom (i.e., displacement and rotation). The evaluation of the first equation gives:

$$
\frac{k_s A G}{4L}(-4u_{2Z} - 2L\varphi_{2Y}) = F_{1Z}^R
\tag{3.267}
$$

$$
\Rightarrow F_{1Z}^R = -q_0 L .
\tag{3.268}
$$

The evaluation of the second equation gives in a similar way:

$$
M_{1Y}^R = \frac{q_0 L^2}{2} .
\tag{3.269}
$$

The internal reactions (i.e., bending moment and shear force) in the element can be obtained from the relations provided in Table 3.10.

$$M_y^e(x) = EI \begin{bmatrix} 1 \\ L \end{bmatrix} \phi_{2Y} = -\frac{3EIq_0L^2}{k_sAGL^2 + 12EI},$$ (3.270)

$$Q_z^e(x) = k_sAG \left(\begin{bmatrix} 1 \\ L \end{bmatrix} u_{2Z} + \begin{bmatrix} x \\ L \end{bmatrix} \phi_{2Y} \right)$$

$$= \frac{q_0L(2k_sAGL^2 + 6EI - 3k_sAGLx)}{k_sAGL^2 + 12EI}.$$ (3.271)

The normal and shear stress distributions can be obtained from Eqs. (2.125) and (2.126):

$$\sigma_x^e(x, z) = \frac{M_y^e(x)}{I} z(x) = -\frac{3Eq_0L^2}{k_sAGL^2 + 12EI} z(x),$$ (3.272)

$$\tau_{xz}^e(x) = \frac{Q_z^e(x)}{k_sA} = \frac{q_0L(2k_sAGL^2 + 6EI - 3k_sAGLx)}{k_sA(k_sAGL^2 + 12EI)}.$$ (3.273)

⑩ Check the global equilibrium between the external loads and the support reactions.

$$\sum_i F_{iz} = 0 \quad \Leftrightarrow \quad \underbrace{(F_{1Z}^R)}_{\text{reaction force}} + \underbrace{(q_0L)}_{\text{external load}} = 0, \checkmark$$ (3.274)

$$\sum_i M_{iY} = 0 \quad \Leftrightarrow \quad \underbrace{(M_{1Y}^R)}_{\text{reaction}} + \underbrace{\left(-\frac{q_0L^2}{2}\right)}_{\text{external load}} = 0. \checkmark$$ (3.275)

Deflection of the right-hand end ($x = L$) as a function of the slenderness ratio $\frac{h}{L}$ for $\nu = 0.0, 0.3$ and 0.5.

The nodal displacement at node 2 is given in Eq. (3.264):

$$u_{2Z} = \frac{2q_0L^2(k_sAGL^2 + 3EI)}{k_sAG(k_sAGL^2 + 12EI)}.$$ (3.276)

Considering that $G = \frac{E}{2(1+\nu)}$, $A = bh$, $I = \frac{bh^3}{12}$, and $k_s = \frac{5}{6}$ allows us to express the last equation as:

$$u_{2Z} = \frac{48(1+\nu)\left(\frac{h}{L}\right)^2 + 80}{25\left(\frac{L}{h}\right)^2 \frac{1}{1+\nu} + 60} \times \frac{q_0L^4}{8EI}.$$ (3.277)

Fig. 3.46 Comparison between the finite element solution for a single linear TIMOSHENKO beam element with analytical integration and the corresponding analytical solutions: **a** general view and **b** magnification for small slenderness ratios

(a)

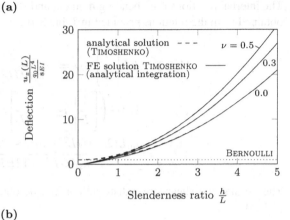

(b)

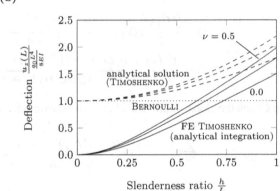

The graphical representation of this equation for different values of POISSON's ratio is given in Fig. 3.46.

3.3.3   Generalized Beam and Frame Elements

3.3.3.1   Generalized Beam Elements

Let us consider a generalized beam element, i.e., a superposition of a rod and a simple beam element, which is composed of two nodes as schematically shown in Fig. 3.47. Each node has three degrees of freedom, i.e., a displacement u_x in the direction of

Fig. 3.47 Superposition of the rod element **a** and the EULER–BERNOULLI beam element **b** to the generalized beam element **c** in the x-z plane

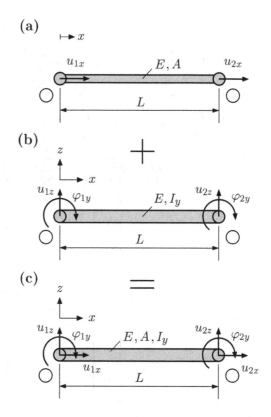

the x-axis, a displacement u_z in the direction of the z-axis (i.e., perpendicular to the principal beam axis), and a rotation φ_y around the y-axis, see Fig. 3.47c. Each node can be loaded by single forces acting in the x- or z-directions or single moments around the y-axis. In the case of distributed loads $p_x(x)$ or $q_z(x)$, a transformation to equivalent nodal loads is required.

The principal finite element equation for the generalized beam can be obtained by combining the expressions for the rod element and the simple beam as given in Eqs. (3.1) and (3.164):

$$
\begin{bmatrix}
\dfrac{EA}{L} & 0 & 0 & -\dfrac{EA}{L} & 0 & 0 \\[2mm]
0 & \dfrac{12EI}{L^3} & \dfrac{6EI}{L^2} & 0 & -\dfrac{12EI}{L^3} & -\dfrac{6EI}{L^2} \\[2mm]
0 & -\dfrac{6EI}{L^2} & \dfrac{4EI}{L} & 0 & \dfrac{6EI}{L^2} & \dfrac{2EI}{L} \\[2mm]
-\dfrac{EA}{L} & 0 & 0 & \dfrac{EA}{L} & 0 & 0 \\[2mm]
0 & -\dfrac{12EI}{L^3} & \dfrac{6EI}{L^2} & 0 & \dfrac{12EI}{L^3} & \dfrac{6EI}{L^2} \\[2mm]
0 & -\dfrac{6EI}{L^2} & \dfrac{2EI}{L} & 0 & \dfrac{6EI}{L^2} & \dfrac{4EI}{L}
\end{bmatrix}
\begin{bmatrix}
u_{1x} \\ u_{1z} \\ \varphi_{1y} \\ u_{2x} \\ u_{2z} \\ \varphi_{2y}
\end{bmatrix}
=
\begin{bmatrix}
F_{1x} \\ F_{1z} \\ M_{1y} \\ F_{2x} \\ F_{2z} \\ M_{2y}
\end{bmatrix}
+
$$

$$
+ \int_0^L
\begin{bmatrix}
N_1 \\ 0 \\ 0 \\ N_2 \\ 0 \\ 0
\end{bmatrix}
p_x(x)\,\mathrm{d}x
+ \int_0^L
\begin{bmatrix}
0 \\ N_{1u} \\ N_{1\varphi} \\ 0 \\ N_{2u} \\ N_{2\varphi}
\end{bmatrix}
q_z(x)\,\mathrm{d}x . \tag{3.278}
$$

The explanation of Eq. (3.278) can be readily taken from the corresponding sections of the rod and simple beam, see Sects. 3.2.1 and 3.3.1.

3.13 Cantilever generalized beam with two point loads

The generalized beam shown in Fig. 3.48 is loaded by two point loads, i.e., a single horizontal force F_0 and a single moment M_0 at its right-hand end. The material constant (E) and the geometrical properties (I, A) are constant and the total length of the beam is equal to L. Model the member with one generalized beam finite element of length L to determine:

Fig. 3.48 Generalized cantilever beam with two point loads

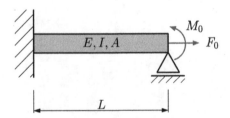

Fig. 3.49 Free-body
diagram of the cantilevered
generalized beam with two
point loads

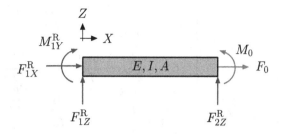

- the unknowns at the nodes,
- the displacement distributions $u_Z = u_Z(X)$ (bending) and $u_X = u_X(X)$ (tension/compression),
- the reactions at the supports,
- the internal reactions (normal force, shear force and bending moment) in the element,
- the strain and stress distributions in the element, and
- the global force and moment equilibrium.

3.13 Solution

The solution will follow the recommended 10 steps outlined on Sect. 3.2.

① Sketch the free-body diagram of the problem, including a global coordinate system, see Fig. 3.49.
② Subdivide the geometry into finite elements. Indicate the node and element numbers, local coordinate systems, and equivalent nodal loads, see Fig. 3.50.
③ Write separately all elemental stiffness matrices expressed in the global coordinate system. Indicate the nodal unknowns on the right-hand sides and over the matrices.

There is only a single element and its stiffness matrix reads:

Fig. 3.50 Free-body
diagram of the discretized
structure

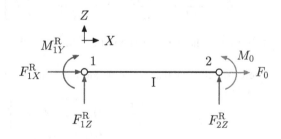

$$
K_I^e = \begin{bmatrix}
\dfrac{EA}{L} & 0 & 0 & -\dfrac{EA}{L} & 0 & 0 \\[2mm]
0 & \dfrac{12EI}{L^3} & -\dfrac{6EI}{L^2} & 0 & -\dfrac{12EI}{L^3} & -\dfrac{6EI}{L^2} \\[2mm]
0 & -\dfrac{6EI}{L^2} & \dfrac{4EI}{L} & 0 & \dfrac{6EI}{L^2} & \dfrac{2EI}{L} \\[2mm]
-\dfrac{EA}{L} & 0 & 0 & \dfrac{EA}{L} & 0 & 0 \\[2mm]
0 & -\dfrac{12EI}{L^3} & \dfrac{6EI}{L^2} & 0 & \dfrac{12EI}{L^3} & \dfrac{6EI}{L^2} \\[2mm]
0 & -\dfrac{6EI}{L^2} & \dfrac{2EI}{L} & 0 & \dfrac{6EI}{L^2} & \dfrac{4EI}{L}
\end{bmatrix}
\begin{matrix} u_{1X} \\ u_{1Z} \\ \varphi_{1Y} \\ u_{2X} \\ u_{2Z} \\ \varphi_{2Y} \end{matrix} \tag{3.279}
$$

where the columns are labelled u_{1X}, u_{1Z}, φ_{1Y}, u_{2X}, u_{2Z}, φ_{2Y}.

Steps ④ to ⑥ can be combined since we have only a single element (Determine the dimensions of the global stiffness matrix and sketch the structure of this matrix with global unknowns on the right-hand side and over the matrix. Insert the values of the elemental stiffness matrices step-by-step into the global stiffness matrix. Add the column matrix of unknowns and external loads to complete the global system of equations).

$$
\begin{bmatrix}
\dfrac{EA}{L} & 0 & 0 & -\dfrac{EA}{L} & 0 & 0 \\[2mm]
0 & \dfrac{12EI}{L^3} & -\dfrac{6EI}{L^2} & 0 & -\dfrac{12EI}{L^3} & -\dfrac{6EI}{L^2} \\[2mm]
0 & -\dfrac{6EI}{L^2} & \dfrac{4EI}{L} & 0 & \dfrac{6EI}{L^2} & \dfrac{2EI}{L} \\[2mm]
-\dfrac{EA}{L} & 0 & 0 & \dfrac{EA}{L} & 0 & 0 \\[2mm]
0 & -\dfrac{12EI}{L^3} & \dfrac{6EI}{L^2} & 0 & \dfrac{12EI}{L^3} & \dfrac{6EI}{L^2} \\[2mm]
0 & -\dfrac{6EI}{L^2} & \dfrac{2EI}{L} & 0 & \dfrac{6EI}{L^2} & \dfrac{4EI}{L}
\end{bmatrix}
\begin{bmatrix} u_{1X} \\ u_{1Z} \\ \varphi_{1Y} \\ u_{2X} \\ u_{2Z} \\ \varphi_{2Y} \end{bmatrix}
=
\begin{bmatrix} F_{1X}^R \\ F_{1Z}^R \\ M_{1Y}^R \\ F_0 \\ F_{2Z}^R \\ -M_0 \end{bmatrix}. \tag{3.280}
$$

⑦ Introduce the boundary conditions to obtain the reduced system of equations.

There are only two degrees of freedom, i.e., the rotation and horizontal displacement at node 2:

$$
\begin{bmatrix} \dfrac{EA}{L} & 0 \\[2mm] 0 & \dfrac{4EI_Y}{L} \end{bmatrix}
\begin{bmatrix} u_{2X} \\ \varphi_{2Y} \end{bmatrix}
=
\begin{bmatrix} F_0 \\ -M_0 \end{bmatrix}. \tag{3.281}
$$

⑧ Solve the reduced system of equations to obtain the unknown nodal deformations.

The solution can be obtained based on the matrix approach $u_p = K^{-1}f$:

$$
\begin{bmatrix} u_{2X} \\ \varphi_{2Y} \end{bmatrix} = \frac{1}{\frac{EA}{L}\frac{4EI_Y}{L} - 0} \begin{bmatrix} \frac{4EI_Y}{L} & 0 \\ 0 & \frac{EA}{L} \end{bmatrix} \begin{bmatrix} F_0 \\ -M_0 \end{bmatrix} = \begin{bmatrix} \dfrac{LF_0}{EA} \\ -\dfrac{LM_0}{4EI_Y} \end{bmatrix}. \tag{3.282}
$$

⑨ Post-computation: determination of reactions, stresses and strains.

The reactions at the supports can be obtained from the non-reduced system of equations as given in step ⑥ under the consideration of the known nodal degrees of freedom (i.e., displacements and rotations). The evaluation of the first equation gives:

$$
-\frac{EA}{L}u_{2X} = -\frac{EA}{L} \times \frac{LF_0}{EA} = F_{1X}^R \quad \Rightarrow \quad F_{1X}^R = -F_0. \tag{3.283}
$$

The other reactions can be obtained in a similar way as:

$$
F_{1Z}^R = \frac{3M_0}{2L}, \tag{3.284}
$$

$$
M_{1Y}^R = -\frac{M_0}{2}, \tag{3.285}
$$

$$
F_{2Z}^R = -\frac{3M_0}{2L}. \tag{3.286}
$$

The internal reactions (i.e., normal force, bending moment, and shear force) in each element can be obtained from the relations provided in Tables 3.2 and 3.8:

$$
N_X^e(X) = \frac{EA}{L}u_{2X} = F_0, \tag{3.287}
$$

$$
Q_Z^e(X) = EI_Y\left(\frac{6}{L^2}\varphi_{2Y}\right) = -\frac{3M_0}{2L}, \tag{3.288}
$$

$$
M_Y^e(X) = EI_Y\left(\left[-\frac{2}{L} + \frac{6X}{L^2}\right]\varphi_{2Y}\right) = \frac{1}{4}\left(\frac{2(L - 3X)}{L^2}\right)LM_0. \tag{3.289}
$$

The total normal stress distribution is a superposition of the contributions from the tensile (N_X) and bending (M_Y) part, see Tables 3.2 and 3.8:

$$\sigma_X^e(X) = \frac{N_X^e(X)}{A} + \frac{M_Y^e(X)}{I_Y}Z = \frac{F_0}{A} + \frac{1}{4}\left(\frac{2(L-3X)}{L^2}\right)\frac{LM_0}{I_Y}Z. \quad (3.290)$$

The strain result from HOOKE's law:

$$\varepsilon_X^e(X, Z) = \frac{\sigma_X^e(X, Z)}{E} = \frac{F_0}{EA} + \frac{1}{4}\left(\frac{2(L-3X)}{L^2}\right)\frac{LM_0}{EI_Y}Z. \quad (3.291)$$

⑩ Check the global equilibrium between the external loads and the support reactions.

$$\sum_i F_{iX} = 0 \quad \Leftrightarrow \quad \underbrace{(F_{1X}^R)}_{\text{reaction force}} + \underbrace{(F_0)}_{\text{external load}} = 0, \checkmark \quad (3.292)$$

$$\sum_i F_{iZ} = 0 \quad \Leftrightarrow \quad \underbrace{(F_{1Z}^R + F_{2Z}^R)}_{\text{reaction force}} + \underbrace{(0)}_{\text{external load}} = 0, \checkmark \quad (3.293)$$

$$\sum_i M_{iY} = 0 \quad \Leftrightarrow \quad \underbrace{(M_{1Y}^R - F_{2Z}^R L)}_{\text{reaction}} + \underbrace{(-M_0)}_{\text{external load}} = 0. \checkmark \quad (3.294)$$

3.14 Generalized cantilever beam with distributed load and end displacement
The generalized beam shown in Fig. 3.51 is loaded by distributed loads p_0 and a vertical displacement u_0 at its right-hand end. The material constant (E) and the geometrical properties (I, A) are constant and the total length of the beam is equal to L. Model the member with one generalized beam finite element of length L to determine:

- the unknowns at the nodes,
- the displacement distributions $u_Z = u_Z(X)$ (bending) and $u_X = u_X(X)$ (tension/compression),
- the reactions at the supports,
- the internal reactions (normal force, shear force and bending moment) in the element,
- the strain and stress distributions in the element, and
- the global force and moment equilibrium.

Fig. 3.51 Generalized cantilever beam with distributed load and end displacement

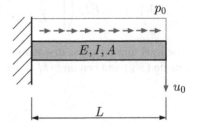

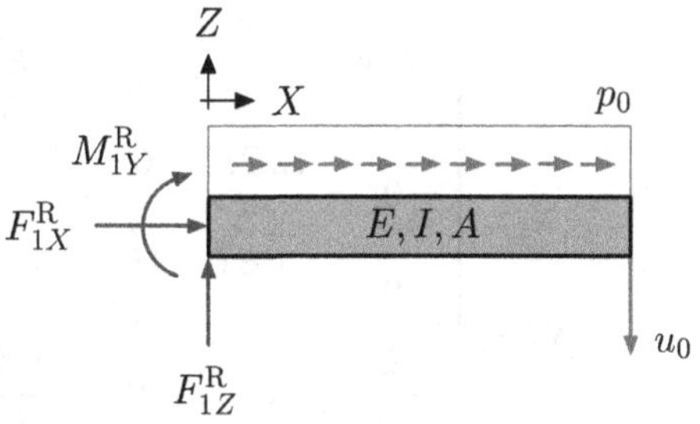

Fig. 3.52 Free-body diagram of the generalized cantilever beam with distributed load and end displacement

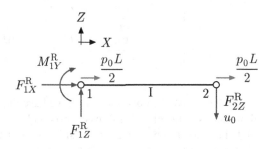

Fig. 3.53 Free-body diagram of the discretized structure with equivalent nodal loads

3.14 Solution

The solution will follow the recommended 10 steps outlined on Sect. 3.2.

① Sketch the free-body diagram of the problem, including a global coordinate system, see Fig. 3.52.
② Subdivide the geometry into finite elements. Indicate the node and element numbers, local coordinate systems, and equivalent nodal loads, see Fig. 3.53.
③ Write separately all elemental stiffness matrices expressed in the global coordinate system. Indicate the nodal unknowns on the right-hand sides and over the matrices.

There is only a single element and its stiffness matrix reads:

$$
K_I^e = \begin{bmatrix}
\dfrac{EA}{L} & 0 & 0 & -\dfrac{EA}{L} & 0 & 0 \\[2mm]
0 & \dfrac{12EI}{L^3} & \dfrac{6EI}{L^2} & 0 & -\dfrac{12EI}{L^3} & \dfrac{6EI}{L^2} \\[2mm]
0 & -\dfrac{6EI}{L^2} & \dfrac{4EI}{L} & 0 & \dfrac{6EI}{L^2} & \dfrac{2EI}{L} \\[2mm]
-\dfrac{EA}{L} & 0 & 0 & \dfrac{EA}{L} & 0 & 0 \\[2mm]
0 & -\dfrac{12EI}{L^3} & \dfrac{6EI}{L^2} & 0 & \dfrac{12EI}{L^3} & \dfrac{6EI}{L^2} \\[2mm]
0 & -\dfrac{6EI}{L^2} & \dfrac{2EI}{L} & 0 & \dfrac{6EI}{L^2} & \dfrac{4EI}{L}
\end{bmatrix}
\begin{matrix} u_{1X} \\ u_{1Z} \\ \varphi_{1Y} \\ u_{2X} \\ u_{2Z} \\ \varphi_{2Y} \end{matrix} \qquad (3.295)
$$

with column headers $u_{1X}\ \ u_{1Z}\ \ \varphi_{1Y}\ \ u_{2X}\ \ u_{2Z}\ \ \varphi_{2Y}$ over the matrix.

Steps ④ to ⑥ can be combined since we have only a single element (Determine the dimensions of the global stiffness matrix and sketch the structure of this matrix with global unknowns on the right-hand side and over the matrix. Insert the values of the elemental stiffness matrices step-by-step into the global stiffness matrix. Add the column matrix of unknowns and external loads to complete the global system of equations):

$$
\begin{bmatrix}
\dfrac{EA}{L} & 0 & 0 & -\dfrac{EA}{L} & 0 & 0 \\[2mm]
0 & \dfrac{12EI}{L^3} & \dfrac{6EI}{L^2} & 0 & -\dfrac{12EI}{L^3} & \dfrac{6EI}{L^2} \\[2mm]
0 & -\dfrac{6EI}{L^2} & \dfrac{4EI}{L} & 0 & \dfrac{6EI}{L^2} & \dfrac{2EI}{L} \\[2mm]
-\dfrac{EA}{L} & 0 & 0 & \dfrac{EA}{L} & 0 & 0 \\[2mm]
0 & -\dfrac{12EI}{L^3} & \dfrac{6EI}{L^2} & 0 & \dfrac{12EI}{L^3} & \dfrac{6EI}{L^2} \\[2mm]
0 & -\dfrac{6EI}{L^2} & \dfrac{2EI}{L} & 0 & \dfrac{6EI}{L^2} & \dfrac{4EI}{L}
\end{bmatrix}
\begin{bmatrix} u_{1X} \\ u_{1Z} \\ \varphi_{1Y} \\ u_{2X} \\ u_{2Z} \\ \varphi_{2Y} \end{bmatrix}
=
\begin{bmatrix} F_{1X}^R + \dfrac{p_0 L}{2} \\[2mm] F_{1Z}^R \\[2mm] M_{1Y}^R \\[2mm] \dfrac{p_0 L}{2} \\[2mm] -F_{2Z}^R \\[2mm] 0 \end{bmatrix}.
$$

$$(3.296)$$

⑦ Introduce the boundary conditions to obtain the reduced system of equations.

Let us first eliminate the degrees of freedom at the left-hand support:

$$\begin{bmatrix} \dfrac{EA}{L} & 0 & 0 \\[2mm] 0 & \dfrac{12EI}{L^3} & \dfrac{6EI}{L^2} \\[2mm] 0 & \dfrac{6EI}{L^2} & \dfrac{4EI}{L} \end{bmatrix} \begin{bmatrix} u_{2X} \\[2mm] u_{2Z} \\[2mm] \varphi_{2Y} \end{bmatrix} = \begin{bmatrix} \dfrac{p_0 L}{2} \\[2mm] -F_{2Z}^{R} \\[2mm] 0 \end{bmatrix} . \qquad (3.297)$$

Let us now multiply the second column of the stiffness matrix with the given displacement $-u_0$:

$$\begin{bmatrix} \dfrac{EA}{L} & 0(-u_0) & 0 \\[2mm] 0 & \dfrac{12EI}{L^3}(-u_0) & \dfrac{6EI}{L^2} \\[2mm] 0 & \dfrac{6EI}{L^2}(-u_0) & \dfrac{4EI}{L} \end{bmatrix} \begin{bmatrix} u_{2X} \\[2mm] u_{2Z} \\[2mm] \varphi_{2Y} \end{bmatrix} = \begin{bmatrix} \dfrac{p_0 L}{2} \\[2mm] -F_{2Z}^{R} \\[2mm] 0 \end{bmatrix} . \qquad (3.298)$$

Rearranging the second column of the stiffness matrix to the right-hand site of the system and canceling the second row gives finally the reduced system of equations:

$$\begin{bmatrix} \dfrac{EA}{L} & 0 \\[2mm] 0 & \dfrac{4EI}{L} \end{bmatrix} \begin{bmatrix} u_{2X} \\[2mm] \varphi_{2Y} \end{bmatrix} = \begin{bmatrix} \dfrac{p_0 L}{2} \\[2mm] 0 \end{bmatrix} + \begin{bmatrix} 0 \\[2mm] \dfrac{6EI}{L^2} \end{bmatrix} u_0 . \qquad (3.299)$$

⑧ Solve the reduced system of equations to obtain the unknown nodal deformations.

The solution can be obtained based on the matrix approach $u_p = K^{-1} f$:

$$\begin{bmatrix} u_{2X} \\[2mm] \varphi_{2Y} \end{bmatrix} = \dfrac{1}{4\frac{EA}{L}\frac{EI}{L} - 0} \begin{bmatrix} \dfrac{4EI}{L} & 0 \\[2mm] 0 & \dfrac{EA}{L} \end{bmatrix} \begin{bmatrix} \dfrac{p_0 L}{2} \\[2mm] \dfrac{6EIu_0}{L^2} \end{bmatrix} = \begin{bmatrix} \dfrac{L}{EA} \times \dfrac{p_0 L}{2} \\[2mm] \dfrac{3u_0}{2L} \end{bmatrix} . \qquad (3.300)$$

⑨ Post-computation: determination of reactions, stresses and strains.

The reactions at the supports can be obtained from the non-reduced system of equations as given in step ⑥ under the consideration of the known nodal degrees of freedom (i.e., displacements and rotations). The evaluation of the first equation gives:

$$-\dfrac{EA}{L} u_{2X} = -\dfrac{EA}{L} \times \dfrac{L}{EA} \dfrac{p_0 L}{2} = F_{1X}^{R} + \dfrac{p_0 L}{2} \implies F_{1X}^{R} = -p_0 L . \qquad (3.301)$$

The other reactions can be obtained in a similar way as:

$$F_{1Z}^{R} = \frac{3EIu_0}{L^3},\tag{3.302}$$

$$M_{1Y}^{R} = -\frac{3EIu_0}{L^2},\tag{3.303}$$

$$F_{2Z}^{R} = \frac{3EIu_0}{L^3}.\tag{3.304}$$

The internal reactions (i.e., normal force, bending moment, and shear force) in each element can be obtained from the relations provided in Tables 3.2 and 3.8:

$$N_X^e(X) = \frac{EA}{L}u_{2X} = \frac{p_0L}{2},\tag{3.305}$$

$$Q_Z^e(X) = EI_Y\left(\frac{12}{L^3}(-u_0) + \frac{6}{L^2}\varphi_{2Y}\right) = -\frac{3EIu_0}{L^3},\tag{3.306}$$

$$M_Y^e(X) = EI_Y\left(\left[-\frac{6}{L^2} + \frac{12X}{L^3}\right](-u_0) + \left[-\frac{2}{L} + \frac{6X}{L^2}\right]\varphi_{2Y}\right)\tag{3.307}$$

$$= -\frac{3EIu_0}{L^3}(X-L).\tag{3.308}$$

The total normal stress distribution is obtained by superposing the contributions from the tensile (N_X) and bending (M_Y) part, see Tables 3.2 and 3.8:

$$\sigma_X^e(X) = \frac{N_X^e(X)}{A} + \frac{M_Y^e(X)}{I_Y}Z = \frac{p_0L}{2A} - \frac{3Eu_0}{L^3}(X-L)\,Z.\tag{3.309}$$

The strains result from HOOKE's law:

$$\varepsilon_X^e(X,Z) = \frac{\sigma_X^e(X,Z)}{E} = \frac{p_0L}{2EA} - \frac{3u_0}{L^3}(X-L)\,Z.\tag{3.310}$$

⑩ Check the global equilibrium between the external loads and the support reactions.

$$\sum_i F_{iX} = 0 \quad \Leftrightarrow \quad \underbrace{(F_{1X}^R)}_{\text{reaction force}} + \underbrace{\left(\frac{p_0L}{2} + \frac{p_0L}{2}\right)}_{\text{external load}} = 0\,, \checkmark\tag{3.311}$$

$$\sum_i F_{iZ} = 0 \quad \Leftrightarrow \quad \underbrace{(F_{1Z}^R - F_{2Z}^R)}_{\text{reaction force}} + \underbrace{(0)}_{\text{external load}} = 0\,, \checkmark\tag{3.312}$$

Fig. 3.54 Generalized
cantilever beam with two
types of distributed loads

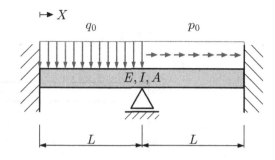

$$\sum_i M_{iY} = 0 \quad \Leftrightarrow \quad \underbrace{(M_{1Y}^{\mathrm{R}} + F_{2Z}^{\mathrm{R}}L)}_{\text{reaction}} + \underbrace{(0)}_{\text{external load}} = 0. \checkmark \qquad (3.313)$$

3.15 Generalized cantilever beam with two types of distributed loads

The generalized beam shown in Fig. 3.54 is loaded by a constant vertical distributed load q_0 in the range $0 \le X \le L$ and a constant horizontal load p_0 in the range $L \le X \le 2L$. The material constant (E) and the geometrical properties (I, A) are constant and the total length of the beam is equal to $2L$. Model the member with two generalized beam finite elements of length L to determine:

- the unknowns at the nodes,
- the displacement distributions $u_Z = u_Z(X)$ (bending) and $u_X = u_X(X)$ (tension/compression),
- the reactions at the supports,
- the internal reactions (normal force, shear force and bending moment) in each element (compare the distributions of the internal reactions with the analytical results),
- the strain and stress distributions in each element, and
- the global force and moment equilibrium.

3.15 Solution

The solution will follow the recommended 10 steps outlined on Sect. 3.2.

① Sketch the free-body diagram of the problem, including a global coordinate system, see Fig. 3.55.
② Subdivide the geometry into finite elements. Indicate the node and element numbers, local coordinate systems, and equivalent nodal loads, see Fig. 3.56.
③ Write separately all elemental stiffness matrices expressed in the global coordinate system. Indicate the nodal unknowns on the right-hand sides and over the matrices:

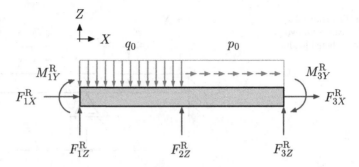

Fig. 3.55 Free-body diagram of the generalized cantilever beam with two types of distributed loads

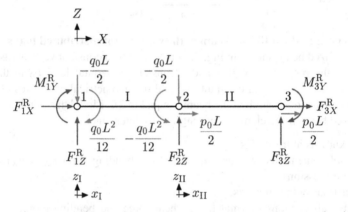

Fig. 3.56 Free-body diagram of the discretized structure with the equivalent nodal loads

$$
\boldsymbol{K}_{\mathrm{I}}^{\mathrm{e}} =
\begin{array}{c}
\begin{array}{cccccc}
u_{1X} & u_{1Z} & \varphi_{1Y} & u_{2X} & u_{2Z} & \varphi_{2Y}
\end{array} \\
\left[
\begin{array}{cccccc}
\dfrac{EA}{L} & 0 & 0 & -\dfrac{EA}{L} & 0 & 0 \\[2.2ex]
0 & \dfrac{12EI}{L^3} & -\dfrac{6EI}{L^2} & 0 & -\dfrac{12EI}{L^3} & -\dfrac{6EI}{L^2} \\[2.2ex]
0 & -\dfrac{6EI}{L^2} & \dfrac{4EI}{L} & 0 & \dfrac{6EI}{L^2} & \dfrac{2EI}{L} \\[2.2ex]
-\dfrac{EA}{L} & 0 & 0 & \dfrac{EA}{L} & 0 & 0 \\[2.2ex]
0 & -\dfrac{12EI}{L^3} & \dfrac{6EI}{L^2} & 0 & \dfrac{12EI}{L^3} & \dfrac{6EI}{L^2} \\[2.2ex]
0 & -\dfrac{6EI}{L^2} & \dfrac{2EI}{L} & 0 & \dfrac{6EI}{L^2} & \dfrac{4EI}{L}
\end{array}
\right]
\begin{array}{c}
u_{1X} \\[2.2ex]
u_{1Z} \\[2.2ex]
\varphi_{1Y} \\[2.2ex]
u_{2X} \\[2.2ex]
u_{2Z} \\[2.2ex]
\varphi_{2Y}
\end{array}
\end{array} ,
\qquad (3.314)
$$

$$
\boldsymbol{K}_{\mathrm{II}}^{\mathrm{e}} =
\begin{array}{c}
\begin{array}{cccccc}
u_{2X} & u_{2Z} & \varphi_{2Y} & u_{3X} & u_{3Z} & \varphi_{3Y}
\end{array} \\
\left[
\begin{array}{cccccc}
\dfrac{EA}{L} & 0 & 0 & -\dfrac{EA}{L} & 0 & 0 \\[2.2ex]
0 & \dfrac{12EI}{L^3} & -\dfrac{6EI}{L^2} & 0 & -\dfrac{12EI}{L^3} & -\dfrac{6EI}{L^2} \\[2.2ex]
0 & -\dfrac{6EI}{L^2} & \dfrac{4EI}{L} & 0 & \dfrac{6EI}{L^2} & \dfrac{2EI}{L} \\[2.2ex]
-\dfrac{EA}{L} & 0 & 0 & \dfrac{EA}{L} & 0 & 0 \\[2.2ex]
0 & -\dfrac{12EI}{L^3} & \dfrac{6EI}{L^2} & 0 & \dfrac{12EI}{L^3} & \dfrac{6EI}{L^2} \\[2.2ex]
0 & -\dfrac{6EI}{L^2} & \dfrac{2EI}{L} & 0 & \dfrac{6EI}{L^2} & \dfrac{4EI}{L}
\end{array}
\right]
\begin{array}{c}
u_{2X} \\[2.2ex]
u_{2Z} \\[2.2ex]
\varphi_{2Y} \\[2.2ex]
u_{3X} \\[2.2ex]
u_{3Z} \\[2.2ex]
\varphi_{3Y}
\end{array}
\end{array}
. \qquad (3.315)
$$

④ Determine the dimensions of the global stiffness matrix and sketch the structure of this matrix with global unknowns on the right-hand side and over the matrix.

The finite element structure is composed of 3 nodes, each having three degrees of freedom (i.e., the vertical and horizontal displacements and the rotation). Thus, the dimensions of the global stiffness matrix are $(3 \times 3) \times (3 \times 3) = (9 \times 9)$:

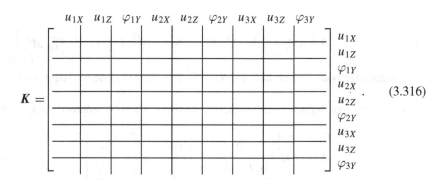

$$
\boldsymbol{K} =
\begin{array}{c}
\begin{array}{ccccccccc}
u_{1X} & u_{1Z} & \varphi_{1Y} & u_{2X} & u_{2Z} & \varphi_{2Y} & u_{3X} & u_{3Z} & \varphi_{3Y}
\end{array} \\
\left[
\begin{array}{ccccccccc}
& & & & & & & & \\
& & & & & & & & \\
& & & & & & & & \\
& & & & & & & & \\
& & & & & & & & \\
& & & & & & & & \\
& & & & & & & & \\
& & & & & & & & \\
& & & & & & & &
\end{array}
\right]
\begin{array}{c}
u_{1X} \\
u_{1Z} \\
\varphi_{1Y} \\
u_{2X} \\
u_{2Z} \\
\varphi_{2Y} \\
u_{3X} \\
u_{3Z} \\
\varphi_{3Y}
\end{array}
\end{array}
. \qquad (3.316)
$$

⑤ Insert the values of the elemental stiffness matrices step-by-step into the global stiffness matrix.

$$K = \begin{array}{c} \\ \begin{array}{ccccccccc} u_{1X} & u_{1Z} & \varphi_{1Y} & u_{2X} & u_{2Z} & \varphi_{2Y} & u_{3X} & u_{3Z} & \varphi_{3Y} \end{array} \\ \begin{bmatrix} \frac{EA}{L} & 0 & 0 & -\frac{EA}{L} & 0 & 0 & 0 & 0 & 0 \\ 0 & \frac{12EI}{L^3} & -\frac{6EI}{L^2} & 0 & -\frac{12EI}{L^3} & -\frac{6EI}{L^2} & 0 & 0 & 0 \\ 0 & -\frac{6EI}{L^2} & \frac{4EI}{L} & 0 & \frac{6EI}{L^2} & \frac{2EI}{L} & 0 & 0 & 0 \\ -\frac{EA}{L} & 0 & 0 & \frac{EA}{L}+\frac{EA}{L} & 0+0 & 0+0 & -\frac{EA}{L} & 0 & 0 \\ 0 & -\frac{12EI}{L^3} & \frac{6EI}{L^2} & 0+0 & \frac{12EI}{L^3}+\frac{12EI}{L^3} & \frac{6EI}{L^2}-\frac{6EI}{L^2} & 0 & -\frac{12EI}{L^3} & -\frac{6EI}{L^2} \\ 0 & -\frac{6EI}{L^2} & \frac{2EI}{L} & 0 & \frac{6EI}{L^2}-\frac{6EI}{L^2} & \frac{4EI}{L}+\frac{4EI}{L} & 0 & \frac{6EI}{L^2} & \frac{2EI}{L} \\ 0 & 0 & 0 & -\frac{EA}{L} & 0 & 0 & \frac{EA}{L} & 0 & 0 \\ 0 & 0 & 0 & 0 & -\frac{12EI}{L^3} & \frac{6EI}{L^2} & 0 & \frac{12EI}{L^3} & \frac{6EI}{L^2} \\ 0 & 0 & 0 & 0 & -\frac{6EI}{L^2} & \frac{2EI}{L} & 0 & \frac{6EI}{L^2} & \frac{4EI}{L} \end{bmatrix} \begin{array}{c} u_{1X} \\ u_{1Z} \\ \varphi_{1Y} \\ u_{2X} \\ u_{2Z} \\ \varphi_{2Y} \\ u_{3X} \\ u_{3Z} \\ \varphi_{3Y} \end{array} \end{array}$$

$$(3.317)$$

⑥ Add the column matrix of unknowns and external loads to complete the global system of equations.

The global system of equations can be written as $K u_{\mathrm{p}} = f$, where the column matrix of nodal unknowns reads

$$u_{\mathrm{p}} = \begin{bmatrix} u_{1X} & u_{1Z} & \varphi_{1Y} & u_{2X} & u_{2Z} & \varphi_{2Y} & u_{3X} & u_{3Z} & \varphi_{3Y} \end{bmatrix}^{\mathrm{T}}, \qquad (3.318)$$

and the column matrix of external loads is given by:

$$f = \begin{bmatrix} F_{1X}^{\mathrm{R}} & F_{1Z}^{\mathrm{R}} - \frac{q_0 L}{2} & M_{1Y}^{\mathrm{R}} + \frac{q_0 L^2}{12} & \frac{p_0 L}{2} & F_{2Z}^{\mathrm{R}} - \frac{q_0 L}{2} & -\frac{q_0 L^2}{12} & F_{3X}^{\mathrm{R}} + \frac{p_0 L}{2} & F_{3Z}^{\mathrm{R}} & M_{3Y}^{\mathrm{R}} \end{bmatrix}^{\mathrm{T}}.$$

$$(3.319)$$

⑦ Introduce the boundary conditions to obtain the reduced system of equations.

There are only two degrees of freedom, i.e., the rotation and horizontal displacement at node 2:

$$\begin{bmatrix} \frac{EA}{L} + \frac{EA}{L} & 0 \\ 0 & \frac{4EI}{L} + \frac{4EI}{L} \end{bmatrix} \begin{bmatrix} u_{2X} \\ \varphi_{2Y} \end{bmatrix} = \begin{bmatrix} \frac{p_0 L}{2} \\ -\frac{q_0 L^2}{12} \end{bmatrix}. \qquad (3.320)$$

⑧ Solve the reduced system of equations to obtain the unknown nodal deformations.

$$\begin{bmatrix} u_{2X} \\ \varphi_{2Y} \end{bmatrix} = \begin{bmatrix} \frac{p_0 L^2}{4EA} \\ -\frac{q_0 L^3}{96 E I_Y} \end{bmatrix}. \qquad (3.321)$$

The obtained nodal unknowns allow to calculate, for example, the elongation and the bending curve based on the nodal approaches provided in Tables 3.2 and 3.7.

Fig. 3.57 Beam elongation along the major axis

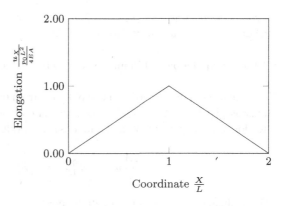

Fig. 3.58 Beam deflection along the major axis

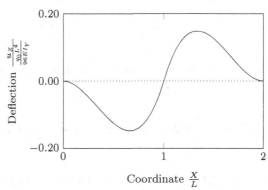

In detail, the elongations in each element can be stated as (see also the graphical representation provided in Fig. 3.57):

$$u_x^e(x_I) = \left[\frac{x_I}{L}\right] u_{2X} = \frac{p_0 L^2}{4EA} \times \frac{x_I}{L}, \tag{3.322}$$

$$u_x^e(x_{II}) = \left[1 - \frac{x_{II}}{L}\right] u_{2X} = \frac{p_0 L^2}{4EA} \times \left(1 - \frac{x_{II}}{L}\right). \tag{3.323}$$

The bending curve for each element reads as follows (see also the graphical representation provided in Fig. 3.58):

$$u_z^e(x_I) = \left[\frac{x_I^2}{L} - \frac{x_I^3}{L^2}\right] \varphi_{2Y} = -\frac{q_0 L^4}{96EI} \left(\left(\frac{x_I}{L}\right)^2 - \left(\frac{x_I}{L}\right)^3\right), \tag{3.324}$$

$$u_z^e(x_{II}) = \left[-x_{II} + \frac{2x_{II}^2}{L} - \frac{x_{II}^3}{L^2}\right] \varphi_{2Y} = -\frac{q_0 L^4}{96EI} \left(-\frac{x_{II}}{L} + 2\left(\frac{x_{II}}{L}\right)^2 - \left(\frac{x_{II}}{L}\right)^3\right). \tag{3.325}$$

It can be seen from Figs. 3.57 and 3.58 that all the support conditions in regards to the displacements and rotations are fulfilled.

⑨ Post-computation: determination of reactions, stresses and strains.

The reactions at the supports can be obtained from the non-reduced system of equations as given in step ⑥ under the consideration of the known nodal degrees of freedom (i.e., displacements and rotations). The evaluation of the first equation gives:

$$-\frac{EA}{L}u_{2X} = F_{1X}^R \quad \Rightarrow \quad F_{1X}^R = -\frac{p_0 L}{4}. \tag{3.326}$$

The evaluation of the second equation gives:

$$-\frac{6EI_Y}{L^2}\varphi_{2Y} = F_{1Z}^R - \frac{q_0 L}{2} \quad \Rightarrow \quad F_{1Z}^R = \frac{9}{16}q_0 L. \tag{3.327}$$

In a similar way, the evaluation of the third, fifth, seventh, eighth and ninth equation gives:

$$M_{1Y}^R = -\frac{5}{48}q_0 L,\ F_{2Z}^R = \frac{q_0 L}{2},\ F_{3X}^R = -\frac{3}{4}p_0 L,\ F_{3Z}^R = -\frac{q_0 L}{16},\ M_{3Y}^R = -\frac{q_0 L^2}{48}. \tag{3.328}$$

The internal reactions (i.e., normal force, bending moment, and shear force) in each element can be obtained from the relations provided in Tables 3.2 and 3.8:

$$N_x^e(x_\mathrm{I}) = \frac{p_0 L}{4}, \tag{3.329}$$

$$N_x^e(x_\mathrm{II}) = -\frac{p_0 L}{4}, \tag{3.330}$$

$$Q_z^e(x_\mathrm{I}) = -\frac{q_0 L}{16}, \tag{3.331}$$

$$Q_z^e(x_\mathrm{II}) = -\frac{q_0 L}{16}, \tag{3.332}$$

$$M_y^e(x_\mathrm{I}) = -\frac{q_0 L^2}{96}\left(-2 + 6\left(\frac{x_\mathrm{I}}{L}\right)\right), \tag{3.333}$$

$$M_y^e(x_\mathrm{II}) = -\frac{q_0 L^2}{96}\left(-4 + 6\left(\frac{x_\mathrm{II}}{L}\right)\right). \tag{3.334}$$

The graphical representations of the internal reactions are shown in Fig. 3.59. It can be seen that the finite element approach gives the correct reactions (as well as displacements and rotations) at the nodes but the distributions of the internal reactions are not correctly represented, especially in the sections with distributed loads. Pay attention to the fact that the internal reactions (analytical solution) are exactly balancing the reactions at the supports.

The analytical solutions for the internal reactions are shown in Fig. 3.60.

The total normal stress distribution is a superposition of the contributions from the tensile (N_X) and bending (M_Y) part, see Tables 3.2 and 3.8:

$$\sigma_x^e(x_\mathrm{I}) = \frac{N_x^e(x_\mathrm{I})}{A} + \frac{M_y^e(x_\mathrm{I})}{I_Y}z = \frac{p_0 L}{4A} - \frac{q_0 L^2}{96 I_Y}\left(-2 + 6\left(\frac{x_\mathrm{I}}{L}\right)\right)z, \qquad (3.335)$$

$$\sigma_x^e(x_\mathrm{II}) = \frac{N_x^e(x_\mathrm{II})}{A} + \frac{M_y^e(x_\mathrm{II})}{I_Y}z = -\frac{p_0 L}{4A} - \frac{q_0 L^2}{96 I_Y}\left(-4 + 6\left(\frac{x_\mathrm{II}}{L}\right)\right)z. \quad (3.336)$$

The strains result from HOOKE's law:

$$\varepsilon_x^e(x_\mathrm{I}, z) = \frac{\sigma_x^e(x_\mathrm{I}, z)}{E} = \frac{p_0 L}{4EA} - \frac{q_0 L^2}{96 E I_Y}\left(-2 + 6\left(\frac{x_\mathrm{I}}{L}\right)\right)z, \qquad (3.337)$$

$$\varepsilon_x^e(x_\mathrm{II}, z) = \frac{\sigma_x^e(x_\mathrm{II}, z)}{E} = -\frac{p_0 L}{4EA} - \frac{q_0 L^2}{96 E I_Y}\left(-4 + 6\left(\frac{x_\mathrm{II}}{L}\right)\right)z. \qquad (3.338)$$

⑩ Check the global equilibrium between the external loads and the support reactions.

$$\sum_i F_{iX} = 0 \quad \Leftrightarrow \quad \underbrace{(F_{1X}^\mathrm{R} + F_{3X}^\mathrm{R})}_{\text{reaction force}} + \underbrace{\left(\tfrac{p_0 L}{2} + \tfrac{p_0 L}{2}\right)}_{\text{external load}} = 0, \ \checkmark \qquad (3.339)$$

$$\sum_i F_{iZ} = 0 \quad \Leftrightarrow \quad \underbrace{(F_{1Z}^\mathrm{R} + F_{2Z}^\mathrm{R} + F_{3Z}^\mathrm{R})}_{\text{reaction force}} + \underbrace{\left(-\tfrac{q_0 L}{2} - \tfrac{q_0 L}{2}\right)}_{\text{external load}} = 0, \ \checkmark \qquad (3.340)$$

$$\sum_i M_{iY} = 0 \quad \Leftrightarrow \quad \underbrace{(M_{1Y}^\mathrm{R} + M_{3Y}^\mathrm{R} - F_{2Z}^\mathrm{R}L - F_{3Z}^\mathrm{R}2L)}_{\text{reaction}} + \underbrace{\left(\tfrac{q_0 L^2}{12} - \tfrac{q_0 L^2}{12} + \tfrac{q_0 L}{2}L\right)}_{\text{external load}} = 0. \ \checkmark$$

$$(3.341)$$

Fig. 3.59 Finite element
solution: **a** Normal force
distribution, **b** shear force
distribution and **c** bending
moment distribution

(a)

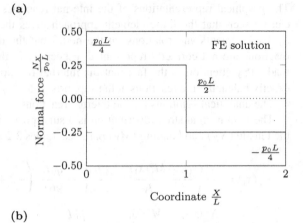

(b)

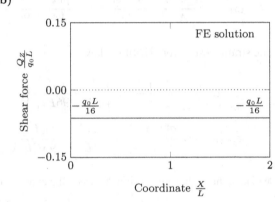

(c)

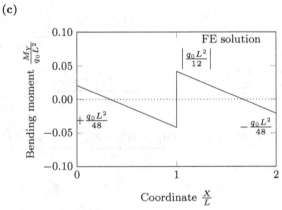

Fig. 3.60 Analytical
solution: **a** Normal force
distribution
$(N_X = -\int p_X dX + c)$,
b shear force distribution
$(Q_Z = -\int q_Z dX + c)$, and
c bending moment
distribution
$(M_Y = \int Q_Z dX + c)$

(a)

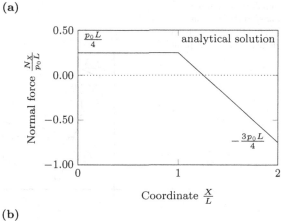

(b)

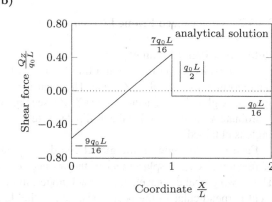

(c)

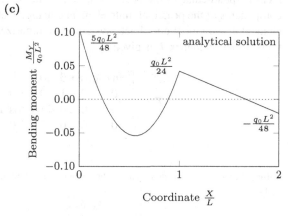

Fig. 3.61 Rotational transformation of an EULER–BERNOULLI beam element in the X-Z plane

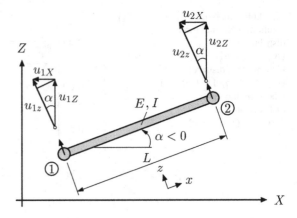

3.3.3.2 Generalized Frame Elements

Rotation of Beam Elements

Let us consider in the following a thin (EULER–BERNOULLI) beam element which can deform in the global X-Z plane. The local x-coordinate is rotated by an angle α against the global coordinate system (X, Z), see Fig. 3.61. If the rotation of the global coordinate system to the local coordinate system is clockwise, a positive rotational angle is obtained.

Each node has now in the global coordinate system two displacement degrees of freedom, i.e., a displacement in the X- and a displacement in the Z-direction. These two global displacements at each node can be used to calculate the displacement perpendicular to the beam axis, i.e., in the direction of the local z-axis. The components of the principal finite element equation can be transformed between the elemental and global coordinate system as summarized in Table 3.11 in which the transformation matrix T is given by

$$T = \begin{bmatrix} \sin\alpha & \cos\alpha & 0 & 0 & 0 & 0 \\ 0 & 0 & 1 & 0 & 0 & 0 \\ 0 & 0 & 0 & \sin\alpha & \cos\alpha & 0 \\ 0 & 0 & 0 & 0 & 0 & 1 \end{bmatrix} . \tag{3.342}$$

Table 3.11 Transformation of matrices between the elemental (x, z) and global coordinate (X, Z) system

Stiffness matrix	
$K_{xz}^{\mathrm{e}} = T K_{XZ}^{\mathrm{e}} T^{\mathrm{T}}$,	$K_{XZ}^{\mathrm{e}} = T^{\mathrm{T}} K_{xz}^{\mathrm{e}} T$
Column matrix of nodal unknowns	
$u_{xz}^{\mathrm{e}} = T u_{XZ}^{\mathrm{e}}$,	$u_{XZ}^{\mathrm{e}} = T^{\mathrm{T}} u_{xz}^{\mathrm{e}}$
Column matrix of external loads	
$f_{xz}^{\mathrm{e}} = T f_{XZ}^{\mathrm{e}}$,	$f_{XZ}^{\mathrm{e}} = T^{\mathrm{T}} f_{xz}^{\mathrm{e}}$

The triple matrix product for the stiffness matrix results in the following formulation for a rotated EULER–BERNOULLI beam element:

$$
\frac{EI_y}{L^3}
\begin{bmatrix}
12s^2\alpha & 12s\alpha c\alpha & -6Ls\alpha & -12s^2\alpha & -12s\alpha c\alpha & -6Ls\alpha \\
12s\alpha c\alpha & 12c^2\alpha & -6Lc\alpha & -12s\alpha c\alpha & -12c^2\alpha & -6Lc\alpha \\
-6Ls\alpha & -6Lc\alpha & 4L^2 & 6Ls\alpha & 6Lc\alpha & 2L^2 \\
-12s^2\alpha & -12s\alpha c\alpha & 6Ls\alpha & 12s^2\alpha & 12s\alpha c\alpha & 6Ls\alpha \\
-12s\alpha c\alpha & -12c^2\alpha & 6Lc\alpha & 12s\alpha c\alpha & 12c^2\alpha & 6Lc\alpha \\
-6Ls\alpha & -6Lc\alpha & 2L^2 & 6Ls\alpha & 6Lc\alpha & 4L^2
\end{bmatrix}
\begin{bmatrix}
u_{1X} \\ u_{1Z} \\ \varphi_{1Y} \\ u_{2X} \\ u_{2Z} \\ \varphi_{2Y}
\end{bmatrix}
=
\begin{bmatrix}
F_{1X} \\ F_{1Z} \\ M_{1Y} \\ F_{2X} \\ F_{2Z} \\ M_{2Y}
\end{bmatrix}.
$$

$$(3.343)$$

The sines ('$s\alpha$') and cosines ('$c\alpha$') values of the rotation angle α can be calculated through the global node coordinates via

$$
s\alpha \overset{\wedge}{=} \sin\alpha = -\frac{Z_2 - Z_1}{L} \quad \text{or} \quad c\alpha \overset{\wedge}{=} \cos\alpha = \frac{X_2 - X_1}{L}, \tag{3.344}
$$

where the element length L results from the global node coordinates as:

$$
L = \sqrt{(X_2 - X_1)^2 + (Z_2 - Z_1)^2}. \tag{3.345}
$$

To simplify the solution of simple beam structures, Table 3.12 collects expressions of the global stiffness matrix for some common angles α.

The results for the transformation of the matrices given in Table 3.11 can be combined with the relationships for the post-processing of nodal values in Tables 3.7 and 3.8 to express the distributions in global coordinates, see Tables 3.13 and 3.14.

Rotation of Generalized Beam Elements

Let us consider in the following a generalized beam element which can deform in the global X-Z plane. Such an element is also called a plane frame element. The local x-coordinate is rotated by an angle α with respect to the global coordinate system (X, Z), see Fig. 3.62. If the rotation of the global coordinate system to the local coordinate system is clockwise, a positive rotational angle is obtained.

Each node has in the global coordinate system two displacement degrees of freedom, i.e., a displacement in the X- and a displacement in the Z-direction. These two global displacements at each node can be used to calculate the displacements in the directions of the local x- and z-axes. The components of the principal finite element equation can be transformed between the elemental and global coordinate system as summarized in Table 3.15 in which the transformation matrix T is given by

$$
T =
\left[
\begin{array}{ccc|ccc}
\cos\alpha & -\sin\alpha & 0 & 0 & 0 & 0 \\
\sin\alpha & \cos\alpha & 0 & 0 & 0 & 0 \\
0 & 0 & 1 & 0 & 0 & 0 \\
\hline
0 & 0 & 0 & \cos\alpha & -\sin\alpha & 0 \\
0 & 0 & 0 & \sin\alpha & \cos\alpha & 0 \\
0 & 0 & 0 & 0 & 0 & 1
\end{array}
\right]. \tag{3.346}
$$

Table 3.12 Elemental stiffness matrices for EULER–BERNOULLI beam elements given for different rotation angles α, cf. Eq. (3.343)

0°

$$\frac{EI_y}{L^3}\begin{bmatrix} 0 & 0 & 0 & 0 & 0 & 0 \\ 0 & 12 & -6L & 0 & -12 & -6L \\ 0 & -6L & 4L^2 & 0 & 6L & 2L^2 \\ 0 & 0 & 0 & 0 & 0 & 0 \\ 0 & -12 & 6L & 0 & 12 & 6L \\ 0 & -6L & 2L^2 & 0 & 6L & 4L^2 \end{bmatrix}$$

180°

$$\frac{EI_y}{L^3}\begin{bmatrix} 0 & 0 & 0 & 0 & 0 & 0 \\ 0 & 12 & 6L & 0 & -12 & 6L \\ 0 & 6L & 4L^2 & 0 & -6L & 2L^2 \\ 0 & 0 & 0 & 0 & 0 & 0 \\ 0 & -12 & -6L & 0 & 12 & -6L \\ 0 & 6L & 2L^2 & 0 & -6L & 4L^2 \end{bmatrix}$$

−30°

$$\frac{EI_y}{L^3}\begin{bmatrix} 3 & -3\sqrt3 & 3L & -3 & 3\sqrt3 & 3L \\ -3\sqrt3 & 9 & -3\sqrt3L & 3\sqrt3 & -9 & -3\sqrt3L \\ 3L & -3\sqrt3L & 4L^2 & -3L & 3\sqrt3L & 2L^2 \\ -3 & 3\sqrt3 & -3L & 3 & -3\sqrt3 & -3L \\ 3\sqrt3 & -9 & 3\sqrt3L & -3\sqrt3 & 9 & 3\sqrt3L \\ 3L & -3\sqrt3L & 2L^2 & -3L & 3\sqrt3L & 4L^2 \end{bmatrix}$$

30°

$$\frac{EI_y}{L^3}\begin{bmatrix} 3 & 3\sqrt3 & -3L & -3 & -3\sqrt3 & -3L \\ 3\sqrt3 & 9 & -3\sqrt3L & -3\sqrt3 & -9 & -3\sqrt3L \\ -3L & -3\sqrt3L & 4L^2 & 3L & 3\sqrt3L & 2L^2 \\ -3 & -3\sqrt3 & 3L & 3 & 3\sqrt3 & 3L \\ -3\sqrt3 & -9 & 3\sqrt3L & 3\sqrt3 & 9 & 3\sqrt3L \\ -3L & -3\sqrt3L & 2L^2 & 3L & 3\sqrt3L & 4L^2 \end{bmatrix}$$

−45°

$$\frac{EI_y}{L^3}\begin{bmatrix} 6 & -6 & 3\sqrt2L & -6 & 6 & 3\sqrt2L \\ -6 & 6 & -3\sqrt2L & 6 & -6 & -3\sqrt2L \\ 3\sqrt2L & -3\sqrt2L & 4L^2 & -3\sqrt2L & 3\sqrt2L & 2L^2 \\ -6 & 6 & -3\sqrt2L & 6 & -6 & -3\sqrt2L \\ 6 & -6 & 3\sqrt2L & -6 & 6 & 3\sqrt2L \\ 3\sqrt2L & -3\sqrt2L & 2L^2 & -3\sqrt2L & 3\sqrt2L & 4L^2 \end{bmatrix}$$

45°

$$\frac{EI_y}{L^3}\begin{bmatrix} 6 & 6 & -3\sqrt2L & -6 & -6 & -3\sqrt2L \\ 6 & 6 & -3\sqrt2L & -6 & -6 & -3\sqrt2L \\ -3\sqrt2L & -3\sqrt2L & 4L^2 & 3\sqrt2L & 3\sqrt2L & 2L^2 \\ -6 & -6 & 3\sqrt2L & 6 & 6 & 3\sqrt2L \\ -6 & -6 & 3\sqrt2L & 6 & 6 & 3\sqrt2L \\ -3\sqrt2L & -3\sqrt2L & 2L^2 & 3\sqrt2L & 3\sqrt2L & 4L^2 \end{bmatrix}$$

−90°

$$\frac{EI_y}{L^3}\begin{bmatrix} 12 & 0 & 6L & -12 & 0 & 6L \\ 0 & 0 & 0 & 0 & 0 & 0 \\ 6L & 0 & 4L^2 & -6L & 0 & 2L^2 \\ -12 & 0 & -6L & 12 & 0 & -6L \\ 0 & 0 & 0 & 0 & 0 & 0 \\ 6L & 0 & 2L^2 & -6L & 0 & 4L^2 \end{bmatrix}$$

90°

$$\frac{EI_y}{L^3}\begin{bmatrix} 12 & 0 & -6L & -12 & 0 & -6L \\ 0 & 0 & 0 & 0 & 0 & 0 \\ -6L & 0 & 4L^2 & 6L & 0 & 2L^2 \\ -12 & 0 & 6L & 12 & 0 & 6L \\ 0 & 0 & 0 & 0 & 0 & 0 \\ -6L & 0 & 2L^2 & 6L & 0 & 4L^2 \end{bmatrix}$$

Table 3.13 Post-processing quantities (part 1) for a rotated EULER–BERNOULLI beam element given as being dependent on the global nodal values as a function of the physical coordinate $0 \le x \le L$ and natural coordinate $-1 \le \xi \le 1$. Bending occurs in the X-Z plane

Vertical displacement (Deflection) u_z

$$u_z^e(x) = \left[1 - 3\left(\frac{x}{L}\right)^2 + 2\left(\frac{x}{L}\right)^3 \right] (u_{1X}\sin\alpha + u_{1Z}\cos\alpha) + \left[-x + \frac{2x^2}{L} - \frac{x^3}{L^2} \right]\varphi_{1Y}$$

$$+ \left[3\left(\frac{x}{L}\right)^2 - 2\left(\frac{x}{L}\right)^3 \right] (u_{2X}\sin\alpha + u_{2Z}\cos\alpha) + \left[+\frac{x^2}{L} - \frac{x^3}{L^2} \right]\varphi_{2Y}$$

$$u_z^e(\xi) = \frac{1}{4}\left[2 - 3\xi + \xi^3 \right] (u_{1X}\sin\alpha + u_{1Z}\cos\alpha) - \frac{1}{4}\left[1 - \xi - \xi^2 + \xi^3 \right]\frac{L}{2}\varphi_{1Y}$$

$$+ \frac{1}{4}\left[2 + 3\xi - \xi^3 \right] (u_{2X}\sin\alpha + u_{2Z}\cos\alpha) - \frac{1}{4}\left[-1 - \xi + \xi^2 + \xi^3 \right]\frac{L}{2}\varphi_{2Y}$$

Rotation (Slope) $\varphi_y = -\dfrac{du_z}{dx} = -\dfrac{2}{L}\dfrac{du_z}{d\xi}$

$$\varphi_y^e(x) = \left[+\frac{6x}{L^2} - \frac{6x^2}{L^3} \right] (u_{1X}\sin\alpha + u_{1Z}\cos\alpha) + \left[1 - \frac{4x}{L} + \frac{3x^2}{L^2} \right]\varphi_{1Y}$$

$$+ \left[-\frac{6x}{L^2} + \frac{6x^2}{L^3} \right] (u_{2X}\sin\alpha + u_{2Z}\cos\alpha) + \left[-\frac{2x}{L} + \frac{3x^2}{L^2} \right]\varphi_{2Y}$$

$$\varphi_y^e(\xi) = \frac{1}{2L}\left[+3 - 3\xi^2 \right] (u_{1X}\sin\alpha + u_{1Z}\cos\alpha) + \frac{1}{4}\left[-1 - 2\xi + 3\xi^2 \right]\varphi_{1Y}$$

$$+ \frac{1}{2L}\left[-3 + 3\xi^2 \right] (u_{2X}\sin\alpha + u_{2Z}\cos\alpha) + \frac{1}{4}\left[-1 + 2\xi + 3\xi^2 \right]\varphi_{2Y}$$

Curvature $\kappa_y = -\dfrac{d^2 u_z}{dx^2} = -\dfrac{4}{L^2}\dfrac{d^2 u_z}{d\xi^2}$

$$\kappa_y^e(x) = \left[+\frac{6}{L^2} - \frac{12x}{L^3} \right] (u_{1X}\sin\alpha + u_{1Z}\cos\alpha) + \left[-\frac{4}{L} + \frac{6x}{L^2} \right]\varphi_{1Y}$$

$$+ \left[-\frac{6}{L^2} + \frac{12x}{L^3} \right] (u_{2X}\sin\alpha + u_{2Z}\cos\alpha) + \left[-\frac{2}{L} + \frac{6x}{L^2} \right]\varphi_{2Y}$$

$$\kappa_y^e(\xi) = \frac{6}{L^2}[-\xi] (u_{1X}\sin\alpha + u_{1Z}\cos\alpha) + \frac{1}{L}[-1 + 3\xi]\varphi_{1Y}$$

$$+ \frac{6}{L^2}[\xi] (u_{2X}\sin\alpha + u_{2Z}\cos\alpha) + \frac{1}{L}[1 + 3\xi]\varphi_{2Y}$$

The triple matrix product for the stiffness matrix results in the following formulation for a rotated generalized beam element, see Eq. (3.347).

Table 3.14 Post-processing quantities (part 2) for a rotated EULER–BERNOULLI beam element given as being dependent on the global nodal values as a function of the physical coordinate $0 \leq x \leq L$ and natural coordinate $-1 \leq \xi \leq 1$. Bending occurs in the X-Z plane

$$\text{Bending moment } M_y = -EI_y \frac{\mathrm{d}^2 u_z}{\mathrm{d}x^2} = -\frac{4}{L^2} EI_y \frac{\mathrm{d}^2 u_z}{\mathrm{d}\xi^2}$$

$$M_y^e(x) = EI_y \left(\left[+\frac{6}{L^2} - \frac{12x}{L^3} \right] (u_{1X} \sin\alpha + u_{1Z} \cos\alpha) + \left[-\frac{4}{L} + \frac{6x}{L^2} \right] \varphi_{1Y} \right.$$

$$\left. + \left[-\frac{6}{L^2} + \frac{12x}{L^3} \right] (u_{2X} \sin\alpha + u_{2Z} \cos\alpha) + \left[-\frac{2}{L} + \frac{6x}{L^2} \right] \varphi_{2Y} \right)$$

$$M_y^e(\xi) = EI_y \left(\frac{6}{L^2}[-\xi](u_{1X} \sin\alpha + u_{1Z} \cos\alpha) + \frac{1}{L}[-1 + 3\xi]\varphi_{1Y} \right.$$

$$\left. + \frac{6}{L^2}[\xi](u_{2X} \sin\alpha + u_{2Z} \cos\alpha) + \frac{1}{L}[1 + 3\xi]\varphi_{2Y} \right)$$

$$\text{Shear force } Q_z = -EI_y \frac{\mathrm{d}^3 u_z}{\mathrm{d}x^3} = -\frac{8}{L^3} EI_y \frac{\mathrm{d}^3 u_z}{\mathrm{d}\xi^3}$$

$$Q_z^e(x) = EI_y \left(\left[-\frac{12}{L^3} \right] (u_{1X} \sin\alpha + u_{1Z} \cos\alpha) + \left[+\frac{6}{L^2} \right] \varphi_{1Y} \right.$$

$$\left. + \left[+\frac{12}{L^3} \right] (u_{2X} \sin\alpha + u_{2Z} \cos\alpha) + \left[+\frac{6}{L^2} \right] \varphi_{2Y} \right)$$

$$Q_z^e(\xi) = EI_y \left(\frac{12}{L^3}[-1](u_{1X} \sin\alpha + u_{1Z} \cos\alpha) + \frac{2}{L^2}[+3]\varphi_{1Y} \right.$$

$$\left. \left(+\frac{12}{L^3}[1](u_{2X} \sin\alpha + u_{2Z} \cos\alpha) + \frac{2}{L^2}[+3]\varphi_{2Y} \right) \right.$$

$$\text{Normal strain } \varepsilon_x^e(x, z) = -z \frac{\mathrm{d}^2 u_z^e(x)}{\mathrm{d}x^2} = -z \frac{4}{L^2} \frac{\mathrm{d}^2 u_z}{\mathrm{d}\xi^2}$$

$$\varepsilon_x^e(x, z) = \left(\left[+\frac{6}{L^2} - \frac{12x}{L^3} \right] (u_{1X} \sin\alpha + u_{1Z} \cos\alpha) + \left[-\frac{4}{L} + \frac{6x}{L^2} \right] \varphi_{1Y} \right.$$

$$\left. + \left[-\frac{6}{L^2} + \frac{12x}{L^3} \right] (u_{2X} \sin\alpha + u_{2Z} \cos\alpha) + \left[-\frac{2}{L} + \frac{6x}{L^2} \right] \varphi_{2Y} \right) z$$

$$\varepsilon_x^e(\xi, z) = \left(\frac{6}{L^2}[-\xi](u_{1X} \sin\alpha + u_{1Z} \cos\alpha) + \frac{1}{L}[-1 + 3\xi]\varphi_{1Y} \right.$$

$$\left. + \frac{6}{L^2}[\xi](u_{2X} \sin\alpha + u_{2Z} \cos\alpha) + \frac{1}{L}[1 + 3\xi]\varphi_{2Y} \right) z$$

$$\text{Normal stress } \sigma_x^e(x, z) = E\varepsilon_x^e(x, z) = E\varepsilon_x^e(\xi, z) \quad \left(= \frac{M_y}{I_y} z \right)$$

$$\sigma_x^e(x, z) = E \left(\left[+\frac{6}{L^2} - \frac{12x}{L^3} \right] (u_{1X} \sin\alpha + u_{1Z} \cos\alpha) + \left[-\frac{4}{L} + \frac{6x}{L^2} \right] \varphi_{1Y} \right.$$

$$\left. + \left[-\frac{6}{L^2} + \frac{12x}{L^3} \right] (u_{2X} \sin\alpha + u_{2Z} \cos\alpha) + \left[-\frac{2}{L} + \frac{6x}{L^2} \right] \varphi_{2Y} \right) z$$

$$\sigma_x^e(\xi, z) = E \left(\frac{6}{L^2}[-\xi](u_{1X} \sin\alpha + u_{1Z} \cos\alpha) + \frac{1}{L}[-1 + 3\xi]\varphi_{1Y} \right.$$

$$\left. + \frac{6}{L^2}[\xi](u_{2X} \sin\alpha + u_{2Z} \cos\alpha) + \frac{1}{L}[1 + 3\xi]\varphi_{2Y} \right) z$$

$$
\begin{pmatrix} F_{1X} \\ F_{1Z} \\ M_{1Y} \\ F_{2X} \\ F_{2Z} \\ M_{2Y} \end{pmatrix}
=
E
\begin{bmatrix}
\frac{12I}{L^3}\sin^2\alpha + \frac{A}{L}\cos^2\alpha & \left(\frac{12I}{L^3} - \frac{A}{L}\right)\sin\alpha\cos\alpha & -\frac{6I}{L^2}\sin\alpha & -\frac{12I}{L^3}\sin^2\alpha - \frac{A}{L}\cos^2\alpha & \left(-\frac{12I}{L^3} + \frac{A}{L}\right)\sin\alpha\cos\alpha & -\frac{6I}{L^2}\sin\alpha \\[6pt]
\left(\frac{12I}{L^3} - \frac{A}{L}\right)\sin\alpha\cos\alpha & \frac{12I}{L^3}\cos^2\alpha + \frac{A}{L}\sin^2\alpha & \frac{6I}{L^2}\cos\alpha & \left(-\frac{12I}{L^3} + \frac{A}{L}\right)\sin\alpha\cos\alpha & -\frac{12I}{L^3}\cos^2\alpha - \frac{A}{L}\sin^2\alpha & \frac{6I}{L^2}\cos\alpha \\[6pt]
-\frac{6I}{L^2}\sin\alpha & \frac{6I}{L^2}\cos\alpha & \frac{4I}{L} & \frac{6I}{L^2}\sin\alpha & -\frac{6I}{L^2}\cos\alpha & \frac{2I}{L} \\[6pt]
-\frac{12I}{L^3}\sin^2\alpha - \frac{A}{L}\cos^2\alpha & \left(-\frac{12I}{L^3} + \frac{A}{L}\right)\sin\alpha\cos\alpha & \frac{6I}{L^2}\sin\alpha & \frac{12I}{L^3}\sin^2\alpha + \frac{A}{L}\cos^2\alpha & \left(\frac{12I}{L^3} - \frac{A}{L}\right)\sin\alpha\cos\alpha & \frac{6I}{L^2}\sin\alpha \\[6pt]
\left(-\frac{12I}{L^3} + \frac{A}{L}\right)\sin\alpha\cos\alpha & -\frac{12I}{L^3}\cos^2\alpha - \frac{A}{L}\sin^2\alpha & -\frac{6I}{L^2}\cos\alpha & \left(\frac{12I}{L^3} - \frac{A}{L}\right)\sin\alpha\cos\alpha & \frac{12I}{L^3}\cos^2\alpha + \frac{A}{L}\sin^2\alpha & -\frac{6I}{L^2}\cos\alpha \\[6pt]
-\frac{6I}{L^2}\sin\alpha & \frac{6I}{L^2}\cos\alpha & \frac{2I}{L} & \frac{6I}{L^2}\sin\alpha & -\frac{6I}{L^2}\cos\alpha & \frac{4I}{L}
\end{bmatrix}
\begin{pmatrix} u_{1X} \\ u_{1Z} \\ \varphi_{1Y} \\ u_{2X} \\ u_{2Z} \\ \varphi_{2Y} \end{pmatrix}
\tag{3.347}
$$

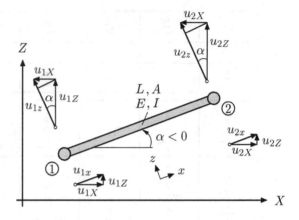

Fig. 3.62 Rotational transformation of a generalized beam element in the X-Z plane

Table 3.15 Transformation of matrices between the elemental (x, z) and global coordinate (X, Z) system

Stiffness matrix	
$K^{\mathrm{e}}_{xz} = T K^{\mathrm{e}}_{XZ} T^{\mathrm{T}}$,	$K^{\mathrm{e}}_{XZ} = T^{\mathrm{T}} K^{\mathrm{e}}_{xz} T$
Column matrix of nodal unknowns	
$u^{\mathrm{e}}_{xz} = T u^{\mathrm{e}}_{XZ}$,	$u^{\mathrm{e}}_{XZ} = T^{\mathrm{T}} u^{\mathrm{e}}_{xz}$
Column matrix of external loads	
$f^{\mathrm{e}}_{xz} = T f^{\mathrm{e}}_{XZ}$,	$f^{\mathrm{e}}_{XZ} = T^{\mathrm{T}} f^{\mathrm{e}}_{xz}$

To simplify the solution of simple beam structures, Tables 3.16 and 3.17 collect expressions for the global stiffness matrix of some common angles α.

3.16 Triangular shaped plane frame structure composed of generalized beam elements

The plane frame structure shown in Fig. 3.63 is composed of generalized beams which are arranged in triangular shape. The structure is loaded by a single horizontal force F_0 at the right-hand corner of the structure and a vertical displacement $-u_0$ at the same location. The material constant (E) and the geometrical properties (I, A) are constant and the horizontal length of the beam is equal to L while the vertical dimension is equal to L. Model the structure with two generalized beam finite elements to determine:

- the unknowns at the nodes,
- the displacement distributions in each member,
- the reactions at the supports,
- the internal reactions (normal force, shear force, and bending moment) in each element, and
- the global force and moment equilibrium.

Table 3.16 Elemental stiffness matrices for plane frame elements given for different rotation angles α in the X-Z plane, see Eq. (3.347)

$0°$

$$E\begin{bmatrix} \frac{A}{L} & 0 & 0 & -\frac{A}{L} & 0 & 0 \\ 0 & \frac{12I}{L^3} & -\frac{6I}{L^2} & 0 & -\frac{12I}{L^3} & -\frac{6I}{L^2} \\ 0 & -\frac{6I}{L^2} & \frac{4I}{L} & 0 & \frac{6I}{L^2} & \frac{2I}{L} \\ -\frac{A}{L} & 0 & 0 & \frac{A}{L} & 0 & 0 \\ 0 & -\frac{12I}{L^3} & \frac{6I}{L^2} & 0 & \frac{12I}{L^3} & \frac{6I}{L^2} \\ 0 & -\frac{6I}{L^2} & \frac{2I}{L} & 0 & \frac{6I}{L^2} & \frac{4I}{L} \end{bmatrix}$$

$180°$

$$E\begin{bmatrix} \frac{A}{L} & 0 & 0 & -\frac{A}{L} & 0 & 0 \\ 0 & \frac{12I}{L^3} & \frac{6I}{L^2} & 0 & -\frac{12I}{L^3} & -\frac{6I}{L^2} \\ 0 & \frac{6I}{L^2} & \frac{4I}{L} & 0 & -\frac{6I}{L^2} & \frac{2I}{L} \\ -\frac{A}{L} & 0 & 0 & \frac{A}{L} & 0 & 0 \\ 0 & -\frac{12I}{L^3} & -\frac{6J}{L^2} & 0 & \frac{12I}{L^3} & -\frac{6I}{L^2} \\ 0 & -\frac{6I}{L^2} & \frac{2I}{L} & 0 & -\frac{6I}{L^2} & \frac{4I}{L} \end{bmatrix}$$

$-90°$

$$E\begin{bmatrix} \frac{12I}{L^3} & 0 & \frac{6I}{L^2} & -\frac{12I}{L^3} & 0 & \frac{6I}{L^2} \\ 0 & \frac{A}{L} & 0 & 0 & -\frac{A}{L} & 0 \\ \frac{6I}{L^2} & 0 & \frac{4I}{L} & -\frac{6I}{L^2} & 0 & \frac{2I}{L} \\ -\frac{12I}{L^3} & 0 & -\frac{6I}{L^2} & \frac{12I}{L^3} & 0 & -\frac{6I}{L^2} \\ 0 & -\frac{A}{L} & 0 & 0 & \frac{A}{L} & 0 \\ \frac{6I}{L^2} & 0 & \frac{2I}{L} & -\frac{6I}{L^2} & 0 & \frac{4I}{L} \end{bmatrix}$$

$90°$

$$E\begin{bmatrix} \frac{12I}{L^3} & 0 & -\frac{6I}{L^2} & -\frac{12I}{L^3} & 0 & -\frac{6I}{L^2} \\ 0 & \frac{A}{L} & 0 & 0 & -\frac{A}{L} & 0 \\ -\frac{6I}{L^2} & 0 & \frac{4I}{L} & \frac{6I}{L^2} & 0 & \frac{2I}{L} \\ -\frac{12I}{L^3} & 0 & \frac{6I}{L^2} & \frac{12I}{L^3} & 0 & \frac{6I}{L^2} \\ 0 & -\frac{A}{L} & 0 & 0 & \frac{A}{L} & 0 \\ -\frac{6I}{L^2} & 0 & \frac{2I}{L} & \frac{6I}{L^2} & 0 & \frac{4I}{L} \end{bmatrix}$$

$-45°$

$$E\begin{bmatrix} \frac{6I}{L^3}+\frac{1}{2}\frac{A}{L} & -\frac{6I}{L^3}+\frac{1}{2}\frac{A}{L} & \frac{3I\sqrt{2}}{L^2} & -\frac{6I}{L^3}-\frac{1}{2}\frac{A}{L} & +\frac{6I}{L^3}-\frac{1}{2}\frac{A}{L} & \frac{3I\sqrt{2}}{L^2} \\ -\frac{6I}{L^3}+\frac{1}{2}\frac{A}{L} & \frac{6I}{L^3}+\frac{1}{2}\frac{A}{L} & -\frac{3I\sqrt{2}}{L^2} & +\frac{6I}{L^3}-\frac{1}{2}\frac{A}{L} & -\frac{6I}{L^3}-\frac{1}{2}\frac{A}{L} & -\frac{3I\sqrt{2}}{L^2} \\ \frac{3I\sqrt{2}}{L^2} & -\frac{3I\sqrt{2}}{L^2} & \frac{4I}{L} & -\frac{3I\sqrt{2}}{L^2} & +\frac{3I\sqrt{2}}{L^2} & \frac{2I}{L} \\ -\frac{6I}{L^3}-\frac{1}{2}\frac{A}{L} & \frac{6I}{L^3}-\frac{1}{2}\frac{A}{L} & -\frac{3I\sqrt{2}}{L^2} & \frac{6I}{L^3}+\frac{1}{2}\frac{A}{L} & -\frac{6I}{L^3}+\frac{1}{2}\frac{A}{L} & -\frac{3I\sqrt{2}}{L^2} \\ +\frac{6I}{L^3}-\frac{1}{2}\frac{A}{L} & -\frac{6I}{L^3}-\frac{1}{2}\frac{A}{L} & +\frac{3I\sqrt{2}}{L^2} & -\frac{6I}{L^3}+\frac{1}{2}\frac{A}{L} & \frac{6I}{L^3}+\frac{1}{2}\frac{A}{L} & \frac{3I\sqrt{2}}{L^2} \\ \frac{3I\sqrt{2}}{L^2} & -\frac{3I\sqrt{2}}{L^2} & \frac{2I}{L} & -\frac{3I\sqrt{2}}{L^2} & \frac{3I\sqrt{2}}{L^2} & \frac{4I}{L} \end{bmatrix}$$

$45°$

$$E\begin{bmatrix} \frac{6I}{L^3}+\frac{1}{2}\frac{A}{L} & \frac{6I}{L^3}-\frac{1}{2}\frac{A}{L} & -\frac{3I\sqrt{2}}{L^2} & -\frac{6I}{L^3}-\frac{1}{2}\frac{A}{L} & -\frac{6I}{L^3}+\frac{1}{2}\frac{A}{L} & -\frac{3I\sqrt{2}}{L^2} \\ \frac{6I}{L^3}-\frac{1}{2}\frac{A}{L} & \frac{6I}{L^3}+\frac{1}{2}\frac{A}{L} & -\frac{3I\sqrt{2}}{L^2} & -\frac{6I}{L^3}+\frac{1}{2}\frac{A}{L} & -\frac{6I}{L^3}-\frac{1}{2}\frac{A}{L} & -\frac{3I\sqrt{2}}{L^2} \\ -\frac{3I\sqrt{2}}{L^2} & -\frac{3I\sqrt{2}}{L^2} & \frac{4I}{L} & \frac{3I\sqrt{2}}{L^2} & \frac{3I\sqrt{2}}{L^2} & \frac{2I}{L} \\ -\frac{6I}{L^3}-\frac{1}{2}\frac{A}{L} & -\frac{6I}{L^3}+\frac{1}{2}\frac{A}{L} & \frac{3I\sqrt{2}}{L^2} & \frac{6I}{L^3}+\frac{1}{2}\frac{A}{L} & \frac{6I}{L^3}-\frac{1}{2}\frac{A}{L} & \frac{3I\sqrt{2}}{L^2} \\ -\frac{6I}{L^3}+\frac{1}{2}\frac{A}{L} & -\frac{6I}{L^3}-\frac{1}{2}\frac{A}{L} & \frac{3I\sqrt{2}}{L^2} & \frac{6I}{L^3}-\frac{1}{2}\frac{A}{L} & \frac{6I}{L^3}+\frac{1}{2}\frac{A}{L} & \frac{3I\sqrt{2}}{L^2} \\ -\frac{3I\sqrt{2}}{L^2} & -\frac{3I\sqrt{2}}{L^2} & \frac{2I}{L} & \frac{3I\sqrt{2}}{L^2} & \frac{3I\sqrt{2}}{L^2} & \frac{4I}{L} \end{bmatrix}$$

Table 3.17 Elemental stiffness matrices for plane frame elements given for different rotation angles α in the X-Z plane, see Eq. (3.347)

$-30°$

$$E\begin{bmatrix}
\frac{3I}{L^3}+\frac{3}{4}\frac{A}{L} & \frac{\sqrt{3}}{4}\left(-\frac{12I}{L^3}+\frac{A}{L}\right) & \frac{3I}{L^2} & -\frac{3I}{L^3}-\frac{3}{4}\frac{A}{L} & -\frac{\sqrt{3}}{4}\left(-\frac{12I}{L^3}+\frac{A}{L}\right) & \frac{3I}{L^2} \\[4pt]
\frac{\sqrt{3}}{4}\left(-\frac{12I}{L^3}+\frac{A}{L}\right) & \frac{9I}{L^3}+\frac{1}{4}\frac{A}{L} & -\frac{3I\sqrt{3}}{L^2} & \frac{\sqrt{3}}{4}\left(\frac{12I}{L^3}-\frac{A}{L}\right) & -\frac{9I}{L^3}-\frac{1}{4}\frac{A}{L} & -\frac{3I\sqrt{3}}{L^2} \\[4pt]
\frac{3I}{L^2} & -\frac{3I\sqrt{3}}{L^2} & \frac{4I}{L} & -\frac{3I}{L^2} & \frac{3I\sqrt{3}}{L^2} & \frac{2I}{L} \\[4pt]
-\frac{3I}{L^3}-\frac{3}{4}\frac{A}{L} & \frac{\sqrt{3}}{4}\left(\frac{12I}{L^3}-\frac{A}{L}\right) & -\frac{3I}{L^2} & \frac{3I}{L^3}+\frac{3}{4}\frac{A}{L} & \frac{\sqrt{3}}{4}\left(-\frac{12I}{L^3}+\frac{A}{L}\right) & -\frac{3I}{L^2} \\[4pt]
-\frac{\sqrt{3}}{4}\left(-\frac{12I}{L^3}+\frac{A}{L}\right) & -\frac{9I}{L^3}-\frac{1}{4}\frac{A}{L} & \frac{3I\sqrt{3}}{L^2} & \frac{\sqrt{3}}{4}\left(-\frac{12I}{L^3}+\frac{A}{L}\right) & \frac{9I}{L^3}+\frac{1}{4}\frac{A}{L} & \frac{3I\sqrt{3}}{L^2} \\[4pt]
\frac{3I}{L^2} & -\frac{3I\sqrt{3}}{L^2} & \frac{2I}{L} & -\frac{3I}{L^2} & \frac{3I\sqrt{3}}{L^2} & \frac{4I}{L}
\end{bmatrix}$$

$30°$

$$E\begin{bmatrix}
\frac{3I}{L^3}+\frac{3}{4}\frac{A}{L} & -\frac{\sqrt{3}}{4}\left(-\frac{12I}{L^3}+\frac{A}{L}\right) & -\frac{3I}{L^2} & -\frac{3I}{L^3}-\frac{3}{4}\frac{A}{L} & -\frac{\sqrt{3}}{4}\left(\frac{12I}{L^3}-\frac{A}{L}\right) & -\frac{3I}{L^2} \\[4pt]
-\frac{\sqrt{3}}{4}\left(-\frac{12I}{L^3}+\frac{A}{L}\right) & \frac{9I}{L^3}+\frac{1}{4}\frac{A}{L} & \frac{3I\sqrt{3}}{L^2} & -\frac{\sqrt{3}}{4}\left(\frac{12I}{L^3}-\frac{A}{L}\right) & -\frac{9I}{L^3}-\frac{1}{4}\frac{A}{L} & \frac{3I\sqrt{3}}{L^2} \\[4pt]
-\frac{3I}{L^2} & \frac{3I\sqrt{3}}{L^2} & \frac{4I}{L} & \frac{3I}{L^2} & -\frac{3I\sqrt{3}}{L^2} & \frac{2I}{L} \\[4pt]
-\frac{3I}{L^3}-\frac{3}{4}\frac{A}{L} & -\frac{\sqrt{3}}{4}\left(\frac{12I}{L^3}-\frac{A}{L}\right) & \frac{3I}{L^2} & \frac{3I}{L^3}+\frac{3}{4}\frac{A}{L} & \frac{\sqrt{3}}{4}\left(-\frac{12I}{L^3}+\frac{A}{L}\right) & \frac{3I}{L^2} \\[4pt]
-\frac{\sqrt{3}}{4}\left(\frac{12I}{L^3}-\frac{A}{L}\right) & -\frac{9I}{L^3}-\frac{1}{4}\frac{A}{L} & -\frac{3I\sqrt{3}}{L^2} & \frac{\sqrt{3}}{4}\left(-\frac{12I}{L^3}+\frac{A}{L}\right) & \frac{9I}{L^3}+\frac{1}{4}\frac{A}{L} & -\frac{3I\sqrt{3}}{L^2} \\[4pt]
-\frac{3I}{L^2} & \frac{3I\sqrt{3}}{L^2} & \frac{2I}{L} & \frac{3I}{L^2} & -\frac{3I\sqrt{3}}{L^2} & \frac{4I}{L}
\end{bmatrix}$$

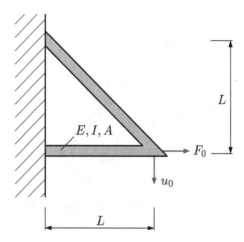

Fig. 3.63 Triangular shaped plane frame structure composed of generalized beam elements

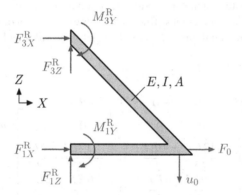

Fig. 3.64 Free-body diagram of the triangular shaped plane frame structure composed of generalized beam elements

3.16 Solution

The solution will follow the recommended 10 steps outlined on Sect. 3.2.

① Sketch the free-body diagram of the problem, including a global coordinate system (see Fig. 3.64).

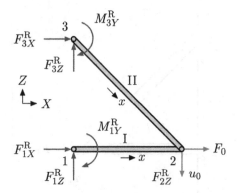

Fig. 3.65 Free-body diagram of the discretized structure with nodal loads

② Subdivide the geometry into finite elements. Indicate the node and element numbers, local coordinate systems, and equivalent nodal loads, see Fig. 3.65.
③ Write separately all elemental stiffness matrices expressed in the global coordinate system. Indicate the nodal unknowns on the right-hand sides and over the matrices.

Element I: $\alpha_I = 0°$, $L_I = L$

$$
K_I^e =
\begin{array}{c}
\begin{array}{cccccc} u_{1X} & u_{1Z} & \varphi_{1Y} & u_{2X} & u_{2Z} & \varphi_{2Y} \end{array} \\
\begin{bmatrix}
\dfrac{EA}{L} & 0 & 0 & -\dfrac{EA}{L} & 0 & 0 \\[2ex]
0 & \dfrac{12EI}{L^3} & -\dfrac{6EI}{L^2} & 0 & -\dfrac{12EI}{L^3} & -\dfrac{6EI}{L^2} \\[2ex]
0 & -\dfrac{6EI}{L^2} & \dfrac{4EI}{L} & 0 & \dfrac{6EI}{L^2} & \dfrac{2EI}{L} \\[2ex]
-\dfrac{EA}{L} & 0 & 0 & \dfrac{EA}{L} & 0 & 0 \\[2ex]
0 & -\dfrac{12EI}{L^3} & \dfrac{6EI}{L^2} & 0 & \dfrac{12EI}{L^3} & \dfrac{6EI}{L^2} \\[2ex]
0 & -\dfrac{6EI}{L^2} & \dfrac{2EI}{L} & 0 & \dfrac{6EI}{L^2} & \dfrac{4EI}{L}
\end{bmatrix}
\begin{array}{c} u_{1X} \\[2ex] u_{1Z} \\[2ex] \varphi_{1Y} \\[2ex] u_{2X} \\[2ex] u_{2Z} \\[2ex] \varphi_{2Y} \end{array}
\end{array}
\quad . \qquad (3.348)
$$

Element II: $\alpha_{II} = 45°$, $L_{II} = \sqrt{2}L$

$$K^e_{II} = E \times \qquad\qquad\qquad\qquad\qquad\qquad\qquad\qquad\qquad\qquad (3.349)$$

$$
\begin{array}{cccccc}
u_{3X} & u_{3Z} & \varphi_{3Y} & u_{2X} & u_{2Z} & \varphi_{2Y}
\end{array}
$$

$$
\begin{bmatrix}
\frac{6I}{(\sqrt{2}L)^3} + \frac{1}{2}\frac{A}{(\sqrt{2}L)} & \frac{6I}{(\sqrt{2}L)^3} - \frac{1}{2}\frac{A}{(\sqrt{2}L)} & -\frac{3I\sqrt{2}}{(\sqrt{2}L)^2} & -\frac{6I}{(\sqrt{2}L)^3} - \frac{1}{2}\frac{A}{(\sqrt{2}L)} & -\frac{6I}{(\sqrt{2}L)^3} + \frac{1}{2}\frac{A}{(\sqrt{2}L)} & -\frac{3I\sqrt{2}}{(\sqrt{2}L)^2} \\[6pt]
\frac{6I}{(\sqrt{2}L)^3} - \frac{1}{2}\frac{A}{(\sqrt{2}L)} & \frac{6I}{(\sqrt{2}L)^3} + \frac{1}{2}\frac{A}{(\sqrt{2}L)} & -\frac{3I\sqrt{2}}{(\sqrt{2}L)^2} & -\frac{6I}{(\sqrt{2}L)^3} + \frac{1}{2}\frac{A}{(\sqrt{2}L)} & -\frac{6I}{(\sqrt{2}L)^3} - \frac{1}{2}\frac{A}{(\sqrt{2}L)} & -\frac{3I\sqrt{2}}{(\sqrt{2}L)^2} \\[6pt]
-\frac{3I\sqrt{2}}{(\sqrt{2}L)^2} & -\frac{3I\sqrt{2}}{(\sqrt{2}L)^2} & \frac{4I}{(\sqrt{2}L)} & \frac{3I\sqrt{2}}{(\sqrt{2}L)^2} & \frac{3I\sqrt{2}}{(\sqrt{2}L)^2} & \frac{2I}{(\sqrt{2}L)} \\[6pt]
-\frac{6I}{(\sqrt{2}L)^3} - \frac{1}{2}\frac{A}{(\sqrt{2}L)} & -\frac{6I}{(\sqrt{2}L)^3} + \frac{1}{2}\frac{A}{(\sqrt{2}L)} & \frac{3I\sqrt{2}}{(\sqrt{2}L)^2} & \frac{6I}{(\sqrt{2}L)^3} + \frac{1}{2}\frac{A}{(\sqrt{2}L)} & \frac{6I}{(\sqrt{2}L)^3} - \frac{1}{2}\frac{A}{(\sqrt{2}L)} & \frac{3I\sqrt{2}}{(\sqrt{2}L)^2} \\[6pt]
-\frac{6I}{(\sqrt{2}L)^3} + \frac{1}{2}\frac{A}{(\sqrt{2}L)} & -\frac{6I}{(\sqrt{2}L)^3} - \frac{1}{2}\frac{A}{(\sqrt{2}L)} & \frac{3I\sqrt{2}}{(\sqrt{2}L)^2} & \frac{6I}{(\sqrt{2}L)^3} - \frac{1}{2}\frac{A}{(\sqrt{2}L)} & \frac{6I}{(\sqrt{2}L)^3} + \frac{1}{2}\frac{A}{(\sqrt{2}L)} & \frac{3I\sqrt{2}}{(\sqrt{2}L)^2} \\[6pt]
-\frac{3I\sqrt{2}}{(\sqrt{2}L)^2} & -\frac{3I\sqrt{2}}{(\sqrt{2}L)^2} & \frac{2I}{(\sqrt{2}L)} & \frac{3I\sqrt{2}}{(\sqrt{2}L)^2} & \frac{3I\sqrt{2}}{(\sqrt{2}L)^2} & \frac{4I}{(\sqrt{2}L)}
\end{bmatrix}
\begin{array}{l}
u_{3X} \\[6pt] u_{3Z} \\[6pt] \varphi_{3Y} \\[6pt] u_{2X} \\[6pt] u_{2Z} \\[6pt] \varphi_{2Y}
\end{array}
$$

④ Determine the dimensions of the global stiffness matrix and sketch the structure of this matrix with global unknowns on the right-hand side and over the matrix.

The finite element structure is composed of 3 nodes, each having three degrees of freedom (i.e., the vertical and horizontal displacements and the rotation). Thus, the dimensions of the global stiffness matrix are $(3 \times 3) \times (3 \times 3) = (9 \times 9)$:

$$
\boldsymbol{K} =
\begin{array}{cccccccccl}
u_{1X} & u_{1Z} & \varphi_{1Y} & u_{2X} & u_{2Z} & \varphi_{2Y} & u_{3X} & u_{3Z} & \varphi_{3Y} & \\
\end{array}
$$

$$
\boldsymbol{K} =
\left[\begin{array}{c|c|c|c|c|c|c|c|c}
 & & & & & & & & \\ \hline
 & & & & & & & & \\ \hline
 & & & & & & & & \\ \hline
 & & & & & & & & \\ \hline
 & & & & & & & & \\ \hline
 & & & & & & & & \\ \hline
 & & & & & & & & \\ \hline
 & & & & & & & & \\ \hline
 & & & & & & & & \\
\end{array}\right]
\begin{array}{l}
u_{1X} \\ u_{1Z} \\ \varphi_{1Y} \\ u_{2X} \\ u_{2Z} \\ \varphi_{2Y} \\ u_{3X} \\ u_{3Z} \\ \varphi_{3Y}
\end{array} \qquad (3.350)
$$

⑤ Insert the values of the elemental stiffness matrices step-by-step into the global stiffness matrix.

$$K = E \times$$

$$
\begin{array}{ccccccccc}
u_{1X} & u_{1Z} & \varphi_{1Y} & u_{2X} & u_{2Z} & \varphi_{2Y} & u_{3X} & u_{3Z} & \varphi_{3Y} \\
\frac{A}{L} & 0 & 0 & -\frac{A}{L} & 0 & 0 & 0 & 0 & 0 \\[4pt]
0 & \frac{12I}{L^3} & -\frac{6I}{L^2} & 0 & -\frac{12I}{L^3} & -\frac{6I}{L^2} & 0 & 0 & 0 \\[4pt]
0 & -\frac{6I}{L^2} & \frac{4I}{L} & 0 & \frac{6I}{L^2} & \frac{2I}{L} & 0 & 0 & 0 \\[4pt]
-\frac{A}{L} & 0 & 0 & \frac{A}{L}+\frac{3I\sqrt{2}}{2L^3}+\frac{A\sqrt{2}}{4L} & \frac{3I\sqrt{2}}{2L^3}-\frac{A\sqrt{2}}{4L} & \frac{3I\sqrt{2}}{2L^2} & -\frac{3I\sqrt{2}}{2L^3}-\frac{A\sqrt{2}}{4L} & -\frac{3I\sqrt{2}}{2L^3}+\frac{A\sqrt{2}}{4L} & \frac{3I\sqrt{2}}{2L^2} \\[4pt]
0 & -\frac{12I}{L^3} & \frac{6I}{L^2} & \frac{3I\sqrt{2}}{2L^3}-\frac{A\sqrt{2}}{4L} & \frac{12I}{L^3}+\frac{3I\sqrt{2}}{2L^3}+\frac{A\sqrt{2}}{4L} & \frac{6I}{L^2}+\frac{3I\sqrt{2}}{2L^2} & -\frac{3I\sqrt{2}}{2L^3}+\frac{A\sqrt{2}}{4L} & -\frac{3I\sqrt{2}}{2L^3}-\frac{A\sqrt{2}}{4L} & \frac{3I\sqrt{2}}{2L^2} \\[4pt]
0 & -\frac{6I}{L^2} & \frac{2I}{L} & \frac{3I\sqrt{2}}{2L^2} & \frac{6I}{L^2}+\frac{3I\sqrt{2}}{2L^2} & \frac{4I}{L}+\frac{2I\sqrt{2}}{L} & -\frac{3I\sqrt{2}}{2L^2} & -\frac{3I\sqrt{2}}{2L^2} & \frac{I\sqrt{2}}{L} \\[4pt]
0 & 0 & 0 & -\frac{3I\sqrt{2}}{2L^3}-\frac{A\sqrt{2}}{4L} & -\frac{3I\sqrt{2}}{2L^3}+\frac{A\sqrt{2}}{4L} & -\frac{3I\sqrt{2}}{2L^2} & \frac{3I\sqrt{2}}{2L^3}+\frac{A\sqrt{2}}{4L} & \frac{3I\sqrt{2}}{2L^3}-\frac{A\sqrt{2}}{4L} & -\frac{3I\sqrt{2}}{2L^2} \\[4pt]
0 & 0 & 0 & -\frac{3I\sqrt{2}}{2L^3}+\frac{A\sqrt{2}}{4L} & -\frac{3I\sqrt{2}}{2L^3}-\frac{A\sqrt{2}}{4L} & -\frac{3I\sqrt{2}}{2L^2} & \frac{3I\sqrt{2}}{2L^3}-\frac{A\sqrt{2}}{4L} & \frac{3I\sqrt{2}}{2L^3}+\frac{A\sqrt{2}}{4L} & -\frac{3I\sqrt{2}}{2L^2} \\[4pt]
0 & 0 & 0 & \frac{3I\sqrt{2}}{2L^2} & \frac{3I\sqrt{2}}{2L^2} & \frac{I\sqrt{2}}{L} & -\frac{3I\sqrt{2}}{2L^2} & -\frac{3I\sqrt{2}}{2L^2} & \frac{2I\sqrt{2}}{L}
\end{array}
\begin{array}{l}
u_{1X} \\[4pt]
u_{1Z} \\[4pt]
\varphi_{1Y} \\[4pt]
u_{2X} \\[4pt]
u_{2Z} \\[4pt]
\varphi_{2Y} \\[4pt]
u_{3X} \\[4pt]
u_{3Z} \\[4pt]
\varphi_{3Y}
\end{array}
$$

$$(3.351)$$

⑥ Add the column matrix of unknowns and external loads to complete the global system of equations.

The global system of equations can be written as $K u_p = f$, where the column matrix of the nodal unknowns reads

$$u_p = \begin{bmatrix} u_{1X} & u_{1Z} & \varphi_{1Y} & u_{2X} & u_{2Z} & \varphi_{2Y} & u_{3X} & u_{3Z} & \varphi_{3Y} \end{bmatrix}^T, \qquad (3.352)$$

and the column matrix of the external loads is given by:

$$f = \begin{bmatrix} F_{1X}^R & F_{1Z}^R & M_{1Y}^R & F_0 & -F_{2Z}^R & 0 & F_{3X}^R & F_{3Z}^R & M_{3Y}^R \end{bmatrix}^T. \qquad (3.353)$$

⑦ Introduce the boundary conditions to obtain the reduced system of equations.

There are only two degrees of freedom, i.e., the rotation and horizontal displacement at node 2:

$$E \begin{bmatrix} \dfrac{A}{L} + \dfrac{3\sqrt{2}I}{2L^3} + \dfrac{\sqrt{2}A}{4L} & \dfrac{3\sqrt{2}I}{2L^2} \\[3mm] \dfrac{3\sqrt{2}I}{2L^2} & \dfrac{4I}{L} + \dfrac{2\sqrt{2}I}{L} \end{bmatrix} \begin{bmatrix} u_{2X} \\[3mm] \varphi_{2Y} \end{bmatrix} = \begin{bmatrix} F_0 + u_0 E \left(\dfrac{3\sqrt{2}I}{2L^3} - \dfrac{A\sqrt{2}}{4L} \right) \\[3mm] u_0 E \left(\dfrac{6I}{L^2} + \dfrac{3\sqrt{2}I}{2L^2} \right) \end{bmatrix}.$$

$$(3.354)$$

⑧ Solve the reduced system of equations to obtain the unknown nodal deformations.

The solution can be obtained based on the matrix approach $u_p = K^{-1} f$:

$$\begin{bmatrix} u_{2X} \\[3mm] \varphi_{2Y} \end{bmatrix} = \dfrac{\dfrac{2L^4}{EI}}{10AL^2 + 6A\sqrt{2}L^2 + 3I + 12I\sqrt{2}} \begin{bmatrix} \dfrac{4I}{L} + \dfrac{2\sqrt{2}I}{L} & -\dfrac{3\sqrt{2}I}{2L^2} \\[3mm] -\dfrac{3\sqrt{2}I}{2L^2} & \dfrac{A}{L} + \dfrac{3\sqrt{2}I}{2L^3} + \dfrac{\sqrt{2}A}{4L} \end{bmatrix}$$

$$\times \begin{bmatrix} F_0 + u_0 E \left(\dfrac{3\sqrt{2}I}{2L^3} - \dfrac{\sqrt{2}A}{4L} \right) \\[3mm] u_0 E \left(\dfrac{6I}{L^2} + \dfrac{3\sqrt{2}I}{2L^3} \right) \end{bmatrix}, \qquad (3.355)$$

or after the multiplication:

$$
\begin{bmatrix} u_{2X} \\ \\ \varphi_{2Y} \end{bmatrix} = \begin{bmatrix} -\dfrac{2\sqrt{2}EAL^2u_0 + 2EAL^2u_0 - 4\sqrt{2}F_0L^3 + 6\sqrt{2}EIu_0 - 8F_0L^3 - 3EIu_0}{E\left(10AL^2 + 6\sqrt{2}AL^2 + 3I + 12\sqrt{2}I\right)} \\ \\ 3 \times \dfrac{2\sqrt{2}EAL^2u_0 + 5EAL^2u_0 - \sqrt{2}F_0L^3 + 6\sqrt{2}EIu_0}{E\left(10AL^2 + 6\sqrt{2}AL^2 + 3I + 12\sqrt{2}I\right)L} \end{bmatrix}.
$$

$$(3.356)$$

If we approximate $\sqrt{2}$ by its numerical value and consider only decimals with a precision of 2, we get the following simplified expression:

$$
\begin{bmatrix} u_{2X} \\ \\ \varphi_{2Y} \end{bmatrix} \approx \begin{bmatrix} -\dfrac{4.83EAL^2u_0 - 13.66F_0L^3 + 5.49EIu_0}{E\left(18.49AL^2 + 19.97I\right)} \\ \\ 3 \times \dfrac{7.83EAL^2u_0 - 1.41F_0L^3 + 8.49EIu_0}{E\left(18.49AL^2 + 19.97I\right)L} \end{bmatrix}.
$$

$$(3.357)$$

The nodal deformations at node 2 allow the calculation of the displacement distributions in local z- (bending) and x-direction (axial) in each element. Based on the relationships in Tables 3.13 and 3.5, one gets:

$$
u_{z_I}^e(x_I) = \left[3\left(\frac{x_I}{L}\right)^2 - 2\left(\frac{x_I}{L}\right)^3\right]u_{2z_I} + \left[\frac{x_I^2}{L} - \frac{x_I^3}{L^2}\right]\varphi_{2y_I}, \tag{3.358}
$$

$$
u_{z_{II}}^e(x_{II}) = \left[3\left(\frac{x_{II}}{\sqrt{2}L}\right)^2 - 2\left(\frac{x_{II}}{\sqrt{2}L}\right)^3\right]u_{2z_{II}} + \left[\frac{x_{II}^2}{\sqrt{2}L} - \frac{x_{II}^3}{(\sqrt{2}L)^2}\right]\varphi_{2y_{II}}.
$$

$$(3.359)$$

The deformations at node 2, expressed in the local coordinate systems (x_I, y_I, z_I) and (x_{II}, y_{II}, z_{II}), can be calculated from the global values based on the relationships (consider $\alpha_I = 0°$, $\alpha_{II} = +45°$, $u_{2Z} = -u_0$) given in Table 3.15:

$$
u_{2z_I} = \sin(\alpha_I)u_{2X} + \cos(\alpha_I)u_{2Z}, \tag{3.360}
$$

$$
\varphi_{2y_I} = \varphi_{2Y}, \tag{3.361}
$$

$$
u_{2z_{II}} = \sin(\alpha_{II})u_{2X} + \cos(\alpha_{II})u_{2Z}, \tag{3.362}
$$

$$
\varphi_{2y_{II}} = \varphi_{2Y}. \tag{3.363}
$$

Thus, the displacement distributions (bending) can be approximated as:

$$u_{z_I}^e(x_I) \approx - \left[\frac{x_I^2}{L^2} - 2\frac{x_I^3}{L^3}\right] u_0 +$$

$$3 \times \frac{7.83EAL^2u_0 - 1.41F_0L^3 + 8.49EIu_0}{EL\left(18.49AL^2 + 19.97I\right)} \left[\frac{x_I^2}{L} - \frac{x_I^3}{L^2}\right], \qquad (3.364)$$

$$u_{z_{II}}^e(x_{II}) \approx \left[1.5\frac{x_{II}^2}{L^2} - 0.71\frac{x_{II}^3}{L^3}\right] \times$$

$$\left(-0.71\frac{4.83EAL^2u_0 - 13.66F_0L^3 + 5.49EIu_0}{E\left(18.49AL^2 + 19.97I\right)} - 0.71u_0\right)$$

$$+ 3.0 \times \frac{7.83EAL^2u_0 - 1.41F_0L^3 + 8.49EIu_0}{EL\left(18.49AL^2 + 19.97I\right)}$$

$$\times \left[0.71\frac{x_{II}^2}{L} - 0.5\frac{x_{II}^3}{L^2}\right]. \qquad (3.365)$$

The axial displacement distributions can be obtained in a similar manner:

$$u_{x_I}^e(x_I) = \left[\frac{x_I}{L}\right] u_{2x_I} = \left[\frac{x_I}{L}\right](\cos(\alpha_I)u_{2X} - \sin(\alpha_I)u_{2Z}), \qquad (3.366)$$

$$u_{x_{II}}^e(x_{II}) = \left[\frac{x_{II}}{\sqrt{2}L}\right] u_{2x_{II}} = \left[\frac{x_{II}}{\sqrt{2}L}\right](\cos(\alpha_{II})u_{2X} - \sin(\alpha_{II})u_{2Z}), \qquad (3.367)$$

or based on the given values:

$$u_{x_I}^e(x_I) \approx -\frac{4.83EAL^2u_0 - 13.66F_0L^3 + 5.49EIu_0}{E\left(18.49AL^2 + 19.97I\right)} \left[\frac{x_I}{L}\right], \qquad (3.368)$$

$$u_{x_{II}}^e(x_{II}) \approx 0.50 \left[\frac{x_{II}}{L}\right]\left(-\frac{4.83EAL^2u_0 - 13.66F_0L^3 + 5.49EIu_0}{E\left(18.49AL^2 + 19.97I\right)} + u_0\right).$$

$$(3.369)$$

⑨ Post-computation: determination of reactions, stresses and strains.

The reactions at the supports can be obtained from the non-reduced system of equations as given in step ⑥ under the consideration of the known nodal degrees of freedom (i.e., displacements and rotations). The evaluation of the first equation gives:

$$0 - \frac{EA}{L}u_{2X} + 0 = F^{R}_{1X},$$ (3.370)

or

$$F^{R}_{1X} = \frac{A\left(2\sqrt{2}EAL^{2}u_{0} + 2EAL^{2}u_{0} - 4\sqrt{2}F_{0}L^{3} + 6\sqrt{2}EIu_{0} - 8F_{0}L^{3} - 3EIu_{0}\right)}{L\left(10AL^{2} + 6\sqrt{2}AL^{2} + 3I + 12\sqrt{2}I\right)}$$

$$\approx \frac{A\left(4.83EAL^{2}u_{0} - 13.66F_{0}L^{3} + 5.49EIu_{0}\right)}{L\left(18.49AL^{2} + 19.97I\right)}.$$ (3.371)

In a similar way, the evaluation of the remaining equations gives:

$$F^{R}_{1Z} = 6 \times \frac{I\left(6\sqrt{2}EAL^{2}u_{0} + 5EAL^{2}u_{0} + 3\sqrt{2}F_{0}L^{3} + 6\sqrt{2}EIu_{0} + 6EIu_{0}\right)}{L^{3}\left(10AL^{2} + 6\sqrt{2}AL^{2} + 3I + 12\sqrt{2}I\right)}$$

$$\approx 6 \times \frac{I\left(13.49EAL^{2}u_{0} + 4.24F_{0}L^{3} + 14.49EIu_{0}\right)}{L^{3}\left(18.49AL^{2} + 19.97I\right)},$$ (3.372)

$$M^{R}_{1Y} = -6 \times \frac{I\left(4\sqrt{2}EAL^{2}u_{0} + 5EAL^{2}u_{0} + \sqrt{2}F_{0}L^{3} + 6\sqrt{2}EIu_{0} + 3EIu_{0}\right)}{\left(10AL^{2} + 6\sqrt{2}AL^{2} + 3I + 12\sqrt{2}I\right)L^{2}}$$

$$\approx -6 \times \frac{I\left(10.66EAL^{2}u_{0} + 1.41F_{0}L^{3} + 11.49EIu_{0}\right)}{\left(18.49AL^{2} + 19.97I\right)L^{2}},$$ (3.373)

$$F^{R}_{2Z} \approx \frac{4.83EA^{2}L^{4}u_{0} + 4.83AF_{0}L^{5} + 85.67EAIL^{2}u_{0} + 5.49F_{0}IL^{3} + 86.91EI^{2}u_{0}}{L^{3}\left(18.49AL^{2} + 19.97I\right)},$$ (3.374)

$$F^{R}_{3X} \approx -0.71 \times \frac{6.83EA^{2}L^{2}u_{0} + 6.83AF_{0}L^{3} + 7.76EAIu_{0} + 28.24F_{0}IL}{L\left(18.49AL^{2} + 19.97I\right)},$$ (3.375)

$$F^{R}_{3Z} \approx 0.71 \times \frac{6.83EA^{2}L^{2}u_{0} + 6.83AF_{0}L^{3} + 6.73EAIu_{0} - 28.24F_{0}IL}{L\left(18.49AL^{2} + 19.97I\right)},$$ (3.376)

$$M_{3Y}^{R} \approx -4.24 \times \frac{I\left(3.83EAL^2 u_0 - 5.41 F_0 L^3 + 4.24 EI u_0\right)}{\left(18.49 AL^2 + 19.97I\right) L^2}. \quad (3.377)$$

The internal reactions (i.e., bending moment, shear force, and normal force) in each element can be obtained from the relations provided in Tables 3.14 and 3.5.

Bending moment distribution:

$$M_{y_I}^{e}(x_I) = EI\left(\left[-\frac{6}{L^2} + \frac{12x_I}{L^3}\right] u_{2z_I} + \left[-\frac{2}{L} + \frac{6x_I}{L^2}\right] \varphi_{2y_I}\right), \quad (3.378)$$

$$M_{y_{II}}^{e}(x_{II}) = EI\left(\left[-\frac{6}{(\sqrt{2}L)^2} + \frac{12x_{II}}{(\sqrt{2}L)^3}\right] u_{2z_{II}} + \left[-\frac{2}{\sqrt{2}L} + \frac{6x_{II}}{(\sqrt{2}L)^2}\right] \varphi_{2y_{II}}\right), \quad (3.379)$$

or based on the given values:

$$M_{y_I}^{e}(x_I) \approx EI\left(\left[\frac{6}{L^2} - \frac{12x_I}{L^3}\right] u_0 + 3 \times \frac{7.83EAL^2 u_0 - 1.41 F_0 L^3 + 8.49 EI u_0}{EL\left(18.49 AL^2 + 19.97I\right)}\right.$$
$$\left. \times \left[-\frac{2}{L} + \frac{6x_I}{L^2}\right]\right), \quad (3.380)$$

$$M_{y_{II}}^{e}(x_I) \approx EI\left(\left[-\frac{3}{L^2} + 4.25\frac{x_{II}}{L^3}\right] \times \right.$$
$$\left(-0.71\frac{4.83EAL^2 u_0 - 13.66 F_0 L^3 + 5.49 EI u_0}{E\left(18.49 AL^2 + 19.97I\right)} - 0.71 u_0\right)$$
$$\left. +3\frac{7.83EAL^2 u_0 - 1.41 F_0 L^3 + 8.49 EI u_0}{EL\left(18.49 AL^2 + 19.97I\right)}\left[-\frac{1.41}{L} + \frac{3x_{II}}{L^2}\right]\right). \quad (3.381)$$

Shear force distribution:

$$Q_{z_I}^{e}(x_I) = EI\left(\left[\frac{12}{L^3}\right] u_{2z_I} + \left[\frac{6}{L^2}\right] \varphi_{2y_I}\right), \quad (3.382)$$

$$Q_{z_{II}}^{e}(x_{II}) = EI\left(\left[\frac{12}{(\sqrt{2}L)^3}\right] u_{2z_I} + \left[\frac{6}{\sqrt{2}L^2}\right] \varphi_{2y_{II}}\right), \quad (3.383)$$

or based on the given values:

$$Q_{z_{\mathrm{I}}}^{\mathrm{e}}(x_{\mathrm{I}}) \approx EI\left(-\frac{12u_0}{L^3} + 18 \times \frac{7.83EAL^2u_0 - 1.41F_0L^3 + 8.49EIu_0}{EL^3\left(18.49AL^2 + 19.97I\right)}\right),$$

$$(3.384)$$

$$Q_{z_{\mathrm{II}}}^{\mathrm{e}}(x_{\mathrm{II}}) \approx EI\left(\frac{4.24}{L^3}\left(-0.71\frac{4.83EAL^2u_0 - 13.66F_0L^3 + 5.49EIu_0}{E\left(18.49AL^2 + 19.97I\right)} - 0.71u_0\right)\right.$$

$$\left.+9 \times \frac{7.83EAL^2u_0 - 1.41F_0L^3 + 8.49EIu_0}{EL^3\left(18.49AL^2 + 19.97I\right)}\right).$$

$$(3.385)$$

Normal force distribution:

$$N_{x_{\mathrm{I}}}^{\mathrm{e}}(x_{\mathrm{I}}) = \frac{EA}{L}u_{2x_{\mathrm{I}}},$$

$$(3.386)$$

$$N_{x_{\mathrm{II}}}^{\mathrm{e}}(x_{\mathrm{II}}) = \frac{EA}{\sqrt{2}L}u_{2x_{\mathrm{II}}},$$

$$(3.387)$$

or based on the given values:

$$N_{x_{\mathrm{I}}}^{\mathrm{e}}(x_{\mathrm{I}}) \approx -\frac{A\left(4.83EAL^2u_0 - 13.66F_0L^3 + 5.49EIu_0\right)}{L\left(18.49AL^2 + 19.97I\right)},$$

$$(3.388)$$

$$N_{x_{\mathrm{II}}}^{\mathrm{e}}(x_{\mathrm{II}}) \approx 0.71\frac{AE}{L}\left(-0.71\frac{4.83EAL^2u_0 - 13.66F_0L^3 + 5.49EIu_0}{E\left(18.49AL^2 + 19.97I\right)} + 0.71u_0\right).$$

$$(3.389)$$

⑩ Check the global equilibrium between the external loads and the support reactions.

$$\sum_i F_{iX} = 0 \quad \Leftrightarrow \quad \underbrace{(F_{1X}^{\mathrm{R}} + F_{3X}^{\mathrm{R}})}_{\text{reaction force}} + \underbrace{(F_0)}_{\text{external load}} = 0, \ \checkmark \qquad (3.390)$$

$$\sum_i F_{iZ} = 0 \quad \Leftrightarrow \quad \underbrace{(F_{1Z}^{\mathrm{R}} - F_{2Z}^{\mathrm{R}} + F_{3Z}^{\mathrm{R}})}_{\text{reaction force}} + \underbrace{(0)}_{\text{external load}} = 0, \ \checkmark \qquad (3.391)$$

$$\sum_i M_{iY} = 0 \quad \Leftrightarrow \quad \underbrace{(M_{1Y}^{\mathrm{R}} + M_{3Y}^{\mathrm{R}} + F_{2Z}^{\mathrm{R}}L + F_{3X}^{\mathrm{R}}L)}_{\text{reaction}} + \underbrace{(0)}_{\text{external load}} = 0. \ \checkmark \quad (3.392)$$

3.17 Plane frame structure composed of generalized beam elements

The plane frame structure shown in Fig. 3.66 is composed of generalized beams which are arranged in a T-shape formation. The structure is loaded by a single force F_0 in the middle of the structure. The material constant (E) and the geometrical properties (I, A) are constant and the horizontal length of the beam is equal to L

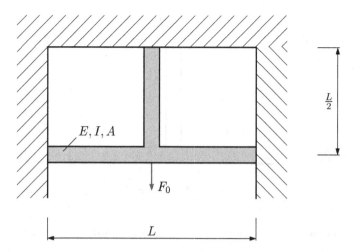

Fig. 3.66 Plane frame structure composed of generalized beam elements

while the vertical dimension is equal to $\frac{L}{2}$. Model the structure with three generalized beam finite elements of length $\frac{L}{2}$ to determine:

- the unknowns at the nodes,
- the displacement distributions in each member,
- the reactions at the supports,
- the internal reactions (normal force, shear force and bending moment) in each element,
- the strain and stress distributions in the elements, and
- the global force and moment equilibrium.

3.17 Solution

The solution will follow the recommended 10 steps outlined on Sect. 3.2.

① Sketch the free-body diagram of the problem, including a global coordinate system, see Fig. 3.67.
② Subdivide the geometry into finite elements. Indicate the node and element numbers, local coordinate systems, and equivalent nodal loads, see Fig. 3.68.
③ Write separately all elemental stiffness matrices expressed in the global coordinate system. Indicate the nodal unknowns on the right-hand sides and over the matrices.

Element I: $\alpha_I = 0°$, $L_I = \frac{L}{2}$

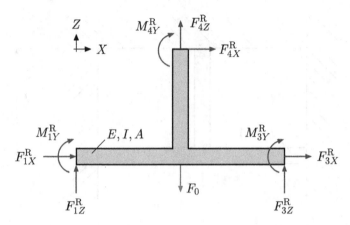

Fig. 3.67 Free-body diagram of the plane frame structure composed of generalized beam elements

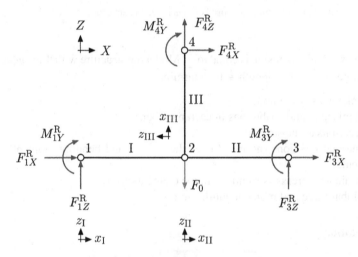

Fig. 3.68 Free-body diagram of the discretized structure with nodal loads

$$
K_I^e = E
\begin{array}{c}
\begin{array}{cccccc} u_{1X} & u_{1Z} & \varphi_{1Y} & u_{2X} & u_{2Z} & \varphi_{2Y} \end{array} \\
\begin{bmatrix}
\dfrac{A}{L_I} & 0 & 0 & -\dfrac{A}{L_I} & 0 & 0 \\[2mm]
0 & \dfrac{12I}{L_I^3} & -\dfrac{6I}{L_I^2} & 0 & -\dfrac{12I}{L_I^3} & -\dfrac{6I}{L_I^2} \\[2mm]
0 & -\dfrac{6I}{L_I^2} & \dfrac{4I}{L_I} & 0 & \dfrac{6I}{L_I^2} & \dfrac{2I}{L_I} \\[2mm]
-\dfrac{A}{L_I} & 0 & 0 & \dfrac{A}{L_I} & 0 & 0 \\[2mm]
0 & -\dfrac{12I}{L_I^3} & \dfrac{6I}{L_I^2} & 0 & \dfrac{12I}{L_I^3} & \dfrac{6I}{L_I^2} \\[2mm]
0 & -\dfrac{6I}{L_I^2} & \dfrac{2I}{L_I} & 0 & \dfrac{6I}{L_I^2} & \dfrac{4I}{L_I}
\end{bmatrix}
\begin{array}{l}
u_{1X} \\[2mm] u_{1Z} \\[2mm] \varphi_{1Y} \\[2mm] u_{2X} \\[2mm] u_{2Z} \\[2mm] \varphi_{2Y}
\end{array}
\end{array}
. \qquad (3.393)
$$

Element II: $\alpha_{\mathrm{II}} = 0°$, $L_{\mathrm{II}} = \frac{L}{2}$

$$
\boldsymbol{K}_{\mathrm{II}}^{\mathrm{e}} = E
\begin{array}{c}
\begin{array}{cccccc}
u_{2X} & u_{2Z} & \varphi_{2Y} & u_{3X} & u_{3Z} & \varphi_{3Y}
\end{array} \\
\left[
\begin{array}{cccccc}
\dfrac{A}{L_{\mathrm{II}}} & 0 & 0 & -\dfrac{A}{L_{\mathrm{II}}} & 0 & 0 \\[2.5ex]
0 & \dfrac{12I}{L_{\mathrm{II}}^{3}} & \dfrac{6I}{L_{\mathrm{II}}^{2}} & 0 & -\dfrac{12I}{L_{\mathrm{II}}^{3}} & \dfrac{6I}{L_{\mathrm{II}}^{2}} \\[2.5ex]
0 & -\dfrac{6I}{L_{\mathrm{II}}^{2}} & \dfrac{4I}{L_{\mathrm{II}}} & 0 & \dfrac{6I}{L_{\mathrm{II}}^{2}} & \dfrac{2I}{L_{\mathrm{II}}} \\[2.5ex]
-\dfrac{A}{L_{\mathrm{II}}} & 0 & 0 & \dfrac{A}{L_{\mathrm{II}}} & 0 & 0 \\[2.5ex]
0 & -\dfrac{12I}{L_{\mathrm{II}}^{3}} & \dfrac{6I}{L_{\mathrm{II}}^{2}} & 0 & \dfrac{12I}{L_{\mathrm{II}}^{3}} & \dfrac{6I}{L_{\mathrm{II}}^{2}} \\[2.5ex]
0 & -\dfrac{6I}{L_{\mathrm{II}}^{2}} & \dfrac{2I}{L_{\mathrm{II}}} & 0 & \dfrac{6I}{L_{\mathrm{II}}^{2}} & \dfrac{4I}{L_{\mathrm{II}}}
\end{array}
\right]
\begin{array}{c}
u_{2X} \\[2.5ex] u_{2Z} \\[2.5ex] \varphi_{2Y} \\[2.5ex] u_{3X} \\[2.5ex] u_{3Z} \\[2.5ex] \varphi_{3Y}
\end{array}
\end{array}
\tag{3.394}
$$

Element III: $\alpha_{\mathrm{III}} = 90°$, $L_{\mathrm{III}} = \frac{L}{2}$

$$
\boldsymbol{K}_{\mathrm{III}}^{\mathrm{e}} = E
\begin{array}{c}
\begin{array}{cccccc}
u_{2X} & u_{2Z} & \varphi_{2Y} & u_{4X} & u_{4Z} & \varphi_{4Y}
\end{array} \\
\left[
\begin{array}{cccccc}
\dfrac{12I}{L_{\mathrm{III}}^{3}} & 0 & -\dfrac{6I}{L_{\mathrm{III}}^{2}} & -\dfrac{12I}{L_{\mathrm{III}}^{3}} & 0 & -\dfrac{6I}{L_{\mathrm{III}}^{2}} \\[2.5ex]
0 & \dfrac{A}{L_{\mathrm{III}}} & 0 & 0 & -\dfrac{A}{L_{\mathrm{III}}} & 0 \\[2.5ex]
-\dfrac{6I}{L_{\mathrm{III}}^{2}} & 0 & \dfrac{4I}{L_{\mathrm{III}}} & \dfrac{6I}{L_{\mathrm{III}}^{2}} & 0 & \dfrac{2I}{L_{\mathrm{III}}} \\[2.5ex]
-\dfrac{12I}{L_{\mathrm{III}}^{3}} & 0 & \dfrac{6I}{L_{\mathrm{III}}^{2}} & \dfrac{12I}{L_{\mathrm{III}}^{3}} & 0 & \dfrac{6I}{L_{\mathrm{III}}^{2}} \\[2.5ex]
0 & -\dfrac{A}{L_{\mathrm{III}}} & 0 & 0 & \dfrac{A}{L_{\mathrm{III}}} & 0 \\[2.5ex]
-\dfrac{6I}{L_{\mathrm{III}}^{2}} & 0 & \dfrac{2I}{L_{\mathrm{III}}} & \dfrac{6I}{L_{\mathrm{III}}^{2}} & 0 & \dfrac{4I}{L_{\mathrm{III}}}
\end{array}
\right]
\begin{array}{c}
u_{2X} \\[2.5ex] u_{2Z} \\[2.5ex] \varphi_{2Y} \\[2.5ex] u_{4X} \\[2.5ex] u_{4Z} \\[2.5ex] \varphi_{4Y}
\end{array}
\end{array}
\tag{3.395}
$$

④ Determine the dimensions of the global stiffness matrix and sketch the structure of this matrix with global unknowns on the right-hand side and over the matrix.

The finite element structure is composed of four nodes, each having three degrees of freedom (i.e., the vertical and horizontal displacements and the rotation). Thus, the dimensions of the global stiffness matrix are $(4 \times 3) \times (4 \times 3) = (12 \times 12)$:

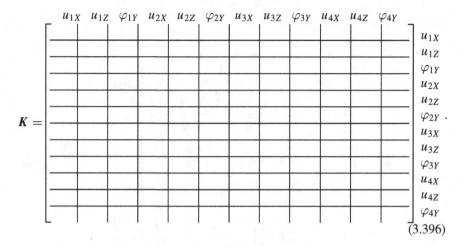

$$\boldsymbol{K} = \qquad\qquad\qquad\qquad\qquad\qquad\qquad\qquad (3.396)$$

⑤ Insert the values of the elemental stiffness matrices step-by-step into the global stiffness matrix:

$$\boldsymbol{K} = E \times \qquad\qquad\qquad\qquad\qquad\qquad\qquad\qquad (3.397)$$

$$
\begin{array}{cccccccccccc}
u_{1X} & u_{1Z} & \varphi_{1Y} & u_{2X} & u_{2Z} & \varphi_{2Y} & u_{3X} & u_{3Z} & \varphi_{3Y} & u_{4X} & u_{4Z} & \varphi_{4Y}
\end{array}
$$

$$
\begin{bmatrix}
\frac{2A}{L} & 0 & 0 & -\frac{2A}{L} & 0 & 0 & 0 & 0 & 0 & 0 & 0 & 0 \\
0 & \frac{96I}{L^3} & -\frac{24I}{L^2} & 0 & -\frac{96I}{L^3} & -\frac{24I}{L^2} & 0 & 0 & 0 & 0 & 0 & 0 \\
0 & -\frac{24I}{L^2} & \frac{8I}{L} & 0 & \frac{24I}{L^2} & \frac{4I}{L} & 0 & 0 & 0 & 0 & 0 & 0 \\
-\frac{2A}{L} & 0 & 0 & \left(\frac{4A}{L}+\frac{96I}{L^3}\right) & 0 & -\frac{24I}{L^2} & -\frac{2A}{L} & 0 & 0 & -\frac{96I}{L^3} & 0 & -\frac{24I}{L^2} \\
0 & -\frac{96I}{L^3} & \frac{24I}{L^2} & 0 & \left(\frac{192I}{L^3}+\frac{2A}{L}\right) & 0 & 0 & -\frac{96I}{L^3} & -\frac{24I}{L^2} & 0 & -\frac{2A}{L} & 0 \\
0 & -\frac{24I}{L^2} & \frac{4I}{L} & -\frac{24I}{L^2} & 0 & \frac{24I}{L} & 0 & \frac{24I}{L^2} & \frac{4I}{L} & \frac{24I}{L^2} & 0 & \frac{4I}{L} \\
0 & 0 & 0 & -\frac{2A}{L} & 0 & 0 & \frac{2A}{L} & 0 & 0 & 0 & 0 & 0 \\
0 & 0 & 0 & 0 & -\frac{96I}{L^3} & \frac{24I}{L^2} & 0 & \frac{96I}{L^3} & \frac{24I}{L^2} & 0 & 0 & 0 \\
0 & 0 & 0 & 0 & -\frac{24I}{L^2} & \frac{4I}{L} & 0 & \frac{24I}{L^2} & \frac{8I}{L} & 0 & 0 & 0 \\
0 & 0 & 0 & -\frac{96I}{L^3} & 0 & \frac{24I}{L^2} & 0 & 0 & 0 & \frac{96I}{L^3} & 0 & \frac{24I}{L^2} \\
0 & 0 & 0 & 0 & -\frac{2A}{L} & 0 & 0 & 0 & 0 & 0 & \frac{2A}{L} & 0 \\
0 & 0 & 0 & -\frac{24I}{L^2} & 0 & \frac{4I}{L} & 0 & 0 & 0 & \frac{24I}{L^2} & 0 & \frac{8I}{L}
\end{bmatrix}
\begin{array}{l}
u_{1X} \\ u_{1Z} \\ \varphi_{1Y} \\ u_{2X} \\ u_{2Z} \\ \varphi_{2Y} \\ u_{3X} \\ u_{3Z} \\ \varphi_{3Y} \\ u_{4X} \\ u_{4Z} \\ \varphi_{4Y}
\end{array}
$$

⑥ Add the column matrix of unknowns and external loads to complete the global system of equations.

The global system of equations can be written as $\boldsymbol{K}\boldsymbol{u}_{\mathrm{p}} = \boldsymbol{f}$, where the column matrix of the nodal unknowns reads

$$\boldsymbol{u}_\mathrm{p} = \begin{bmatrix} u_{1X} & u_{1Z} & \varphi_{1Y} & u_{2X} & u_{2Z} & \varphi_{2Y} & u_{3X} & u_{3Z} & \varphi_{3Y} & u_{4X} & u_{4Z} & \varphi_{4Y} \end{bmatrix}^\mathrm{T} , \tag{3.398}$$

and the column matrix of the external loads is given by:

$$\boldsymbol{f} = \begin{bmatrix} F_{1X}^\mathrm{R} & F_{1Z}^\mathrm{R} & M_{1Y}^\mathrm{R} & 0 & -F_0 & 0 & F_{3X}^\mathrm{R} & F_{3Z}^\mathrm{R} & M_{3Y}^\mathrm{R} & F_{4X}^\mathrm{R} & F_{4Z}^\mathrm{R} & M_{4Y}^\mathrm{R} \end{bmatrix}^\mathrm{T} . \tag{3.399}$$

⑦ Introduce the boundary conditions to obtain the reduced system of equations.

There are only three degrees of freedom, i.e., the rotation and displacements at node 2:

$$E \begin{bmatrix} \left(\dfrac{4A}{L} + \dfrac{96I}{L^3} \right) & 0 & -\dfrac{24I}{L^2} \\[2ex] 0 & \left(\dfrac{192I}{L^3} + \dfrac{2A}{L} \right) & 0 \\[2ex] -\dfrac{24I}{L^2} & 0 & \dfrac{24I}{L} \end{bmatrix} \begin{bmatrix} u_{2X} \\[2ex] u_{2Z} \\[2ex] \varphi_{2Y} \end{bmatrix} = \begin{bmatrix} 0 \\[2ex] -F_0 \\[2ex] 0 \end{bmatrix} . \tag{3.400}$$

⑧ Solve the reduced system of equations to obtain the unknown nodal deformations.

The solution can be obtained based on the matrix approach $\boldsymbol{u}_\mathrm{p} = \boldsymbol{K}^{-1}\boldsymbol{f}$:

$$\begin{bmatrix} u_{2X} \\[2ex] u_{2Z} \\[2ex] \varphi_{2Y} \end{bmatrix} = \begin{bmatrix} 0 \\[2ex] -\dfrac{1}{2} \times \dfrac{L^3 F_0}{E(AL^2 + 96I)} \\[2ex] 0 \end{bmatrix} . \tag{3.401}$$

The nodal deformation at node 2 allows the calculation of the displacement distributions in local z- (bending) and x-direction (axial) in each element. Based on the relationships in Tables 3.13 and 3.5, one gets:

$$\begin{aligned} u_{z_\mathrm{I}}^\mathrm{e}(x_\mathrm{I}) &= \left[3\left(\frac{x_\mathrm{I}}{L}\right)^2 - 2\left(\frac{x_\mathrm{I}}{L}\right)^3 \right] \underbrace{u_{2Z} \cos(\alpha_\mathrm{I})}_{1} \\ &= -\frac{1}{2} \times \frac{L^3 F_0}{E\left(AL^2 + 96I\right)} \left[\frac{12x_\mathrm{I}^2}{L^2} - \frac{16x_\mathrm{I}^3}{L^3} \right] , \\ u_{z_{\mathrm{II}}}^\mathrm{e}(x_{\mathrm{II}}) &= \left[1 - 3\left(\frac{x_{\mathrm{II}}}{L}\right)^2 + 2\left(\frac{x_{\mathrm{II}}}{L}\right)^3 \right] \underbrace{u_{2Z} \cos(\alpha_{\mathrm{II}})}_{1} \end{aligned} \tag{3.402}$$

$$= -\frac{1}{2} \times \frac{L^3 F_0}{E\left(AL^2 + 96I\right)} \left[1 - \frac{12x_{\mathrm{II}}^2}{L^2} + \frac{16x_{\mathrm{II}}^3}{L^3}\right],$$ (3.403)

$$u_{z_{\mathrm{III}}}^e(x_{\mathrm{III}}) = \left[1 - 3\left(\frac{x_{\mathrm{III}}}{L}\right)^2 + 2\left(\frac{x_{\mathrm{III}}}{L}\right)^3\right] u_{2Z} \underbrace{\cos(\alpha_{\mathrm{III}})}_{0}$$

$$= 0.$$ (3.404)

The axial displacement distributions can be obtained in a similar manner:

$$u_{x_{\mathrm{I}}}^e(x_{\mathrm{I}}) = -\left[\frac{x_{\mathrm{I}}}{L}\right] u_{2Z} \underbrace{\sin(\alpha_{\mathrm{I}})}_{0} = 0,$$ (3.405)

$$u_{x_{\mathrm{II}}}^e(x_{\mathrm{II}}) = -\left[1 - \frac{x_{\mathrm{II}}}{L}\right] u_{2Z} \underbrace{\sin(\alpha_{\mathrm{II}})}_{0} = 0,$$ (3.406)

$$u_{x_{\mathrm{III}}}^e(x_{\mathrm{III}}) = -\left[\frac{x_{\mathrm{III}}}{L}\right] u_{2Z} \underbrace{\sin(\alpha_{\mathrm{III}})}_{1} = 0,$$

$$= \frac{1}{2} \times \frac{L^3 F_0}{E\left(AL^2 + 96I\right)} \left[1 - \frac{2x_{\mathrm{III}}}{L}\right].$$ (3.407)

⑨ Post-computation: determination of reactions, stresses and strains.

The reactions at the supports can be obtained from the non-reduced system of equations as given in step ⑥ under the consideration of the known nodal degrees of freedom (i.e., displacements and rotations). The evaluation of the first equation gives:

$$\frac{2A}{L} u_{1X} - \frac{2A}{L} u_{2X} = F_{1X}^{\mathrm{R}} \quad \Rightarrow \quad F_{1X}^{\mathrm{R}} = 0.$$ (3.408)

In a similar way, the evaluation of the remaining equations gives:

$$F_{1Z}^{\mathrm{R}} = \frac{48I F_0}{AL^2 + 96I},$$ (3.409)

$$M_{1Y}^{\mathrm{R}} = -\frac{12ILF_0}{AL^2 + 96I},$$ (3.410)

$$F_{3X}^{\mathrm{R}} = 0,$$ (3.411)

$$F_{3Z}^{\mathrm{R}} = \frac{48I F_0}{AL^2 + 96I},$$ (3.412)

$$M_{3Y}^R = \frac{12ILF_0}{AL^2 + 96I}, \tag{3.413}$$

$$F_{4X}^R = 0, \tag{3.414}$$

$$F_{4Z}^R = \frac{AL^2 F_0}{AL^2 + 96I}, \tag{3.415}$$

$$M_{4Y}^R = 0. \tag{3.416}$$

The internal reactions (i.e., bending moment, shear force, and normal force) in each element can be obtained from the relations provided in Tables 3.14 and 3.5.

Bending moment distribution:

$$
\begin{aligned}
M_{y_I}^e(x_I) &= EI\left(\left[-\frac{6}{L_I^2} + \frac{12x_I}{L_I^3}\right] u_{2Z}\cos(\alpha_I)\right) \\
&= \frac{1}{2} \times \frac{I}{AL^2 + 96I}\left(-\left[-24\left(1 - 4\frac{x_I}{L}\right)\right]LF_0\right),
\end{aligned} \tag{3.417}
$$

$$
\begin{aligned}
M_{y_{II}}^e(x_{II}) &= EI\left(\left[\frac{6}{L_{II}^2} - \frac{12x_{II}}{L_{II}^3}\right] u_{2Z}\cos(\alpha_{II})\right) \\
&= \frac{1}{2} \times \frac{I}{AL^2 + 96I}\left(-\left[24\left(1 - 4\frac{x_{II}}{L}\right)\right]LF_0\right),
\end{aligned} \tag{3.418}
$$

$$
\begin{aligned}
M_{y_{III}}^e(x_{III}) &= EI\left(\left[\frac{6}{L_{III}^2} - \frac{12x_{III}}{L_{III}^3}\right] u_{2Z}\cos(\alpha_{III})\right) \\
&= 0.
\end{aligned} \tag{3.419}
$$

Shear force distribution:

$$
\begin{aligned}
Q_{z_I}^e(x_I) &= EI\left(\left[\frac{12}{L_I^3}\right] u_{2Z}\cos(\alpha_I)\right) \\
&= -\frac{48I F_0}{AL^2 + 96I},
\end{aligned} \tag{3.420}
$$

$$
\begin{aligned}
Q_{z_{II}}^e(x_{II}) &= EI\left(\left[-\frac{12}{L_{II}^3}\right] u_{2Z}\cos(\alpha_{II})\right) \\
&= \frac{48I F_0}{AL^2 + 96I},
\end{aligned} \tag{3.421}
$$

$$
\begin{aligned}
Q_{z_{III}}^e(x_{III}) &= EI\left(\left[-\frac{12}{L_{III}^3}\right] u_{2Z}\cos(\alpha_{III})\right) \\
&= 0.
\end{aligned} \tag{3.422}
$$

Normal force distribution:

$$N^e_{x_I}(x_I) = -\frac{EA}{L_I} u_Z \sin(\alpha_I)$$
$$= 0, \tag{3.423}$$

$$N^e_{x_{II}}(x_{II}) = \frac{EA}{L_{II}} u_Z \sin(\alpha_{II})$$
$$= 0, \tag{3.424}$$

$$N^e_{x_{III}}(x_{III}) = \frac{EA}{L_{III}} u_Z \sin(\alpha_{III})$$
$$= -\frac{AL^2 F_0}{AL^2 + 96I}. \tag{3.425}$$

The graphical representation of the internal reactions is shown in Fig. 3.69.

⑩ Check the global equilibrium between the external loads and the support reactions.

$$\sum_i F_{iX} = 0 \quad \Leftrightarrow \quad \underbrace{(F^R_{1X} + F^R_{3X} + F^R_{3X})}_{\text{reaction force}} + \underbrace{(0)}_{\text{external load}} = 0, \checkmark \tag{3.426}$$

$$\sum_i F_{iZ} = 0 \quad \Leftrightarrow \quad \underbrace{(F^R_{1Z} + F^R_{3Z} + F^R_{4Z})}_{\text{reaction force}} + \underbrace{(-F_0)}_{\text{external load}} = 0, \checkmark \tag{3.427}$$

$$\sum_i M_{iY} = 0 \quad \Leftrightarrow \tag{3.428}$$

$$\underbrace{(M^R_{1Y} + M^R_{3Y} + M^R_{4Y}L + F_0 L_I - F^R_{3Z}(L_I + L_{II}) - F^R_{4Z}L_I + F^R_{4X}L_{III} +}_{\text{reaction}}$$
$$+ \underbrace{(0)}_{\text{external load}} = 0. \checkmark \tag{3.429}$$

3.18 Plane frame structure representing a crane (computational problem)
The plane frame structure shown in Fig. 3.70 is composed of generalized beams which are arranged to represent a simple crane. The structure is loaded by a single force F_0 at the right-hand end. The material constant (E) and the geometrical properties (I, A) are constant and the horizontal length of the frame structure is equal to $\frac{L}{2}$ while the vertical dimension of the left-hand part is equal to L. Model the structure with three generalized beam finite elements to determine:

• the unknowns at the nodes,
• the reactions at the supports,

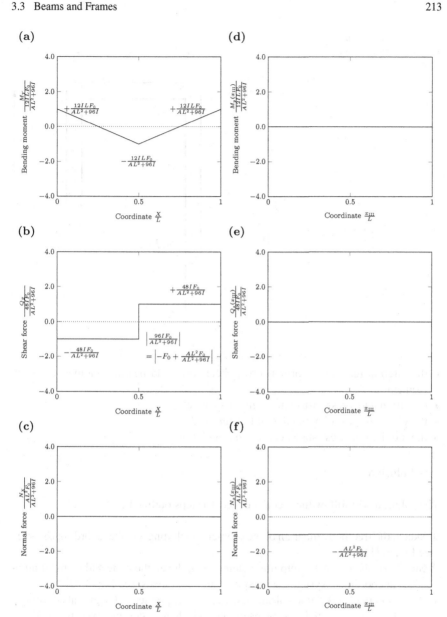

Fig. 3.69 Graphical representation of the internal reactions: **a–c** horizontal beams I and II, and **d–f** vertical beam III

Fig. 3.70 Plane frame
structure representing a
crane

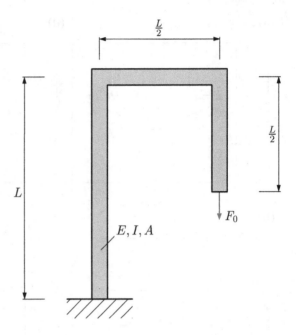

- the internal reactions (normal force, shear force and bending moment) in each element,
- the strain and stress distributions in the elements,
- the global force and moment equilibrium, and
- the multi-axial stress state near to the foundation.

3.18 Solution

The solution will follow the recommended 10 steps outlined on Sect. 3.2.

① Sketch the free-body diagram of the problem, including a global coordinate system (see Fig. 3.71).
② Subdivide the geometry into finite elements. Indicate the node and element numbers, local coordinate systems, and equivalent nodal loads, see Fig. 3.72.
③ Write separately all elemental stiffness matrices expressed in the global coordinate system. Indicate the nodal unknowns on the right-hand sides and over the matrices.

Element I: $\alpha_I = -90°$, $L_I = L$

Fig. 3.71 Free-body diagram of the plane frame structure representing a crane

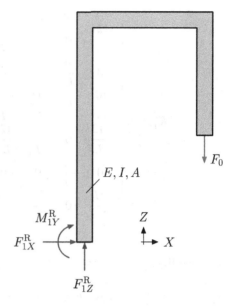

Fig. 3.72 Free-body diagram of the discretized structure

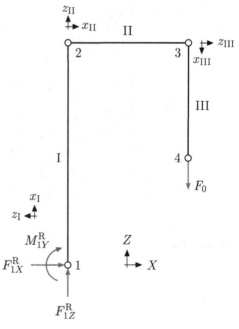

$$
\boldsymbol{K}_{\mathrm{I}}^{\mathrm{e}} = E
\begin{array}{c}
\begin{array}{cccccc}
u_{1X} & u_{1Z} & \varphi_{1Y} & u_{2X} & u_{2Z} & \varphi_{2Y}
\end{array}\\
\left[
\begin{array}{cccccc}
\dfrac{12I}{L_{\mathrm{I}}^{3}} & 0 & \dfrac{6I}{L_{\mathrm{I}}^{2}} & -\dfrac{12I}{L_{\mathrm{I}}^{3}} & 0 & \dfrac{6I}{L_{\mathrm{I}}^{2}} \\[2ex]
0 & \dfrac{A}{L_{\mathrm{I}}} & 0 & 0 & -\dfrac{A}{L_{\mathrm{I}}} & 0 \\[2ex]
\dfrac{6I}{L_{\mathrm{I}}^{2}} & 0 & \dfrac{4I}{L_{\mathrm{I}}} & -\dfrac{6I}{L_{\mathrm{I}}^{2}} & 0 & \dfrac{2I}{L_{\mathrm{I}}} \\[2ex]
-\dfrac{12I}{L_{\mathrm{I}}^{3}} & 0 & -\dfrac{6I}{L_{\mathrm{I}}^{2}} & \dfrac{12I}{L_{\mathrm{I}}^{3}} & 0 & -\dfrac{6I}{L_{\mathrm{I}}^{2}} \\[2ex]
0 & -\dfrac{A}{L_{\mathrm{I}}} & 0 & 0 & \dfrac{A}{L_{\mathrm{I}}} & 0 \\[2ex]
\dfrac{6I}{L_{\mathrm{I}}^{2}} & 0 & \dfrac{2I}{L_{\mathrm{I}}} & -\dfrac{6I}{L_{\mathrm{I}}^{2}} & 0 & \dfrac{4I}{L_{\mathrm{I}}}
\end{array}
\right]
\begin{array}{c}
u_{1X}\\[2ex] u_{1Z}\\[2ex] \varphi_{2Y}\\[2ex] u_{2X}\\[2ex] u_{2Z}\\[2ex] \varphi_{2Y}
\end{array}
\end{array}
\tag{3.430}
$$

Element II: $\alpha_{\mathrm{II}} = 0°$, $L_{\mathrm{II}} = \frac{L}{2}$

$$
\boldsymbol{K}_{\mathrm{II}}^{\mathrm{e}} = E
\begin{array}{c}
\begin{array}{cccccc}
u_{2X} & u_{2Z} & \varphi_{2Y} & u_{3X} & u_{3Z} & \varphi_{3Y}
\end{array}\\
\left[
\begin{array}{cccccc}
\dfrac{A}{L_{\mathrm{II}}} & 0 & 0 & -\dfrac{A}{L_{\mathrm{II}}} & 0 & 0 \\[2ex]
0 & \dfrac{12I}{L_{\mathrm{II}}^{3}} & -\dfrac{6I}{L_{\mathrm{II}}^{2}} & 0 & -\dfrac{12I}{L_{\mathrm{II}}^{3}} & -\dfrac{6I}{L_{\mathrm{II}}^{2}} \\[2ex]
0 & -\dfrac{6I}{L_{\mathrm{II}}^{2}} & \dfrac{4I}{L_{\mathrm{II}}} & 0 & \dfrac{6I}{L_{\mathrm{II}}^{2}} & \dfrac{2I}{L_{\mathrm{II}}} \\[2ex]
-\dfrac{A}{L_{\mathrm{II}}} & 0 & 0 & \dfrac{A}{L_{\mathrm{II}}} & 0 & 0 \\[2ex]
0 & -\dfrac{12I}{L_{\mathrm{II}}^{3}} & \dfrac{6I}{L_{\mathrm{II}}^{2}} & 0 & \dfrac{12I}{L_{\mathrm{II}}^{3}} & \dfrac{6I}{L_{\mathrm{II}}^{2}} \\[2ex]
0 & -\dfrac{6I}{L_{\mathrm{II}}^{2}} & \dfrac{2I}{L_{\mathrm{II}}} & 0 & \dfrac{6I}{L_{\mathrm{II}}^{2}} & \dfrac{4I}{L_{\mathrm{II}}}
\end{array}
\right]
\begin{array}{c}
u_{2X}\\[2ex] u_{2Z}\\[2ex] \varphi_{2Y}\\[2ex] u_{3X}\\[2ex] u_{3Z}\\[2ex] \varphi_{3Y}
\end{array}
\end{array}
\tag{3.431}
$$

Element III: $\alpha_{\mathrm{III}} = 90°$, $L_{\mathrm{III}} = \frac{L}{2}$

$$
K_{III}^e = E
\begin{array}{c}
\begin{array}{cccccc}
u_{3X} & u_{3Z} & \varphi_{3Y} & u_{4X} & u_{4Z} & \varphi_{4Y}
\end{array} \\
\left[
\begin{array}{cccccc}
\dfrac{12I}{L_{III}^{3}} & 0 & -\dfrac{6I}{L_{III}^{2}} & -\dfrac{12I}{L_{III}^{3}} & 0 & -\dfrac{6I}{L_{III}^{2}} \\[10pt]
0 & \dfrac{A}{L_{III}} & 0 & 0 & -\dfrac{A}{L_{III}} & 0 \\[10pt]
-\dfrac{6I}{L_{III}^{2}} & 0 & \dfrac{4I}{L_{III}} & \dfrac{6I}{L_{III}^{2}} & 0 & \dfrac{2I}{L_{III}} \\[10pt]
-\dfrac{12I}{L_{III}^{3}} & 0 & \dfrac{6I}{L_{III}^{2}} & \dfrac{12I}{L_{III}^{3}} & 0 & \dfrac{6I}{L_{III}^{2}} \\[10pt]
0 & -\dfrac{A}{L_{III}} & 0 & 0 & \dfrac{A}{L_{III}} & 0 \\[10pt]
-\dfrac{6I}{L_{III}^{2}} & 0 & \dfrac{2I}{L_{III}} & \dfrac{6I}{L_{III}^{2}} & 0 & \dfrac{4I}{L_{III}}
\end{array}
\right]
\begin{array}{c}
u_{3X} \\[10pt]
u_{3Z} \\[10pt]
\varphi_{3Y} \\[10pt]
u_{4X} \\[10pt]
u_{4Z} \\[10pt]
\varphi_{4Y}
\end{array}
\end{array}
. \tag{3.432}
$$

④ Determine the dimensions of the global stiffness matrix and sketch the structure of this matrix with global unknowns on the right-hand side and over the matrix.

The finite element structure is composed of four nodes, each having three degrees of freedom (i.e., the vertical and horizontal displacements and the rotation). Thus, the dimensions of the global stiffness matrix are $(4 \times 3) \times (4 \times 3) = (12 \times 12)$:

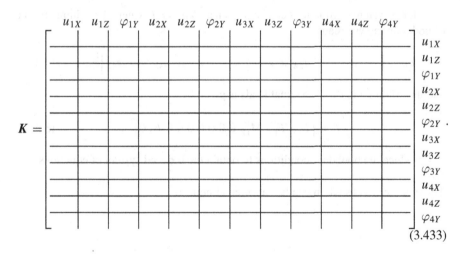

$$ \tag{3.433} $$

⑤ Insert the values of the elemental stiffness matrices step-by-step into the global stiffness matrix.

$$K = E\times \tag{3.434}$$

$$
\begin{array}{cccccccccccc}
u_{1X} & u_{1Z} & \varphi_{1Y} & u_{2X} & u_{2Z} & \varphi_{2Y} & u_{3X} & u_{3Z} & \varphi_{3Y} & u_{4X} & u_{4Z} & \varphi_{4Y}
\end{array}
$$

$$
\left[
\begin{array}{cccccccccccc}
\frac{12I}{L^3} & 0 & \frac{6I}{L^2} & -\frac{12I}{L^3} & 0 & \frac{6I}{L^2} & 0 & 0 & 0 & 0 & 0 & 0 \\[4pt]
0 & \frac{A}{L} & 0 & 0 & -\frac{A}{L} & 0 & 0 & 0 & 0 & 0 & 0 & 0 \\[4pt]
\frac{6I}{L^2} & 0 & \frac{4I}{L} & -\frac{6I}{L^2} & 0 & \frac{2I}{L} & 0 & 0 & 0 & 0 & 0 & 0 \\[4pt]
-\frac{12I}{L^3} & 0 & -\frac{6I}{L^2} & \left(\frac{12I}{L^3}+\frac{2A}{L}\right) & 0 & -\frac{6I}{L^2} & -\frac{2A}{L} & 0 & 0 & 0 & 0 & 0 \\[4pt]
0 & -\frac{A}{L} & 0 & 0 & \left(\frac{A}{L}+\frac{96I}{L^3}\right) & -\frac{24I}{L^2} & 0 & -\frac{96I}{L^3} & -\frac{24I}{L^2} & 0 & 0 & 0 \\[4pt]
\frac{6I}{L^2} & 0 & \frac{2I}{L} & -\frac{6I}{L^2} & -\frac{24I}{L^2} & \frac{12I}{L} & 0 & \frac{24I}{L^2} & \frac{4I}{L} & 0 & 0 & 0 \\[4pt]
0 & 0 & 0 & -\frac{2A}{L} & 0 & 0 & \left(\frac{2A}{L}+\frac{96I}{L^3}\right) & 0 & -\frac{24I}{L^2} & -\frac{96I}{L^3} & 0 & -\frac{24I}{L^2} \\[4pt]
0 & 0 & 0 & 0 & -\frac{96I}{L^3} & \frac{24I}{L^2} & 0 & \left(\frac{2A}{L}+\frac{96I}{L^3}\right) & \frac{24I}{L^2} & 0 & -\frac{2A}{L} & 0 \\[4pt]
0 & 0 & 0 & 0 & -\frac{24I}{L^2} & \frac{4I}{L} & -\frac{24I}{L^2} & \frac{24I}{L^2} & \frac{16I}{L} & \frac{24I}{L^2} & 0 & \frac{4I}{L} \\[4pt]
0 & 0 & 0 & 0 & 0 & 0 & -\frac{96I}{L^3} & 0 & \frac{24I}{L^2} & \frac{96I}{L^3} & 0 & \frac{24I}{L^2} \\[4pt]
0 & 0 & 0 & 0 & 0 & 0 & 0 & -\frac{2A}{L} & 0 & 0 & \frac{2A}{L} & 0 \\[4pt]
0 & 0 & 0 & 0 & 0 & 0 & -\frac{24I}{L^2} & 0 & \frac{4I}{L} & \frac{24I}{L^2} & 0 & \frac{8I}{L}
\end{array}
\right]
\begin{array}{l}
u_{1X} \\[4pt]
u_{1Z} \\[4pt]
\varphi_{1Y} \\[4pt]
u_{2X} \\[4pt]
u_{2Z} \\[4pt]
\varphi_{2Y} \\[4pt]
u_{3X} \\[4pt]
u_{3Z} \\[4pt]
\varphi_{3Y} \\[4pt]
u_{4X} \\[4pt]
u_{4Z} \\[4pt]
\varphi_{4Y}
\end{array}
$$

⑥ Add the column matrix of unknowns and external loads to complete the global system of equations.

The global system of equations can be written as $K u_{\mathrm{p}} = f$, where the column matrix of nodal unknowns reads

$$u_{\mathrm{p}} = \begin{bmatrix} u_{1X} & u_{1Z} & \varphi_{1Y} & u_{2X} & u_{2Z} & \varphi_{2Y} & u_{3X} & u_{3Z} & \varphi_{3Y} & u_{4X} & u_{4Z} & \varphi_{4Y} \end{bmatrix}^{\mathrm{T}}, \tag{3.435}$$

and the column matrix of external loads is given by:

$$f = \begin{bmatrix} F_{1X}^{\mathrm{R}} & F_{1Z}^{\mathrm{R}} & M_{1Y}^{\mathrm{R}} & 0 & 0 & 0 & 0 & 0 & 0 & 0 & F_0 & 0 \end{bmatrix}^{\mathrm{T}}. \tag{3.436}$$

⑦ Introduce the boundary conditions to obtain the reduced system of equations.

Three degrees of freedom can be canceled at node 1:

$$
\begin{bmatrix}
\frac{12I}{L^3}+\frac{2A}{L} & 0 & -\frac{6I}{L^2} & -\frac{2A}{L} & 0 & 0 & 0 & 0 & 0 \\[4pt]
0 & \frac{A}{L}+\frac{96I}{L^3} & -\frac{24I}{L^2} & 0 & -\frac{96I}{L^3} & -\frac{24I}{L^2} & 0 & 0 & 0 \\[4pt]
-\frac{6I}{L^2} & -\frac{24I}{L^2} & \frac{12I}{L} & 0 & \frac{24I}{L^2} & \frac{4I}{L} & 0 & 0 & 0 \\[4pt]
-\frac{2A}{L} & 0 & 0 & \frac{2A}{L}+\frac{96I}{L^3} & 0 & -\frac{24I}{L^2} & -\frac{96I}{L^3} & 0 & -\frac{24I}{L^2} \\[4pt]
0 & -\frac{96I}{L^3} & \frac{24I}{L^2} & 0 & \frac{2A}{L}+\frac{96I}{L^3} & \frac{24I}{L^2} & 0 & -\frac{2A}{L} & 0 \\[4pt]
0 & -\frac{24I}{L^2} & \frac{4I}{L} & -\frac{24I}{L^2} & \frac{24I}{L^2} & \frac{16I}{L} & \frac{24I}{L^2} & 0 & \frac{4I}{L} \\[4pt]
0 & 0 & 0 & -\frac{96I}{L^3} & 0 & \frac{24I}{L^2} & \frac{96I}{L^3} & 0 & \frac{24I}{L^2} \\[4pt]
0 & 0 & 0 & 0 & -\frac{2A}{L} & 0 & 0 & \frac{2A}{L} & 0 \\[4pt]
0 & 0 & 0 & -\frac{24I}{L^2} & 0 & \frac{4I}{L} & \frac{24I}{L^2} & 0 & \frac{8I}{L}
\end{bmatrix}
\begin{bmatrix}
u_{2X} \\ u_{2Z} \\ \varphi_{2Y} \\ u_{3X} \\ u_{3Z} \\ \varphi_{3Y} \\ u_{4X} \\ u_{4Z} \\ \varphi_{4Y}
\end{bmatrix}
=
\begin{bmatrix}
0 \\ 0 \\ 0 \\ 0 \\ 0 \\ 0 \\ 0 \\ F_0 \\ 0
\end{bmatrix}.
$$

$$(3.437)$$

⑧ Solve the reduced system of equations to obtain the unknown nodal deformations.

The solution can be obtained based on the matrix approach $u_{\mathrm{p}} = K^{-1} f$:

$$
\begin{bmatrix}
u_{2X} \\ u_{2Z} \\ \varphi_{2Y} \\ u_{3X} \\ u_{3Z} \\ \varphi_{3Y} \\ u_{4X} \\ u_{4Z} \\ \varphi_{4Y}
\end{bmatrix}
=
\begin{bmatrix}
\frac{L^3 F_0}{4EI} \\[4pt]
-\frac{L F_0}{EA} \\[4pt]
\frac{L^2 F_0}{2EI} \\[4pt]
\frac{L^3 F_0}{4EI} \\[4pt]
-\frac{L(7AL^2+24I)F_0}{24IEA} \\[4pt]
\frac{5L^2 F_0}{8EI} \\[4pt]
-\frac{L^3 F_0}{16EI} \\[4pt]
-\frac{L(7AL^2+36I)F_0}{24IEA} \\[4pt]
\frac{5L^2 F_0}{8IEA}
\end{bmatrix}.
$$

$$(3.438)$$

⑨ Post-computation: determination of reactions, stresses and strains.

The reactions at the supports can be obtained from the non-reduced system of equations as given in step ⑥ under the consideration of the known nodal degrees of freedom (i.e., displacements and rotations). The evaluation of the first equation gives:

$$
-\frac{12I}{L_{\mathrm{I}}^3} u_{2X} + \frac{6I}{L_{\mathrm{I}}^2}\varphi_{2Y} = F_{1X}^{\mathrm{R}} \quad\Rightarrow\quad F_{1X}^{\mathrm{R}} = 0. \tag{3.439}
$$

In a similar way, the evaluation of the remaining equations gives:

$$F_{1Z}^R = F_{2X}^R = F_{2Z}^R = F_{3X}^R = F_{3Z}^R = F_{4X}^R = 0, \qquad (3.440)$$

$$F_{1Z}^R = F_0, \qquad (3.441)$$

$$M_{1Y}^R = -\frac{LF_0}{2}, \qquad (3.442)$$

$$M_{2Y}^R = M_{3Y}^R = M_{4Y}^R = 0. \qquad (3.443)$$

The internal reactions (i.e., bending moment, shear force, and normal force) in each element can be obtained from the relations provided in Tables 3.14 and 3.5.

Bending moment distribution:

$$M_{y_I}^e(x_I) = EI \left(\left[-\frac{6}{L^2} + \frac{12x_I}{L^3} \right] (u_{2X}(-1) + \left[-\frac{2}{L} + \frac{6x_I}{L^2} \right] \varphi_{2Y} \right)$$

$$= \frac{LF_0}{2}, \qquad (3.444)$$

$$M_{y_{II}}^e(x_{II}) = EI \left(\left[+\frac{6}{L^2} - \frac{12x_{II}}{L^3} \right] (u_{2Z}(1)) + \left[-\frac{4}{L} + \frac{6x_{II}}{L^2} \right] \varphi_{2Y} \right.$$

$$+ \left[-\frac{6}{L^2} + \frac{12x_{II}}{L^3} \right] (u_{3Z}(1)) + \left[-\frac{2}{L} + \frac{6x_{II}}{L^2} \right] \varphi_{3Y} \right)$$

$$= F_0 \left(\frac{L}{2} - 2x_{II} \right), \qquad (3.445)$$

$$M_{y_{III}}^e(x_{III}) = EI \left(\left[+\frac{6}{L^2} - \frac{12x_{III}}{L^3} \right] (u_{3X}(1)) + \left[-\frac{4}{L} + \frac{6x_{III}}{L^2} \right] \varphi_{3Y} \right.$$

$$+ \left[-\frac{6}{L^2} + \frac{12x_{III}}{L^3} \right] (u_{4X}(1)) + \left[-\frac{2}{L} + \frac{6x_{III}}{L^2} \right] \varphi_{4Y} \right)$$

$$= 0. \qquad (3.446)$$

Shear force distribution

$$Q_{z_I}^e(x_I) = EI_y \left(\left[+\frac{12}{L^3} \right] (u_{2X}(-1)) + \left[+\frac{6}{L^2} \right] \varphi_{2Y} \right)$$

$$= 0, \qquad (3.447)$$

$$Q_{z_{II}}^e(x_{II}) = EI_y \left(\left[-\frac{12}{L^3} \right] (u_{2Z}(1)) + \left[+\frac{6}{L^2} \right] \varphi_{2Y} \right.$$

$$+ \left[+\frac{12}{L^3} \right] (u_{3Z}(1)) + \left[+\frac{6}{L^2} \right] \varphi_{3Y} \Bigg)$$

$$= -F_0 , \tag{3.448}$$

$$Q^e_{z_{\text{III}}}(x_{\text{III}}) = EI_y \left(\left[-\frac{12}{L^3} \right] (u_{3X}(1)) + \left[+\frac{6}{L^2} \right] \varphi_{3Y} \right.$$

$$+ \left[+\frac{12}{L^3} \right] (u_{4X}(1)) + \left[+\frac{6}{L^2} \right] \varphi_{4Y} \Bigg)$$

$$= 0 . \tag{3.449}$$

Normal force distribution:

$$N^e_{x_{\text{I}}}(x_{\text{I}}) = = \frac{EA}{L} \left((-(-1)u_{2Z}) \right)$$

$$= -F_0 \tag{3.450}$$

$$N^e_{x_{\text{II}}}(x_{\text{II}}) = = \frac{EA}{L} \left(((1)u_{3X}) - ((1)u_{2X}) \right)$$

$$= 0 \tag{3.451}$$

$$N^e_{x_{\text{III}}}(x_{\text{III}}) = = \frac{EA}{L} \left((-(1)u_{4Z}) - (-(1)u_{3Z}) \right)$$

$$= F_0 . \tag{3.452}$$

⑩ Check the global equilibrium between the external loads and the support reactions.

$$\sum_i F_{iX} = 0 \quad \Leftrightarrow \quad \underbrace{(F^R_{1X})}_{\text{reaction force}} + \underbrace{(0)}_{\text{external load}} = 0 , \checkmark \tag{3.453}$$

$$\sum_i F_{iZ} = 0 \quad \Leftrightarrow \quad \underbrace{(F^R_{1Z})}_{\text{reaction force}} + \underbrace{(-F_0)}_{\text{external load}} = 0 , \checkmark \tag{3.454}$$

$$\sum_i M_{iY} = 0 \quad \Leftrightarrow \quad \underbrace{(M^R_{1Y})}_{\text{reaction}} + \underbrace{(F_0 L_{\text{II}})}_{\text{external load}} = 0 . \checkmark \tag{3.455}$$

Multi-axial stress state near to the foundation, i.e., $x_{\text{I}} = 0$:

The total normal stress distribution is a superposition of the contributions from the tensile ($N_{x_{\text{I}}}$) and bending ($M_{y_{\text{I}}}$) parts, see Tables 3.2 and 3.8:

$$\sigma^e_{x_{\text{I}}}(x_{\text{I}}) = \frac{N^e_{x_{\text{I}}}(x_{\text{I}})}{A} + \frac{M^e_{y_{\text{I}}}(x_{\text{I}})}{I} z_{\text{I}} = -\frac{F_0}{A} + \frac{L F_0}{2I} \times z_{\text{I}} . \tag{3.456}$$

(a)   (b)

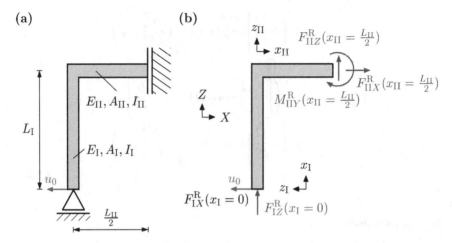

Fig. 3.73 **a** Mechanical model of the sensor under consideration of symmetry; **b** free-body diagram

3.4 Extensometer Analysis

The solution will follow the recommended 10 steps outlined on Sect. 3.2.

① Sketch the free-body diagram of the problem, including a global coordinate system.

It is advantageous to work only with half of the sensor, i.e., to consider the symmetry of the problem, see Fig. 3.73a. The free-body diagrams as outlines in Fig. 3.73b contains also unknown reactions where the displacement boundary condition u_0 is imposed. The reactions at the symmetry line, i.e., the normal force $F_{\mathrm{II}X}^{\mathrm{R}}$, the vertical force[6] $F_{\mathrm{II}Z}^{\mathrm{R}}$, and the bending moment $M_{\mathrm{II}Y}^{\mathrm{R}}$ will serve to calculate the total normal strain in element II.
② Subdivide the geometry into finite elements. Indicate the node and element numbers, local coordinate systems, and equivalent nodal loads.

The finite element approach will be based on two frame elements (see Fig. 3.74), i.e., the superposition of rod and beam elements. Since the horizontal beam is not loaded by a shear force (see Sect. 2.4), we can rely on the EULER–BERNOULLI beam theory. The vertical beam is subjected to a shear force. However, the contribution of this shear force to the deformation can only be estimated if real real numbers are assigned to the design variables, i.e., L_{I}, E_{I}, I_{I}. Thus, we assume at this point of the derivation that the vertical beam is thin.

---

[6]This vertical force could be omitted right from the beginning since the introduced support has a vertical degree of freedom. Thus, there will be no vertical reaction force. From this, one could conclude immediately that $F_{\mathrm{I}Z}^{\mathrm{R}}$ must be zero. Nevertheless, the finite element approach will show this.

Fig. 3.74 Finite element
model of the sensor based on
two elements (I and II)

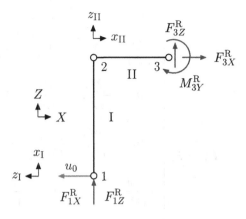

③ Write separately all elemental stiffness matrices expressed in the global coordinate
system. Indicate the nodal unknowns on the right-hand sides and over the matrices.

Element I is rotated by an angle of $\alpha = -90°$ and application of Eq. (3.347) gives:

$$\boldsymbol{K}_{\mathrm{I}}^{\mathrm{e}} = E_{\mathrm{I}}
\begin{array}{c}
\begin{array}{cccccc} u_{1X} & u_{1Z} & \varphi_{1Y} & u_{2X} & u_{2Z} & \varphi_{2Y} \end{array} \\
\begin{bmatrix}
\dfrac{12I_{\mathrm{I}}}{L_{\mathrm{I}}^{3}} & 0 & \dfrac{6I_{\mathrm{I}}}{L_{\mathrm{I}}^{2}} & -\dfrac{12I_{\mathrm{I}}}{L_{\mathrm{I}}^{3}} & 0 & \dfrac{6I_{\mathrm{I}}}{L_{\mathrm{I}}^{2}} \\[3mm]
0 & \dfrac{A_{\mathrm{I}}}{L_{\mathrm{I}}} & 0 & 0 & -\dfrac{A_{\mathrm{I}}}{L_{\mathrm{I}}} & 0 \\[3mm]
\dfrac{6I_{\mathrm{I}}}{L_{\mathrm{I}}^{2}} & 0 & \dfrac{4I_{\mathrm{I}}}{L_{\mathrm{I}}} & -\dfrac{6I_{\mathrm{I}}}{L_{\mathrm{I}}^{2}} & 0 & \dfrac{2I_{\mathrm{I}}}{L_{\mathrm{I}}} \\[3mm]
-\dfrac{12I_{\mathrm{I}}}{L_{\mathrm{I}}^{3}} & 0 & -\dfrac{6I_{\mathrm{I}}}{L_{\mathrm{I}}^{2}} & \dfrac{12I_{\mathrm{I}}}{L_{\mathrm{I}}^{3}} & 0 & -\dfrac{6I_{\mathrm{I}}}{L_{\mathrm{I}}^{2}} \\[3mm]
0 & -\dfrac{A_{\mathrm{I}}}{L_{\mathrm{I}}} & 0 & 0 & \dfrac{A_{\mathrm{I}}}{L_{\mathrm{I}}} & 0 \\[3mm]
\dfrac{6I_{\mathrm{I}}}{L_{\mathrm{I}}^{2}} & 0 & \dfrac{2I_{\mathrm{I}}}{L_{\mathrm{I}}} & -\dfrac{6I_{\mathrm{I}}}{L_{\mathrm{I}}^{2}} & 0 & \dfrac{4I_{\mathrm{I}}}{L_{\mathrm{I}}}
\end{bmatrix}
\begin{array}{l} u_{1X} \\ u_{1Z} \\ \varphi_{1Y} \\ u_{2X} \\ u_{2Z} \\ \varphi_{2Y} \end{array}
\end{array}, \qquad (3.457)$$

Element II does not require any rotation and its elemental stiffness matrix reads:

$$
\boldsymbol{K}_{\mathrm{II}}^{\mathrm{e}} = E_{\mathrm{II}}
\begin{array}{c}
\begin{array}{cccccc}
u_{2X} & u_{2Z} & \varphi_{2Y} & u_{3X} & u_{3Z} & \varphi_{3Y}
\end{array} \\
\left[
\begin{array}{cccccc}
\dfrac{A_{\mathrm{II}}}{\left(\frac{L_{\mathrm{II}}}{2}\right)} & 0 & 0 & -\dfrac{A_{\mathrm{II}}}{\left(\frac{L_{\mathrm{II}}}{2}\right)} & 0 & 0 \\[2ex]
0 & \dfrac{12 I_{\mathrm{II}}}{\left(\frac{L_{\mathrm{II}}}{2}\right)^{3}} & \dfrac{6 I_{\mathrm{II}}}{\left(\frac{L_{\mathrm{II}}}{2}\right)^{2}} & 0 & -\dfrac{12 I_{\mathrm{II}}}{\left(\frac{L_{\mathrm{II}}}{2}\right)^{3}} & \dfrac{6 I_{\mathrm{II}}}{\left(\frac{L_{\mathrm{II}}}{2}\right)^{2}} \\[2ex]
0 & -\dfrac{6 I_{\mathrm{II}}}{\left(\frac{L_{\mathrm{II}}}{2}\right)^{2}} & \dfrac{4 I_{\mathrm{II}}}{\left(\frac{L_{\mathrm{II}}}{2}\right)} & 0 & \dfrac{6 I_{\mathrm{II}}}{\left(\frac{L_{\mathrm{II}}}{2}\right)^{2}} & \dfrac{2 I_{\mathrm{II}}}{\left(\frac{L_{\mathrm{II}}}{2}\right)} \\[2ex]
-\dfrac{A_{\mathrm{II}}}{\left(\frac{L_{\mathrm{II}}}{2}\right)} & 0 & 0 & \dfrac{A_{\mathrm{II}}}{\left(\frac{L_{\mathrm{II}}}{2}\right)} & 0 & 0 \\[2ex]
0 & -\dfrac{12 I_{\mathrm{II}}}{\left(\frac{L_{\mathrm{II}}}{2}\right)^{3}} & \dfrac{6 I_{\mathrm{II}}}{\left(\frac{L_{\mathrm{II}}}{2}\right)^{2}} & 0 & \dfrac{12 I_{\mathrm{II}}}{\left(\frac{L_{\mathrm{II}}}{2}\right)^{3}} & \dfrac{6 I_{\mathrm{II}}}{\left(\frac{L_{\mathrm{II}}}{2}\right)^{2}} \\[2ex]
0 & -\dfrac{6 I_{\mathrm{II}}}{\left(\frac{L_{\mathrm{II}}}{2}\right)^{2}} & \dfrac{2 I_{\mathrm{II}}}{\left(\frac{L_{\mathrm{II}}}{2}\right)} & 0 & \dfrac{6 I_{\mathrm{II}}}{\left(\frac{L_{\mathrm{II}}}{2}\right)^{2}} & \dfrac{4 I_{\mathrm{II}}}{\left(\frac{L_{\mathrm{II}}}{2}\right)}
\end{array}
\right]
\begin{array}{c}
u_{2X} \\[2ex] u_{2Z} \\[2ex] \varphi_{2Y} \\[2ex] u_{3X} \\[2ex] u_{3Z} \\[2ex] \varphi_{3Y}
\end{array}
\end{array}. \quad (3.458)
$$

④ Determine the dimensions of the global stiffness matrix and sketch the structure of this matrix with global unknowns on the right-hand side and over the matrix.

The finite element structure is composed of 3 nodes, each having 3 degrees of freedom (i.e., the horizontal and vertical displacements, and the rotation). Thus, the dimensions of the global stiffness matrix are $(3 \times 3) \times (3 \times 3) = (9 \times 9)$:

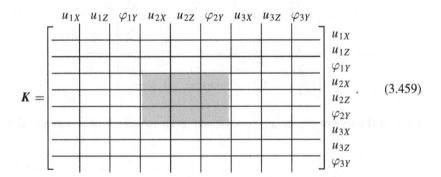

The cells highlighted in gray color relate to the overlap zone, i.e., node 2 which connects elements I and II. These cells combine stiffness contributions from both elements.

⑤ Insert the values of the elemental stiffness matrices step-by-step into the global stiffness matrix.

$$
\mathbf{K} =
\begin{array}{r}
\begin{array}{ccccccccc}
u_{1X} & u_{1Z} & \varphi_{1Y} & u_{2X} & u_{2Z} & \varphi_{2Y} & u_{3X} & u_{3Z} & \varphi_{3Y}
\end{array} \\
\left[
\begin{array}{ccccccccc}
\dfrac{12E_1I_1}{L_1^3} & 0 & \dfrac{6E_1I_1}{L_1^2} & -\dfrac{12E_1I_1}{L_1^3} & 0 & \dfrac{6E_1I_1}{L_1^2} & 0 & 0 & 0 \\[2ex]
0 & \dfrac{E_1A_1}{L_1} & 0 & 0 & -\dfrac{E_1A_1}{L_1} & 0 & 0 & 0 & 0 \\[2ex]
\dfrac{6E_1I_1}{L_1^2} & 0 & \dfrac{4E_1I_1}{L_1} & -\dfrac{6E_1I_1}{L_1^2} & 0 & \dfrac{2E_1I_1}{L_1} & 0 & 0 & 0 \\[2ex]
-\dfrac{12E_1I_1}{L_1^3} & 0 & -\dfrac{6E_1I_1}{L_1^2} & \dfrac{12E_1I_1}{L_1^3}+\dfrac{E_{II}A_{II}}{\left(\frac{L_{II}}{2}\right)} & 0+0 & -\dfrac{6E_1I_1}{L_1^2}+0 & -\dfrac{E_{II}A_{II}}{\left(\frac{L_{II}}{2}\right)} & 0 & 0 \\[2ex]
0 & -\dfrac{E_1A_1}{L_1} & 0 & 0+0 & \dfrac{E_1A_1}{L_1}+\dfrac{12E_{II}I_{II}}{\left(\frac{L_{II}}{2}\right)^3} & 0-\dfrac{6E_{II}I_{II}}{\left(\frac{L_{II}}{2}\right)^2} & 0 & -\dfrac{12E_{II}I_{II}}{\left(\frac{L_{II}}{2}\right)^3} & -\dfrac{6E_{II}I_{II}}{\left(\frac{L_{II}}{2}\right)^2} \\[2ex]
\dfrac{6E_1I_1}{L_1^2} & 0 & \dfrac{2E_1I_1}{L_1} & -\dfrac{6E_1I_1}{L_1^2}+0 & 0-\dfrac{6E_{II}I_{II}}{\left(\frac{L_{II}}{2}\right)^2} & \dfrac{4E_1I_1}{L_1}+\dfrac{4E_{II}I_{II}}{\left(\frac{L_{II}}{2}\right)} & 0 & \dfrac{6E_{II}I_{II}}{\left(\frac{L_{II}}{2}\right)^2} & \dfrac{2E_{II}I_{II}}{\left(\frac{L_{II}}{2}\right)} \\[2ex]
0 & 0 & 0 & \dfrac{E_{II}A_{II}}{\left(\frac{L_{II}}{2}\right)} & 0 & 0 & \dfrac{E_{II}A_{II}}{\left(\frac{L_{II}}{2}\right)} & 0 & 0 \\[2ex]
0 & 0 & 0 & 0 & -\dfrac{12E_{II}I_{II}}{\left(\frac{L_{II}}{2}\right)^3} & \dfrac{6E_{II}I_{II}}{\left(\frac{L_{II}}{2}\right)^2} & 0 & \dfrac{12E_{II}I_{II}}{\left(\frac{L_{II}}{2}\right)^3} & \dfrac{6E_{II}I_{II}}{\left(\frac{L_{II}}{2}\right)^2} \\[2ex]
0 & 0 & 0 & 0 & -\dfrac{6E_{II}I_{II}}{\left(\frac{L_{II}}{2}\right)^2} & \dfrac{2E_{II}I_{II}}{\left(\frac{L_{II}}{2}\right)} & 0 & \dfrac{6E_{II}I_{II}}{\left(\frac{L_{II}}{2}\right)^2} & \dfrac{4E_{II}I_{II}}{\left(\frac{L_{II}}{2}\right)}
\end{array}
\right]
\begin{array}{l}
u_{1X} \\ u_{1Z} \\ \varphi_{1Y} \\ u_{2X} \\ u_{2Z} \\ \varphi_{2Y} \\ u_{3X} \\ u_{3Z} \\ \varphi_{3Y}
\end{array}
\end{array}
$$

$$(3.460)$$

⑥ Add the column matrix of unknowns and external loads to complete the global system of equations.

The global system of equations can be expressed in matrix from as

$$\boldsymbol{K}\boldsymbol{u}_{\mathrm{p}} = \boldsymbol{f} \,, \tag{3.461}$$

where the column matrix of the external loads reads:

$$\boldsymbol{f} = \begin{bmatrix} F_{1X}^{\mathrm{R}} & F_{1Z}^{\mathrm{R}} & 0 & 0 & 0 & 0 & F_{3X}^{\mathrm{R}} & F_{3Z}^{\mathrm{R}} & M_{3Y}^{\mathrm{R}} \end{bmatrix}^{\mathrm{T}} \,. \tag{3.462}$$

⑦ Introduce the boundary conditions to obtain the reduced system of equations.

The consideration of the support conditions $u_{1Z} = u_{3X} = 0$ and $\varphi_{3Y} = 0$ results in the following 6×6 system:

$$
\begin{array}{cccccc}
u_{1X} & \varphi_{1Y} & u_{2X} & u_{2Z} & \varphi_{2Y} & u_{3Z}
\end{array}
$$

$$
\begin{bmatrix}
\dfrac{12E_I I_I}{L_I^3} & \dfrac{6E_I I_I}{L_I^2} & -\dfrac{12E_I I_I}{L_I^3} & 0 & \dfrac{6E_I I_I}{L_I^2} & 0 \\[2ex]
\dfrac{6E_I I_I}{L_I^2} & \dfrac{4E_I I_I}{L_I} & -\dfrac{6E_I I_I}{L_I^2} & 0 & \dfrac{2E_I I_I}{L_I} & 0 \\[2ex]
-\dfrac{12E_I I_I}{L_I^3} & -\dfrac{6E_I I_I}{L_I^2} & \dfrac{12E_I I_I}{L_I^3}+\dfrac{E_{II}A_{II}}{\left(\frac{L_{II}}{2}\right)} & 0+0 & -\dfrac{6E_I I_I}{L_I^2}+0 & 0 \\[3ex]
0 & 0 & 0+0 & \dfrac{E_I A_I}{L_I}+\dfrac{12E_{II}I_{II}}{\left(\frac{L_{II}}{2}\right)^3} & 0-\dfrac{6E_{II}I_{II}}{\left(\frac{L_{II}}{2}\right)^2} & -\dfrac{12E_{II}I_{II}}{\left(\frac{L_{II}}{2}\right)^3} \\[3ex]
\dfrac{6E_I I_I}{L_I^2} & \dfrac{2E_I I_I}{L_I} & -\dfrac{6E_I I_I}{L_I^2}+0 & 0-\dfrac{6E_{II}I_{II}}{\left(\frac{L_{II}}{2}\right)^2} & \dfrac{4E_I I_I}{L_I}+\dfrac{4E_{II}I_{II}}{\left(\frac{L_{II}}{2}\right)} & \dfrac{6E_{II}I_{II}}{\left(\frac{L_{II}}{2}\right)^2} \\[3ex]
0 & 0 & 0 & -\dfrac{12E_{II}I_{II}}{\left(\frac{L_{II}}{2}\right)^3} & \dfrac{6E_{II}I_{II}}{\left(\frac{L_{II}}{2}\right)^2} & \dfrac{12E_{II}I_{II}}{\left(\frac{L_{II}}{2}\right)^3}
\end{bmatrix}
\begin{bmatrix}
u_{1X} \\[2ex] \varphi_{1Y} \\[2ex] u_{2X} \\[3ex] u_{2Z} \\[3ex] \varphi_{2Y} \\[3ex] u_{3Z}
\end{bmatrix}
=
\begin{bmatrix}
-F_{1X}^{\mathrm{R}} \\[2ex] 0 \\[2ex] 0 \\[3ex] 0 \\[3ex] 0 \\[3ex] 0
\end{bmatrix} \,.
$$

$$\tag{3.463}$$

Under consideration of $u_{1X} = -u_0$, one gets:

$$\begin{array}{cccccc}
u_{1X} & \varphi_{1Y} & u_{2X} & u_{2Z} & \varphi_{2Y} & u_{3Z}
\end{array}$$

$$
\begin{bmatrix}
1 & 0 & 0 & 0 & 0 & 0 \\[4pt]
\dfrac{6E_I I_I}{L_I^2} & \dfrac{4E_I I_I}{L_I} & -\dfrac{6E_I I_I}{L_I^2} & 0 & \dfrac{2E_I I_I}{L_I} & 0 \\[8pt]
-\dfrac{12E_I I_I}{L_I^3} & -\dfrac{6E_I I_I}{L_I^2} & \dfrac{12E_I I_I}{L_I^3}+\dfrac{E_{II}A_{II}}{\left(\frac{L_{II}}{2}\right)} & 0+0 & -\dfrac{6E_I I_I}{L_I^2}+0 & 0 \\[8pt]
0 & 0 & 0+0 & \dfrac{E_I A_I}{L_I}+\dfrac{12E_{II}I_{II}}{\left(\frac{L_{II}}{2}\right)^3} & 0-\dfrac{6E_{II}I_{II}}{\left(\frac{L_{II}}{2}\right)^2} & -\dfrac{12E_{II}I_{II}}{\left(\frac{L_{II}}{2}\right)^3} \\[8pt]
\dfrac{6E_I I_I}{L_I^2} & \dfrac{2E_I I_I}{L_I} & -\dfrac{6E_I I_I}{L_I^2}+0 & 0-\dfrac{6E_{II}I_{II}}{\left(\frac{L_{II}}{2}\right)^2} & \dfrac{4E_I I_I}{L_I}+\dfrac{4E_{II}I_{II}}{\left(\frac{L_{II}}{2}\right)} & \dfrac{6E_{II}I_{II}}{\left(\frac{L_{II}}{2}\right)^2} \\[8pt]
0 & 0 & 0 & -\dfrac{12E_{II}I_{II}}{\left(\frac{L_{II}}{2}\right)^3} & \dfrac{6E_{II}I_{II}}{\left(\frac{L_{II}}{2}\right)^2} & \dfrac{12E_{II}I_{II}}{\left(\frac{L_{II}}{2}\right)^3}
\end{bmatrix}
\begin{bmatrix} u_{1X} \\[6pt] \varphi_{1Y} \\[6pt] u_{2X} \\[6pt] u_{2Z} \\[6pt] \varphi_{2Y} \\[6pt] u_{3Z} \end{bmatrix}
=
\begin{bmatrix} -u_0 \\[6pt] 0 \\[6pt] 0 \\[6pt] 0 \\[6pt] 0 \\[6pt] 0 \end{bmatrix}.
$$

$$\tag{3.464}$$

⑧ Solve the reduced system of equations to obtain the unknown nodal deformations.

The solution can be obtained based on the matrix approach $\boldsymbol{u}_{\mathrm{p}} = \boldsymbol{K}^{-1}\boldsymbol{f}$:

$$\varphi_{1Y} = \frac{1+\frac{E_{II}}{E_I}\frac{I_{II}}{I_I}\frac{L_I}{L_{II}}}{1+\frac{2}{3}\frac{E_{II}}{E_I}\frac{I_{II}}{I_I}\frac{L_I^2}{L_I L_{II}}+\frac{1}{12}\frac{h_{II}^2}{L_I^2}} \times \frac{u_0}{L_I}, \tag{3.465}$$

$$u_{2X} = -\frac{1}{1+8\frac{E_{II}}{E_I}\frac{I_{II}}{I_I}\frac{L_I^3}{L_{II}h_{II}^2}+12\frac{L_I^2}{h_{II}^2}} \times u_0, \tag{3.466}$$

$$u_{2Z} = 0, \tag{3.467}$$

$$\varphi_{2Y} = \frac{1}{1+\frac{2}{3}\frac{E_{II}}{E_I}\frac{I_{II}}{I_I}\frac{L_I^2}{L_I L_{II}}+\frac{1}{12}\frac{h_{II}^2}{L_I^2}} \times \frac{u_0}{L_I}, \tag{3.468}$$

$$u_{3Z} = -\frac{3}{4} \times \frac{1}{3\frac{L_I}{L_{II}}+2\frac{E_{II}}{E_I}\frac{I_{II}}{I_I}\frac{L_I^2}{L_{II}^2}+\frac{1}{4}\frac{h_{II}^2}{L_I L_{II}}} \times u_0. \tag{3.469}$$

⑨ Post-computation: determination of reaction forces, stresses and strains.

Take into account the non-reduced system of equations as given in step ⑥ under the consideration of the known nodal displacements and rotations. The first equation of this system gives:

$$F_{1X}^{R} = \frac{u_0}{\frac{1}{3}\frac{L_I^3}{E_I I_I}+\frac{1}{2}\frac{L_I^2 L_{II}}{E_{II}I_{II}}+\frac{1}{24}\frac{L_{II}h_{II}^2}{E_{II}I_{II}}}. \tag{3.470}$$

The evaluation of the corresponding other equations gives:

$$F_{1Z}^R = 0,$$ (3.471)

$$F_{3X}^R = \frac{u_0}{\frac{1}{3}\frac{L_I^3}{E_I I_I} + \frac{1}{2}\frac{L_I^2 L_{II}}{E_{II} I_{II}} + \frac{1}{24}\frac{L_{II} h_{II}^2}{E_{II} I_{II}}},$$ (3.472)

$$F_{3Z}^R = 0,$$ (3.473)

$$M_{3Y}^R = -\frac{u_0}{\frac{1}{3}\frac{L_I^3}{E_I I_I} + \frac{1}{2}\frac{L_I^2 L_{II}}{E_{II} I_{II}} + \frac{1}{24}\frac{L_{II} h_{II}^2}{E_{II} I_{II}}} \times L_I.$$ (3.474)

Based on the reaction force and moment at node 3, the axial strain can be evaluated as outlined in Sect. 2.4. Finally, the same equations are obtained as given in Eqs. (2.253) and (2.254). It should be noted here that this result, i.e., the finite element solution is equal to the analytical solution, cannot be generalized to more complex problems. At least, any further mesh refinement does not increase the accuracy in this specific case.

⑩ Check the global equilibrium between the external loads and the support reactions.

$$\sum_i F_{iX} = 0 \quad \Leftrightarrow \quad \underbrace{(-F_{1X}^R + F_{3X}^R)}_{\text{reaction force}} + \underbrace{0}_{\text{external loads}} = 0, \checkmark$$ (3.475)

$$\sum_i F_{iZ} = 0 \quad \Leftrightarrow \quad \underbrace{(F_{1Z}^R + F_{3Z}^R)}_{\text{reaction force}} + \underbrace{0}_{\text{external load}} = 0. \checkmark$$ (3.476)

$$\sum_i M_{iY} = 0 \quad \Leftrightarrow \quad \underbrace{(F_{1X}^R L_I + F_{1Z}^R \frac{L_{II}}{2} + M_{3Y}^R)}_{\text{reaction moment}} + \underbrace{0}_{\text{external moment}} = 0. \checkmark$$ (3.477)

Let us investigate in the following the influence of different ratios on the sensor sensitivity, see Eq. (2.254):

$$\left.\frac{\varepsilon_{x,II}}{\varepsilon_{sp}}\right|_{z=-\frac{h_{II}}{2}} = \frac{1}{\frac{2}{3}\frac{E_{II}I_{II}L_I}{E_I I_I L_{II}} + 1 + \frac{1}{12}\frac{h_{II}^2}{L_I^2}} \times \frac{h_{II}}{L_I} \times \left(+\frac{1}{2} + \frac{1}{12}\left(\frac{h_{II}}{L_I}\right)\right).$$ (3.478)

The influence of the different fractions in Eq. (3.478) is illustrated in Figs. 3.75 and 3.76. The ratios of stiffness, second moment of area and length have the same influence on the strain ratio (pay attention to the fact that Eq. (3.478) contains the ratio $\frac{L_I}{L_{II}}$ while Fig. 3.75c is plotted as a function of the inverse value). The geometrical ratio $\frac{h_{II}}{L_I}$ has a stronger influence on the strain ratio. However, it must be checked for larger ratios if the thin beam assumption is still valid. In case that more than one

Fig. 3.75 Sensor sensitivity as a function of different ratios: **a** stiffness, **b** second moment of area and **c** length

(a)

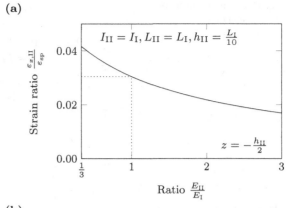

(b)

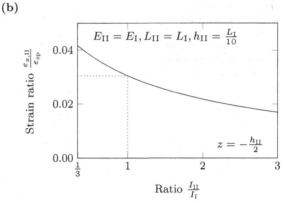

(c)

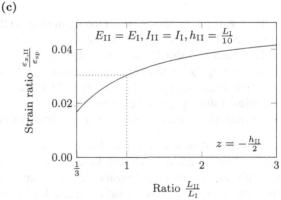

ratio is changed in the corresponding direction, one can expect a stronger influence on the strain ratio, see Fig. 3.77.

The design process, i.e., the choice of the geometrical and material properties of the extensometer, should consider a few limits:

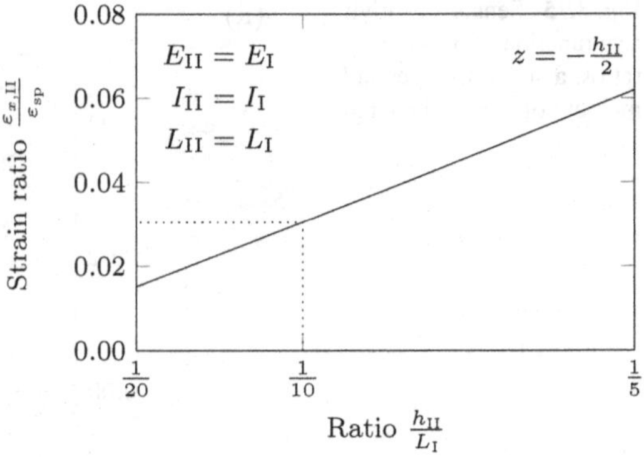

Fig. 3.76 Sensor sensitivity as a function of the geometrical ratio $\frac{h_{II}}{L_I}$

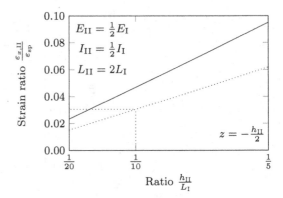

Fig. 3.77 Increased sensor sensitivity as a function of the geometrical ratio $\frac{h_{II}}{L_I}$. The reference curve (dotted line) refers to the configuration which is shown in Fig. 3.76

- The ratio $\varepsilon_{x,II}/\varepsilon_{sp}$ should be not too small to avoid that the strain $\varepsilon_{x,II}$ is below the sensitivity of the selected strain gage.
- The strain $\varepsilon_{x,II}$ should not exceed the upper limit of the selected strain gage to avoid elongation failure (e.g., grid cracking or loss of bond). The limit of high-performance strain gages is typically 1–2% whereas regular self-temperature compensated strain gages are capable to record up to 5–10% strains [9].
- The material of the extensometer should only deform in the elastic range.

In a final step, it is also possible to use a commercial finite element package such as MSC Marc/Mentat [3, 6]. In such packages, it is quite simple to investigate the influence of the mesh density, i.e., the number of nodes per unit length, on the results. In general, only a sufficient number of nodes or elements guarantees a result which is—from a practical point of view—independent from the mesh size. To investigate the mesh dependency, a so-called mesh convergence study must be conducted. Starting from a coarse mesh (see Fig. 3.78a), the mesh is subsequently refined (see Fig. 3.78b, c) and a critical quantity evaluated. This could be in our case the internal bending moment or the normal force. If the difference of this quantity

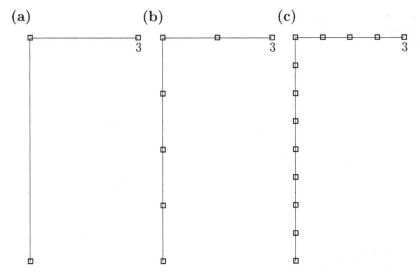

(a)   (b)   (c)

Fig. 3.78 Frame structure with different mesh densities: **a** 3 nodes, **b** 7 nodes, and **c** 13 nodes

Table 3.18 Result of the mesh sensitivity analysis with: $E_I = E_{II} = 70000$, $I_I = I_{II} = 6.75$, $L_I = L_{II} = 30$, square cross section with a side length of 3 (all numerical values in consistent units)

Number of nodes	Bending moment	Normal force
3	1259.37	41.979
7	1259.37	41.979
13	1259.37	41.979
Analytical solution	1259.84	41.995

from one mesh to the next refined configuration is below a certain threshold,[7] one can state that the result is no more mesh dependent.

Let us check this behavior for the internal bending moment at node 3, see Fig. 3.78. The summary in Table 3.18 indicates that the internal bending moment and normal force is not dependent on the mesh density. This result is a special case[8] for the applied thin beam elements and the corresponding support and load conditions.

The original and deformed shapes of the mesh with 13 nodes (see Fig. 3.78c) are shown in Fig. 3.79. It can be seen that all the imposed boundary conditions and the expected mode of deformation are fulfilled.

For a general problem of complex nature, the result of a mesh convergence study may look as schematically shown in Fig. 3.80.

---

[7]Other important factors are the computing time or the size of the result file.

[8]The result of the finite element hand calculation as given in Eq. (3.470) indicated already that the analytical result has been obtained. Thus, the computational approach is for this special case exact and a mesh refinement does not increase the accuracy.

Fig. 3.79 Frame structure with fine mesh: original and deformed shape

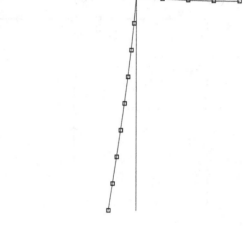

Fig. 3.80 Schematic representation of the result of a mesh convergence study

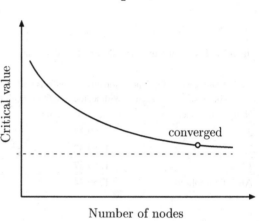

3.5 Supplementary Problems

3.19 Rod structure under dead weight

Given is a rod structure which is deforming under the influence of its dead weight, see Fig. 3.81. The rod is of the original length L, cross-sectional area A, YOUNG's modulus E, and mass density ϱ. The standard gravity is given by g.

Apply two linear rod elements of length $\frac{L}{2}$ to calculate the elongation of the rod due to its dead weight.

3.20 Truss structure with three members

Given is a plane truss structure as shown in Fig. 3.82. The members have a uniform cross-sectional area A and YOUNG's modulus E. The length of each member can be taken from the figure. The structure is fixed at its left-hand sides and loaded by two points loads F_1 and F_2.

Fig. 3.81 Rod loaded under
its dead weight

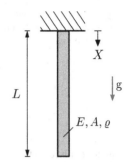

Fig. 3.82 Truss structure
composed of three straight
members

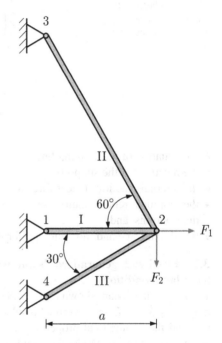

Model the truss structure with three linear finite elements and determine:

- the displacements of the nodes,
- the reaction forces at the supports,
- the strain, stress, and normal force in each element, and
- check the global force equilibrium.

3.21 Simply supported beam partially loaded with distributed load

The beam shown in Fig. 3.83 is loaded by a constant distributed load q_0. The bending stiffness EI is constant and the total length of the beam is equal to $2L$. Model the beam with two finite elements to determine:

- the unknowns at the nodes,

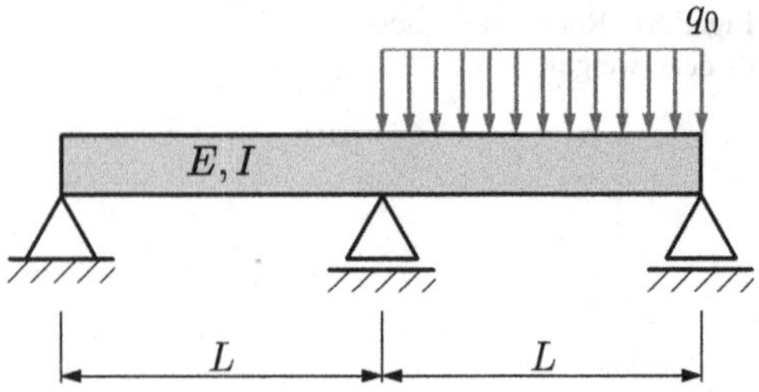

Fig. 3.83 Simply supported
beam partially loaded with
distributed load

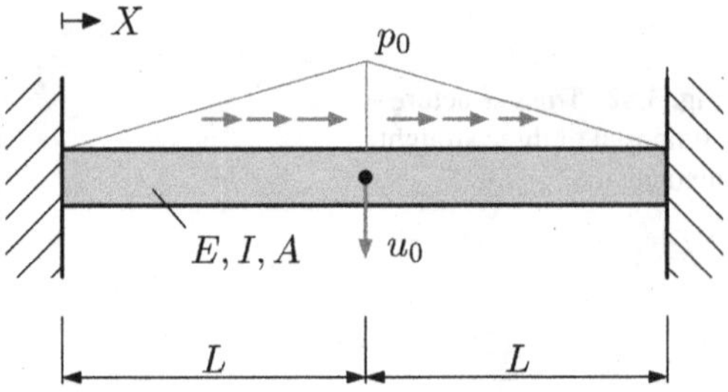

Fig. 3.84 Fixed-end
generalized beam with
distributed load and
displacement boundary
condition

- the equation of the bending line,
- the reactions at the supports,
- the internal reactions (shear force and bending moment) in each element,
- the graphical representations of the deflection, bending moment, and shear force distributions, and
- the global force and moment equilibrium.

3.22 Fixed-end generalized beam with distributed load and displacement boundary condition

The generalized beam shown in Fig. 3.84 is loaded by a distributed load $p(X)$ in the range $0 \leq X \leq 2L$ and a vertical displacement u_0 at $X = L$. The material constant (E) and the geometrical properties (I, A) are constant and the total length of the beam is equal to $2L$. Model the member with two generalized beam finite elements of length L to determine:

- the unknowns at the nodes,
- the displacement distributions $u_Z = u_Z(X)$ (bending) and $u_X = u_X(X)$ (tension/compression), including a graphical representation,
- the reactions at the supports,
- the internal reactions (normal force, shear force and bending moment) in each element, and
- the global force and moment equilibrium.

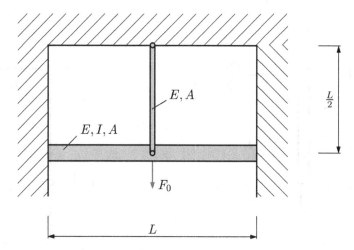

Fig. 3.85 Generalized beam supported by a rod element

3.23 Generalized beam supported by a rod element

The horizontal generalized beam shown in Fig. 3.85 is supported by a vertical rod element. The structure is loaded by a single force F_0 in the middle of the structure. The material property (E) and the geometrical properties (I, A) are constant and the same for beam and rod. The horizontal length of the structure is equal to L while the vertical dimension is equal to $\frac{L}{2}$. Model the structure with two generalized beam finite elements and one rod element of length $\frac{L}{2}$ to determine:

- the unknowns at the nodes,
- the displacement distributions in each member,
- the reactions at the supports,
- the internal reactions (normal force, shear force and bending moment for the beams and normal force for the rod) in each element,
- the strain and stress distributions in the elements, and
- the global force and moment equilibrium.

3.24 Generalized beam supported by a rod element: revised

Consider again the structure from Problem 3.23 (see Fig. 3.85). Replace the generalized beams by a more appropriate element type under consideration of the deformation of the structure. Furthermore, consider the symmetry of the problem to reduce the size of your computational model.

3.25 Extensometer with additional extensions

Derive the calibration curve for the extensometer shown in Fig. 3.86a based on a finite element approach. Use three generalized Euler–Bernoulli beam elements to simulate the extensometer under consideration of the symmetry of the problem.

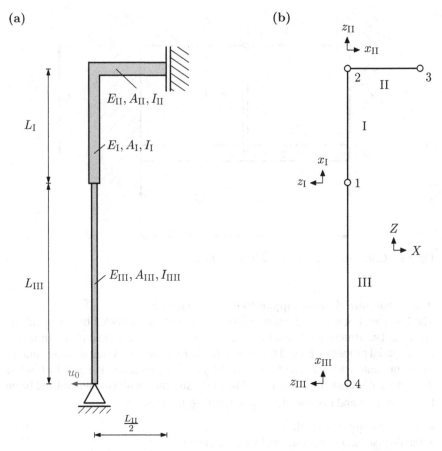

Fig. 3.86 Extensometer with additional extensions (see index 'III'): **a** mechanical model and **b** finite element representation

3.26 Extensometer without rotation at the specimen

Derive the calibration curve for the extensometer shown in Fig. 3.87a based on a finite element approach. Use two generalized Euler–Bernoulli beam elements to simulate the extensometer under consideration of the symmetry of the problem.

3.27 Finite element post-processing of strains

The nodal deformations of the finite element model shown in Fig. 3.74 were obtained in Eqs. (3.465)–(3.469) as follows:

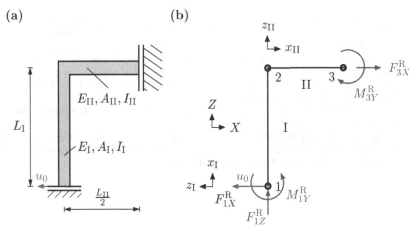

Fig. 3.87 Extensometer without rotation at the specimen: **a** mechanical model and **b** finite element representation

$$\varphi_{1Y} = \frac{1 + \frac{E_{II}}{E_I}\frac{I_{II}}{I_I}\frac{L_I}{L_{II}}}{1 + \frac{2}{3}\frac{E_{II}}{E_I}\frac{I_{II}}{I_I}\frac{L_I^2}{L_I L_{II}} + \frac{1}{12}\frac{h_{II}^2}{L_I^2}} \times \frac{u_0}{L_I}, \tag{3.479}$$

$$u_{2X} = -\frac{1}{1 + 8\frac{E_{II}}{E_I}\frac{I_{II}}{I_I}\frac{L_I^3}{L_{II}h_{II}^2} + 12\frac{L_I^2}{h_{II}^2}} \times u_0, \tag{3.480}$$

$$u_{2Z} = 0, \tag{3.481}$$

$$\varphi_{2Y} = \frac{1}{1 + \frac{2}{3}\frac{E_{II}}{E_I}\frac{I_{II}}{I_I}\frac{L_I^2}{L_I L_{II}} + \frac{1}{12}\frac{h_{II}^2}{L_I^2}} \times \frac{u_0}{L_I}, \tag{3.482}$$

$$u_{3Z} = -\frac{3}{4} \times \frac{1}{3\frac{L_I}{L_{II}} + 2\frac{E_{II}}{E_I}\frac{I_{II}}{I_I}\frac{L_I^2}{L_{II}^2} + \frac{1}{4}\frac{h_{II}^2}{L_I L_{II}}} \times u_0. \tag{3.483}$$

Use Tables 3.2 and 3.8 to calculate the strains, i.e., at the free surfaces, at node 3 and derive the corresponding calibration factors.

References

1. Buchanan GR (1995) Schaum's outline of theory and problems of finite element analysis. McGraw-Hill, New York
2. Cook RD, Malkus DS, Plesha ME, Witt RJ (2002) Concepts and applications of finite element analysis. Wiley, New York
3. Javanbakht Z, Öchsner A (2017) Advanced finite element simulation with MSC Marc: application of user subroutines. Springer, Cham

4. Javanbakht Z, Öchsner A (2018) Computational statics revision course. Springer, Cham
5. Öchsner A, Merkel M (2013) One-dimensional finite elements: an introduction to the FE method. Springer, Berlin
6. Öchsner A, Öchsner M (2016) The finite element analysis program MSC Marc/Mentat. Springer, Singapore
7. Öchsner A (2020) Computational statics and dynamics: an introduction based on the finite element method. Springer, Singapore
8. Reddy JN (1999) On the dynamic behaviour of the Timoshenko beam finite elements. Sadhana-Acad P Eng S 24:175–198
9. Sharpe WN (2008) Springer handbook of experimental solid mechanics. Springer, New York
10. Timoshenko SP (1921) On the correction for shear of the differential equation for transverse vibrations of prismatic bars. Philos Mag 41:744–746

Chapter 4
Outlook: Two- and Three-Dimensional Elements

Abstract This chapter gives a brief outlook on some two- and three-dimensional elements. The similarities between the previously treated one-dimensional elements, i.e., rod and beams, and their multidimensional analogs are presented without going into the mathematical details. The considered elements are restricted to configurations with even node numbers at which nodes are exclusively located at the element corners.

In Chap. 3 introduced one-dimensional elements, i.e., rod and beams, are shown in Fig. 4.1 with their two- and three-dimensional generalizations. The understanding of the rod element can easily be transferred to a four-node plane elasticity element (with plane stress or plane strain behavior) or to the eight-node solid element. The rod element has in its elemental coordinate system just a single degree of freedom per node, i.e the displacement along the rod axis [5]. This concept is extended in the case of plane elasticity elements to two displacement components (u_{ix}, u_{iy}) per node i whereas the solid element possesses three displacement components (u_{ix}, u_{iy}, u_{iz}) per node i. It should be noted here that there are also other element types with, for example, inner nodes or elements with an uneven number for plane cases such as triangular elements.[1] Research in the area of element formulations requires a computer code, which allows that user-written code can be linked to the main program. Some of the commercial packages offer such a functionality [4].

Let us recall again the principal finite element equation for rod elements as introduced in Eq. (3.1). A 2×2 stiffness matrix is multiplied with the column matrix of two nodal unknowns, see Eq. (4.1).

$$\underbrace{\begin{bmatrix} \cdots \cdots \\ \cdots \cdots \end{bmatrix}}_{2 \times 2} \begin{bmatrix} u_{1x} \\ u_{2x} \end{bmatrix} = \begin{bmatrix} F_{1x} \\ F_{2x} \end{bmatrix}. \tag{4.1}$$

---

[1] The theoretical treatment of triangular elements requires the consideration of natural or triangular coordinates which are aligned to the sides of the triangle and these axes do not intersect at right angle. Thus, the approach must be slightly adjusted compared to the elements shown in Fig. 4.1. For example, the linkage between Cartesian and triangular coordinates must be first reviewed.

A. Öchsner, *A Project-Based Introduction to Computational Statics*, https://doi.org/10.1007/978-3-030-58771-0_4

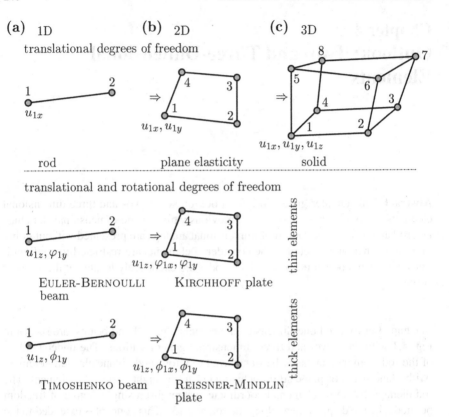

Fig. 4.1 Classification of considered finite elements: **a** one-dimensional, **b** two-dimensional, and **c** three-dimensional elements

Knowing that a plane elasticity element with four nodes has in total eight degrees of freedom (number of nodes times nodal degrees of freedom), the elemental stiffness matrix must have the dimensions 8×8 and the principal finite element equation has the structure as shown in Eq. (4.2). Of course that it must be considered that the mathematical derivations require more work than just considering the structure of the finite element equation [6, 8].

$$
\underbrace{\begin{bmatrix} \cdots & & \cdots \\ & & \\ & & \\ \cdots & & \cdots \end{bmatrix}}_{8 \times 8}
\begin{bmatrix} u_{1x} \\ u_{1y} \\ \vdots \\ u_{4x} \\ u_{4y} \end{bmatrix}
=
\begin{bmatrix} F_{1x} \\ F_{1y} \\ \vdots \\ F_{4x} \\ F_{4y} \end{bmatrix},
\tag{4.2}
$$

The same reasoning can be applied to solid elements. Assuming that eight nodes are forming a hexahedral element, the corresponding stiffness matrix must have the dimensions 24×24 since each node has three translatorial degrees of freedom, see Eq. (4.3).

$$
\underbrace{\begin{bmatrix} \cdots & & & \cdots \\ & & & \\ & & & \\ & & & \\ & & & \\ \cdots & & & \cdots \end{bmatrix}}_{24 \times 24}
\begin{bmatrix} u_{1x} \\ u_{1y} \\ u_{1z} \\ \vdots \\ u_{4x} \\ u_{4y} \\ u_{4z} \end{bmatrix}
=
\begin{bmatrix} F_{1x} \\ F_{1y} \\ F_{1z} \\ \vdots \\ F_{4x} \\ F_{4y} \\ F_{4z} \end{bmatrix}, \tag{4.3}
$$

Furthermore, it should be mentioned that the here introduced beam elements with one rotational and one translatorial degree of freedom for bending in a single plane have their two-dimensional counterparts as thin or thick plates, see Fig. 4.1.

Finally, it must be noted that the numerical simulation or the corresponding homogenization of structured materials such as cellular metals [1, 2, 9] or nanoparticle-reinforced composites [3, 7, 10] is a challenging topic and subject matter of many actual research projects.

References

1. Fiedler T, Sturm B, Öchsner A, Gracio J, Kuhn G (2006) Modelling the mechanical behaviour of adhesively bonded and sintered hollow-sphere structures. Mech Compos Mater 42:559–570
2. Fiedler T, Löffler R, Bernthaler T, Winkler R, Belova IV, Murch GE, Öchsner A (2009) Numerical analyses of the thermal conductivity of random hollow sphere structures. Mater Lett 63:1125–1127
3. Ghavamian A, Öchsner A (2012) Numerical investigation on the influence of defects on the buckling behavior of single-and multi-walled carbon nanotubes. Phys E 46:241–249
4. Javanbakht Z, Öchsner A (2017) Advanced finite element simulation with MSC Marc: application of user subroutines. Springer, Cham
5. Öchsner A, Merkel M (2018) One-dimensional finite elements: an introduction to the FE method. Springer, Cham
6. Öchsner A (2020) Computational statics and dynamics: an introduction based on the finite element method. Springer, Singapore
7. Rahmandoust M, Öchsner A (2012) On finite element modeling of single-and multi-walled carbon nanotubes. J Nanosci Nanotech 12:8129–8136
8. Reddy JN (2006) An introduction to the finite element method. McGraw-Hill, Singapore
9. Vesenjak M, Fiedler T, Ren Z, Öchsner A (2008) Behaviour of syntactic and partial hollow sphere structures under dynamic loading. Adv Eng Mater 10:185–191
10. Yengejeh SI, Kazemi SA, Öchsner A (2016) Advances in mechanical analysis of structurally and atomically modified carbon nanotubes and degenerated nanostructures: a review. Compos Part B-Eng 86:95–107

Chapter 5
Answers to Supplementary Problems

Abstract This chapter provides the short solutions to the supplementary problems given at the end of Chaps. 2 and 3. Where appropriate, some intermediate steps are provided to better understand the solution procedure.

5.1 Problems from Chap. 2

For further problems and explanations, the interested reader may consult reference [3].

2.13 Rod loaded by a single force in its middle
Case (a): Approach based on two sections

$$u_{x_I} = \frac{F_0 L}{2EA}\left(\frac{x_I}{L}\right), \qquad u_{x_{II}} = \frac{F_0 L}{2EA}\left(1 - \frac{x_{II}}{L}\right), \qquad (5.1)$$

$$N_{x_I} = \frac{F_0}{2}, \qquad N_{x_{II}} = -\frac{F_0}{2}. \qquad (5.2)$$

Case (b): Approach based on discontinuous function

$$EA\frac{du_X(X)}{dX} = N_X(X) = -F_0\langle X - L\rangle^0 + c_1, \qquad (5.3)$$

$$EAu_X(X) = -F_0\langle X - L\rangle^1 + c_1 X + c_2, \qquad (5.4)$$

or finally after the determination of the constants of integration:

$$u_X(X) = -\frac{F_0}{EA}\left(\langle X - L\rangle^1 + \frac{X}{2}\right). \qquad (5.5)$$

The displacement and normal force distributions are shown in Fig. 5.1.

Fig. 5.1 Rod loaded by a
single force in its middle:
a displacement distribution,
b normal force

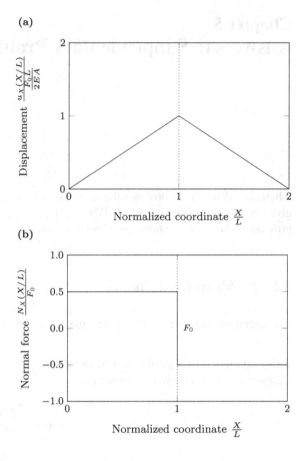

2.14 Rod loaded by a distributed load

PDEs based on two sections:
The general solutions are given as:

$$u_{x_I}(x_I) = \frac{1}{EA}(c_1 x_I + c_2) , \quad u_{x_{II}}(x_{II}) = \frac{1}{EA}\left(-\frac{p_0 x_{II}^2}{2} + c_3 x_{II} + c_4\right) , \quad (5.6)$$

$$N(x_I) = c_1 , \qquad\qquad N(x_{II}) = -p_0 x_{II} + c_3 . \qquad\qquad (5.7)$$

Consideration of the boundary conditions $u_{x_I}(x_I = 0) = 0$ and $u_{x_{II}}(x_{II} = L) = 0$ as
well as the transmission conditions $u_{x_I}(x_I = L) = u_{x_{II}}(x_{II} = 0)$ and $N(x_I = L) =$

$N(x_{II} = 0)$ gives the following constants of integration: $c_1 = \frac{p_0 L}{4}$, $c_2 = 0$, $c_3 = \frac{p_0 L}{4}$, and $c_4 = \frac{p_0 L^2}{4}$.

Thus, we get the following solutions:

$$u_{x_I}(x_I) = \frac{p_0 L}{4EA} x_I, \qquad u_{x_{II}}(x_{II}) = \frac{p_0 L}{4EA}\left(-\frac{2}{L}x_{II}^2 + x_{II} + L\right), \qquad (5.8)$$

$$N(x_I) = \frac{p_0 L}{4}, \qquad N(x_{II}) = \frac{p_0 L}{4}\left(-\frac{4x_{II}}{L} + 1\right). \qquad (5.9)$$

Discontinuous function based on a single section:

$$EA\frac{d^2 u_X(X)}{dX^2} = -p_X = -p_0\langle X - L\rangle^0, \qquad (5.10)$$

$$EA\frac{du_X(X)}{dX} = N(X) = -p_0\langle X - L\rangle^1 + c_1, \qquad (5.11)$$

$$EAu_X(X) = -\frac{p_0}{2}\langle X - L\rangle^2 + c_1 X + c_2. \qquad (5.12)$$

Consideration of the boundary conditions gives: $c_2 = 0$ and $c_1 = \frac{p_0 L}{4}$. Thus, we get as the solutions:

$$u_X(X) = \frac{1}{EA}\left(-\frac{p_0}{2}\langle X - L\rangle^2 + \frac{p_0 L}{4}X\right), \qquad (5.13)$$

$$N(X) = -p_0\langle X - L\rangle^1 + \frac{p_0 L}{4}. \qquad (5.14)$$

2.15 Cantilever beam under the influence of a point or distributed load—rectangular cross section
Case (a):

$$\frac{u_z\left(\frac{x}{L} = 1\right)}{\frac{F_0 L^3}{EI}} = \frac{1}{3} + \frac{1}{5}(1 + v)\left(\frac{h}{L}\right)^2. \qquad (5.15)$$

Case (b):

$$\frac{u_z\left(\frac{x}{L}\right)}{\frac{q_0 L^4}{EI}} = \frac{1}{8} + \frac{1}{10}(1 + v)\left(\frac{h}{L}\right)^2. \qquad (5.16)$$

2.17 Beam-like structure: energy approach

$$u_{x,D} = \frac{F_0 a^3}{8EI},$$ (5.17)

$$u_{z,C} = \frac{F_0 a^3}{3EI}.$$ (5.18)

2.18 Calibration factor for the standard extensometer
A calibration factor equal to 1 requires that:
$$\frac{3}{2} \times \frac{E_I I_I L_{II}}{E_{II} I_{II} L_I} \times \frac{h_{II}}{L_I} \times \left(+\frac{1}{2} + \frac{1}{12}\left(\frac{h_{II}}{L_I}\right)\right) \overset{!}{=} 1.$$ (5.19)

Assuming $h_{II}/L_I = 1/10$, which is based on the classical assumption for thin beams, we obtain:

$$\frac{61}{800} \times \frac{E_I I_I L_{II}}{E_{II} I_{II} L_I} \overset{!}{=} 1.$$ (5.20)

Case (a): $E_I = E_{II}$ and $I_I = I_{II}$, i.e., both parts have the same cross section and are made from the same material: $\rightarrow L_{II} = 13.115 L_I$.

Case (b): $E_I =$ St, $E_{II} =$ Al, i.e., $E_I/E_{II} = 3$, and $I_I = I_{II}$, i.e., both parts have the same cross section: $\rightarrow L_{II} = 4.371 L_I$.

Case (c): $E_I =$ St, $E_{II} =$ Al, i.e., $E_I/E_{II} = 3$, and $I_I = 3I_{II}$: $\rightarrow L_{II} = 1.457 L_I$.

2.19 Elastic limit of the standard extensometer
The critical location is the entire member II. From the calibration factor for the standard extensometer according to Eq. (2.238), together with $\sigma_{II} = E_{II} \varepsilon_{x,II}\big|_{z=-\frac{h_{II}}{2}} = R_{p0.2,II}$ and $\varepsilon_{sp} = u_0/(L_{II}/2)$, one gets:

$$u_{0,max} = \frac{R_{p0.2,II}}{3 \times \dfrac{E_I I_I}{I_{II} L_I} \times \dfrac{h_{II}}{L_I} \times \left(+\dfrac{1}{2} + \dfrac{1}{12}\left(\dfrac{h_{II}}{L_I}\right)\right)}.$$ (5.21)

2.20 Stiffening effect of the extensometer
The force ratio, using the relation $F_{sp}/A_{sp} = E_{sp} u_0/(L_{II}/2)$, is obtained as given in Eq. (5.22) and the numerical values are presented in Table 5.1.

Table 5.1 Stiffening effect of the extensometer

Cross section	Stiffening ratio $\frac{F_0^R}{F_{sp}}$
$b_I = b_{II} = 5$ mm $h_I = h_{II} = 2$ mm	2.3992×10^{-4}
$b_I = b_{II} = 6$ mm $h_I = h_{II} = 3$ mm	9.7130×10^{-4}
$b_I = b_{II} = 7$ mm $h_I = h_{II} = 4$ mm	14.3956×10^{-4}

$$\frac{F_0^R}{F_{sp}} = \frac{1}{E_{sp} A_{sp}} \times \frac{1}{\frac{2L_I^3}{3L_{II}E_I I_I} + \frac{L_I^2}{E_{II}I_{II}} + \frac{h_{II}^2}{12E_{II}I_{II}}}. \tag{5.22}$$

2.21 Calibration factor of the extensometer and corresponding strains
The numerical values are provided in Table 5.2.

2.22 Simple design of a double-cantilever displacement gage
Some intermediate results can be taken from Problem 2.4a, i.e., the general equation of the bending line, slope and moment distribution. It should be noted here that the following solution considers only absolute values and the direction of the vertical coordinate axis is not considered.

• Maximum deflection and rotation at load application point

$$u(x) = \frac{F_0}{6EI}\left(3Lx^2 - x^3\right), \tag{5.23}$$

$$\varphi(x) = \frac{du(x)}{dx} = \frac{F_0}{6EI}\left(6Lx - 3x^2\right). \tag{5.24}$$

Table 5.2 Calibration factor and corresponding strains

Cross section	$\dfrac{\varepsilon_{x,II}\big|_{z=+\frac{h_{II}}{2}}}{\varepsilon_{sp}}$	$\dfrac{\varepsilon_{x,II}\big|_{z=-\frac{h_{II}}{2}}}{\varepsilon_{sp}}$	$\varepsilon_{x,II}\big|_{z=+\frac{h_{II}}{2}}$	$\varepsilon_{x,II}\big|_{z=-\frac{h_{II}}{2}}$
$h_I = h_{II} = 1$ mm	−0.005911	0.005960	−0.001182	0.001192
$h_I = h_{II} = 2$ mm	−0.011772	0.011967	−0.002354	0.002393
$h_I = h_{II} = 3$ mm	−0.017583	0.018022	−0.003517	0.003604

$$u_0 = u(x = L) = \frac{F_0 L^3}{3EI} \Leftrightarrow F_0 = \frac{3EIu_0}{L^3},$$

(5.25)

$$\varphi_0 = \varphi(x = L) = \frac{3F_0 L^2}{6EI} = \frac{3u_0}{2L}.$$

(5.26)

• Limitation of the maximum strain to 75% of the elastic range

$$\varepsilon_{\text{lim}} = \frac{3}{4} \times \frac{R_p}{E} = 0.75 \times 0.008 = 0.006.$$

(5.27)

• Moments at locations O and G

$$M_O = F_0 L, \quad M_G = \alpha F_0 L.$$

(5.28)

• Strains at locations O and G

$$\varepsilon_O = \frac{F_0 Lh}{2EI} = \frac{3hu_0}{2L^2},$$

(5.29)

$$\varepsilon_G = \frac{\alpha F_0 Lh}{2EI} = \alpha \varepsilon_O.$$

(5.30)

• Maximum strains due to elastic limit condition

$$\varepsilon_{\text{lim}} = \frac{3}{4} \times 0.008 = \frac{3hu_{\text{lim}}}{2L^2}$$

(5.31)

$$\Leftrightarrow \quad 2u_{\text{lim}} = 0.008 \frac{L^2}{h}.$$

(5.32)

• Maximum rotation due to elastic limit condition

$$\varphi_{\text{lim}} = \frac{3u_{\text{lim}}}{2L} = \frac{3}{4} 0.008 \frac{L}{h} = 0.006 \frac{L}{h}.$$

(5.33)

• Maximum force per beam width due to elastic limit condition

$$F_{\text{lim}} = \frac{3EIu_{\text{lim}}}{L^3} = \frac{2EI\varepsilon_{\text{lim}}}{Lh} = \frac{2EI}{Lh} \times \frac{3}{4} \times \frac{R_p}{E},$$

(5.34)

$$\frac{F_{\text{lim}}}{b} = \frac{R_p}{8} \times \frac{h^2}{L}.$$

(5.35)

• Sensitivity of the gage

$$\frac{d\varepsilon_G}{du_0} = \frac{3\alpha h}{2L^2},\tag{5.36}$$

where $\alpha = 0.9$ for $L = 31.75\,\text{mm}$ and $\alpha = 0.92$ for $L = 38.10\,\text{mm}$. Taking $L = 31.75\,\text{mm}$ and $h = 1.016\,\text{mm}$, the following characteristics are obtained: $\varphi_{\text{lim}} = 10.74°$, $F_{\text{lim}}/b = 3.36\,\text{N/mm}$, $d\varepsilon_G/du_0 = 1.36\,10^3/\text{mm}$, and $2u_0 = 7.94\,\text{mm}$. Assuming a beam width of $b = 9.525\,\text{mm}$ (3/8 in), a spring force of approximately $32\,\text{N}$ ($\sim 3.3\,\text{kg}$) is necessary for specimen attachment.
The corresponding design curves are shown in Fig. 5.2.

5.2 Problems from Chap. 3

For further problems and explanations, the interested reader may consult reference [2].

3.19 Rod structure under dead weight
The following force acts on a volume element: $dF_g = dmg = \varrho A g dx$. Thus, the distributed load is given by: $p_g = \frac{dF_g}{dx} = \varrho A g$.
• Solution of the reduced system of equations:

$$\begin{bmatrix} u_{2X} \\ u_{3X} \end{bmatrix} = \frac{\varrho g L^2}{8E} \begin{bmatrix} 3 \\ 4 \end{bmatrix}.\tag{5.37}$$

3.20 Truss structure with three members
• Unknown displacements:

$$\begin{bmatrix} u_{2X} \\ u_{2Z} \end{bmatrix} = \frac{a}{3EA(1+\sqrt{3})} \begin{bmatrix} (3+\sqrt{3})F_1 + (3-\sqrt{3})F_2 \\ -(3-\sqrt{3})F_1 - (9+3\sqrt{3})F_2 \end{bmatrix}.\tag{5.38}$$

• Reaction forces:

$$F_{1X}^R = -\frac{1}{3(1+\sqrt{3})}\left((3+\sqrt{3})F_1 + (3-\sqrt{3})F_2\right),\tag{5.39}$$

$$F_{1Z}^R = 0,\tag{5.40}$$

$$F_{3X}^R = -\frac{1}{6(1+\sqrt{3})}\left((\sqrt{3})F_1 + (3+2\sqrt{3})F_2\right),\tag{5.41}$$

$$F_{3Z}^R = -\frac{1}{2(1+\sqrt{3})}\left(F_1 + (2+\sqrt{3})F_2\right),\tag{5.42}$$

$$F_{4X}^R = -\frac{1}{2(1+\sqrt{3})}\left((\sqrt{3})F_1 - (3)F_2\right),\tag{5.43}$$

Fig. 5.2 Design parameters for the double-cantilever displacement gage. Adapted from [1]

(a)

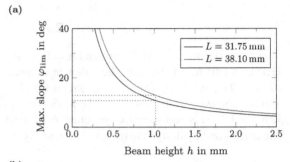

(b)

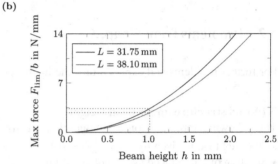

(c)

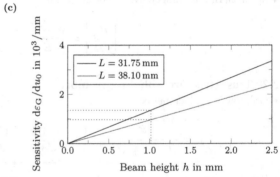

(d)

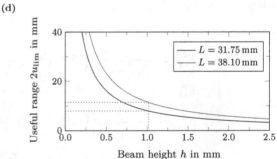

$$F_{4Z}^{R} = -\frac{1}{2(1+\sqrt{3})}\left(F_1 + (-\sqrt{3})F_2\right) . \tag{5.44}$$

- Global force equilibrium:

$$\sum_i F_{iX} = 0 \quad \Leftrightarrow \quad \underbrace{(F_{1X}^{R} + F_{3X}^{R} + F_{4X}^{R})}_{\text{reaction force}} + \underbrace{(F_1)}_{\text{external load}} = 0, \checkmark \tag{5.45}$$

$$\sum_i F_{iZ} = 0 \quad \Leftrightarrow \quad \underbrace{(F_{1Z}^{R} + F_{3Z}^{R} + F_{4Z}^{R})}_{\text{reaction force}} + \underbrace{(-F_2)}_{\text{external load}} = 0. \checkmark \tag{5.46}$$

- Elemental stress, strain and normal force:

$$\sigma_{\mathrm{I}} = \frac{1}{3A(1+\sqrt{3})}\left((3+\sqrt{3})F_1 + (3-\sqrt{3})F_2\right), \tag{5.47}$$

$$\sigma_{\mathrm{II}} = \frac{1}{3A(1+\sqrt{3})}\left((\sqrt{3})F_1 + (3+2\sqrt{3})F_2\right), \tag{5.48}$$

$$\sigma_{\mathrm{III}} = \frac{\sqrt{3}}{3A(1+\sqrt{3})}\left((\sqrt{3})F_1 + (-3)F_2\right). \tag{5.49}$$

$$\varepsilon_{\mathrm{I}} = \frac{1}{3EA(1+\sqrt{3})}\left((3+\sqrt{3})F_1 + (3-\sqrt{3})F_2\right), \tag{5.50}$$

$$\varepsilon_{\mathrm{II}} = \frac{1}{3EA(1+\sqrt{3})}\left((\sqrt{3})F_1 + (3+2\sqrt{3})F_2\right), \tag{5.51}$$

$$\varepsilon_{\mathrm{III}} = \frac{\sqrt{3}}{3EA(1+\sqrt{3})}\left((\sqrt{3})F_1 + (-3)F_2\right). \tag{5.52}$$

$$N_{\mathrm{I}} = \frac{1}{3(1+\sqrt{3})}\left((3+\sqrt{3})F_1 + (3-\sqrt{3})F_2\right), \tag{5.53}$$

$$N_{\mathrm{II}} = \frac{1}{3(1+\sqrt{3})}\left((\sqrt{3})F_1 + (3+2\sqrt{3})F_2\right), \tag{5.54}$$

$$N_{\mathrm{III}} = \frac{\sqrt{3}}{3(1+\sqrt{3})}\left((\sqrt{3})F_1 + (-3)F_2\right). \tag{5.55}$$

3.21 Simply supported beam partially loaded with distributed load

- Unknowns at the nodes:

$$\varphi_{1Y} = -\frac{L^3 q_0}{96EI_Y}, \; \varphi_{2Y} = \frac{L^3 q_0}{48EI_Y}, \; \varphi_{3Y} = -\frac{L^3 q_0}{32EI_Y}. \tag{5.56}$$

• Equation of the bending line:

$$u_{\mathrm{I}Z}^{e} = \frac{L^4 q_0}{96EI_Y}\left(\frac{x_{\mathrm{I}}}{L} - \left(\frac{x_{\mathrm{I}}}{L}\right)^3\right), \tag{5.57}$$

$$u_{\mathrm{II}Z}^{e} = -\frac{L^4 q_0}{96EI_Y}\left(2\frac{x_{\mathrm{II}}}{L} - \left(\frac{x_{\mathrm{II}}}{L}\right)^2 - \left(\frac{x_{\mathrm{II}}}{L}\right)^3\right). \tag{5.58}$$

• Reactions at the supports:

$$F_{1Z}^{R} = -\frac{L q_0}{16}, \; F_{2Z}^{R} = \frac{5 L q_0}{8}, \; F_{3Z}^{R} = \frac{7 L q_0}{16}. \tag{5.59}$$

• Internal reactions in each element:

$$M_{\mathrm{I}Y} = \frac{L^3 q_0}{96}\left(\frac{6 x_{\mathrm{I}}}{L^2}\right), \tag{5.60}$$

$$M_{\mathrm{II}Y} = -\frac{L^3 q_0}{96}\left(\frac{2(L + 3 x_{\mathrm{II}})}{L^2}\right). \tag{5.61}$$

$$Q_{\mathrm{I}Z} = \frac{L q_0}{16}, \; Q_{\mathrm{II}Z} = -\frac{L q_0}{16}. \tag{5.62}$$

• Global force and moment equilibrium:

$$\sum_{i} F_{iZ} = 0 \;\Leftrightarrow\; \underbrace{(F_{1Z}^{R} + F_{2Z}^{R} + F_{3Z}^{R})}_{\text{reaction force}} + \underbrace{\left(-\frac{q_0 L}{2} - \frac{q_0 L}{2}\right)}_{\text{external load}} = 0, \; \checkmark \tag{5.63}$$

Fig. 5.3 **a** Beam deflection along the major axis, **b** bending moment distribution and **c** shear force distribution

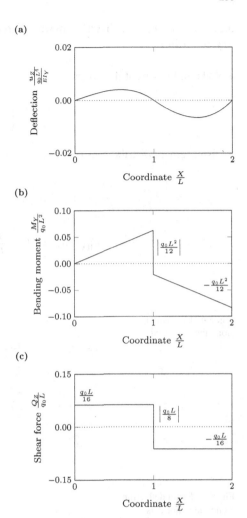

(a)

Deflection $\frac{u_Z}{\frac{q_0 L^4}{EI_Y}}$

Coordinate $\frac{X}{L}$

(b)

Bending moment $\frac{M_Y}{q_0 L^2}$

$\left|\frac{q_0 L^2}{12}\right|$

$-\frac{q_0 L^2}{12}$

Coordinate $\frac{X}{L}$

(c)

Shear force $\frac{Q_Z}{q_0 L}$

$\frac{q_0 L}{16}$

$\left|\frac{q_0 L}{8}\right|$

$-\frac{q_0 L}{16}$

Coordinate $\frac{X}{L}$

$$\sum_i M_{iY}(X=0) = 0 \quad \Leftrightarrow \quad \underbrace{(-F_{2Z}^R L - F_{3Z}^R 2L)}_{\text{reaction}} + \underbrace{\left(\frac{q_0 L^2 3}{2}\right)}_{\text{external load}} = 0. \;\checkmark \qquad (5.64)$$

• Graphical representations of the deflection, bending moment, and shear force distributions

The graphical representation is shown in Fig. 5.3.

3.22 Cantilever generalized beam with distributed load and displacement boundary condition

• Nodal unknowns at the nodes:

$$\begin{bmatrix} u_{2X} \\ \varphi_{2Y} \end{bmatrix} = \begin{bmatrix} \frac{p_0 L^2}{3EA} \\ 0 \end{bmatrix}. \tag{5.65}$$

• Displacement distributions (see Figs. 5.4 and 5.5):

$$u^e_{IZ}(x_I) = \left[3\frac{x_I^2}{L^2} - 2\frac{x_I^3}{L^3} \right](-u_0), \quad u^e_{IIZ}(x_{II}) = \left[1 - 3\frac{x_{II}^2}{L^2} + 2\frac{x_{II}^3}{L^3} \right](-u_0), \tag{5.66}$$

$$u^e_{IX}(x_I) = \frac{x_I}{L} \times \frac{p_0 L^2}{3EA}, \quad u^e_{IIX}(x_{II}) = \left[1 - \frac{x_{II}}{L} \right] \times \frac{p_0 L^2}{3EA}. \tag{5.67}$$

Fig. 5.4 Beam elongation along the major axis

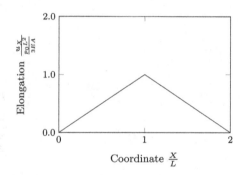

Fig. 5.5 Beam deflection along the major axis

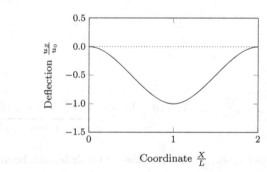

- Reactions at the supports:

$$F^R_{1X} = -\frac{p_0 L}{2}, \quad F^R_{1Z} = \frac{12EIu_0}{L^3}, \quad M^R_{1Y} = -\frac{6EIu_0}{L^2}, \quad F^R_{2Z} = \frac{24EIu_0}{L^3}, \quad (5.68)$$

$$F^R_{3X} = -\frac{p_0 L}{2}, \quad F^R_{3Z} = \frac{12EIu_0}{L^3}, \quad M^R_{3Y} = \frac{6EIu_0}{L^2}. \quad (5.69)$$

- Internal reactions (normal force, shear force and bending moment) in each element:

$$M^e_{IY}(x_I) = EI\left[\frac{6}{L^2} - \frac{12x_I}{L^3}\right]u_0, \quad M^e_{IIY}(x_{II}) = EI\left[-\frac{6}{L^2} + \frac{12x_{II}}{L^3}\right]u_0, \quad (5.70)$$

$$Q^e_{IZ}(x_I) = -EI\left[\frac{12}{L^3}\right]u_0, \quad Q^e_{IIZ}(x_{II}) = EI\left[\frac{12}{L^3}\right]u_0, \quad (5.71)$$

$$N^e_{IX}(x_I) = \frac{p_0 L}{3}, \quad N^e_{IIX}(x_{II}) = -\frac{p_0 L}{3}. \quad (5.72)$$

3.23 Generalized beam supported by a rod element

The reduced system of equations reads:

$$E\begin{bmatrix} \left(\dfrac{A}{L_I} + \dfrac{A}{L_{II}}\right) & 0 & 0 \\[2ex] 0 & \left(\dfrac{12I}{L_I^3} + \dfrac{12I}{L_{II}^3} + \dfrac{A}{L_{III}}\right) & \left(\dfrac{6I}{L_I^2} - \dfrac{6I}{L_{II}^2}\right) \\[2ex] 0 & \left(\dfrac{6I}{L_I^2} - \dfrac{6I}{L_{II}^2}\right) & \left(\dfrac{4I}{L_I} + \dfrac{4I}{L_{II}}\right) \end{bmatrix}\begin{bmatrix} u_{2X} \\[2ex] u_{2Z} \\[2ex] \varphi_{2Y} \end{bmatrix} = \begin{bmatrix} 0 \\[2ex] -F_0 \\[2ex] 0 \end{bmatrix}. \quad (5.73)$$

The solution can be obtained as:

$$\begin{bmatrix} u_{2X} \\[1ex] u_{2Z} \\[1ex] \varphi_{2Y} \end{bmatrix} = \begin{bmatrix} 0 \\[1ex] -\dfrac{1}{2} \times \dfrac{L^3 F_0}{E(AL^2 + 96I)} \\[1ex] 0 \end{bmatrix}. \quad (5.74)$$

Fig. 5.6 Extensometer with
additional extensions: finite
element mesh with reactions

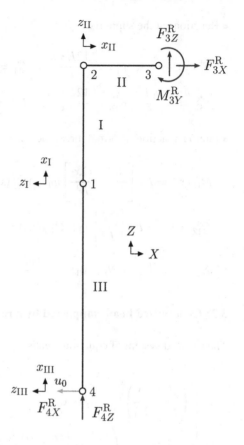

3.24 Generalized beam supported by a rod element: revised
The generalized beams (E, I, A) can be replaced by pure thin beams (E, I) since
there is no elongation of the horizontal structure.

3.25 Extensometer with additional extensions
The finite element representation including the reactions is shown in Fig. 5.6.
The finite element simulation, i.e., the post-processing, must provide the horizontal
reaction force F_{4X}^{R} at node 4 to evaluate the normal force and bending moment at
node 3 according to Eqs. (2.287) and (2.289) or alternatively the reactions F_{3X}^{R} and
M_{3Y}^{R} at node 3.

The reduced stiffness matrix reads:

$$
\begin{bmatrix}
\frac{12E_{III}I_{III}}{L_{III}^3} + \frac{12E_I I_I}{L_I^3} & 0 & \frac{6E_I I_I}{L_I^2} - \frac{6E_{III}I_{III}}{L_{III}^2} & -\frac{12E_I I_I}{L_I^3} & 0 & \frac{6E_I I_I}{L_I^2} & 0 & -\frac{12E_{III}I_{III}}{L_{III}^3} & -\frac{6E_{III}I_{III}}{L_{III}^2} \\[2mm]
0 & \frac{A_{III}}{L_{III}} + \frac{A_I}{L_I} & 0 & 0 & -\frac{A_I}{L_I} & 0 & 0 & 0 & 0 \\[2mm]
\frac{6E_I I_I}{L_I^2} - \frac{6E_{III}I_{III}}{L_{III}^2} & 0 & \frac{4E_{III}I_{III}}{L_{III}} + \frac{4E_I I_I}{L_I} & -\frac{6E_I I_I}{L_I^2} & 0 & \frac{2E_I I_I}{L_I} & 0 & \frac{6E_{III}I_{III}}{L_{III}^2} & \frac{2E_{III}I_{III}}{L_{III}} \\[2mm]
-\frac{12E_I I_I}{L_I^3} & 0 & -\frac{6E_I I_I}{L_I^2} & \frac{2A_{II}E_{II}}{L_{II}} + \frac{12E_I I_I}{L_I^3} & 0 & -\frac{6E_I I_I}{L_I^2} & 0 & 0 & 0 \\[2mm]
0 & -\frac{A_I}{L_I} & 0 & 0 & \frac{96E_{II}I_{II}}{L_{II}^3} + \frac{A_I}{L_I} & -\frac{24E_{II}I_{II}}{L_{II}^2} & -\frac{96E_{II}I_{II}}{L_{II}^3} & 0 & 0 \\[2mm]
\frac{6E_I I_I}{L_I^2} & 0 & \frac{2E_I I_I}{L_I} & -\frac{6E_I I_I}{L_I^2} & -\frac{24E_{II}I_{II}}{L_{II}^2} & \frac{8E_{II}I_{II}}{L_{II}} + \frac{4E_I I_I}{L_I} & \frac{24E_{II}I_{II}}{L_{II}^2} & 0 & 0 \\[2mm]
0 & 0 & 0 & 0 & -\frac{96E_{II}I_{II}}{L_{II}^3} & \frac{24E_{II}I_{II}}{L_{II}^2} & \frac{96E_{II}I_{II}}{L_{II}^3} & 0 & 0 \\[2mm]
0 & 0 & 0 & 0 & 0 & 0 & 0 & 1 & 0 \\[2mm]
-\frac{6E_{III}I_{III}}{L_{III}^2} & 0 & \frac{2E_{III}I_{III}}{L_{III}} & 0 & 0 & 0 & 0 & \frac{6E_{III}I_{III}}{L_{III}^2} & \frac{4E_{III}I_{III}}{L_{III}}
\end{bmatrix},
\qquad (5.75)
$$

from which the horizontal reaction force F_{4X}^R at node 4 is obtained after some calculation as:

$$
F_{4X}^R = \frac{u_0}{\left(\dfrac{L_I\left(\frac{1}{3}L_I^2 + L_I L_{III} + L_{III}^2\right)}{E_I I_I} + \dfrac{L_{II}(L_I + L_{III})^2}{2E_{II}I_{II}} + \dfrac{L_{II}}{2E_{II}A_{II}} + \dfrac{L_{III}^3}{3E_{III}I_{III}} \right)},
\qquad (5.76)
$$

which is identical to Eq. (2.291).

3.26 Extensometer without rotation at the specimen
The reduced system of equations reads

$$
\begin{bmatrix}
1 & 0 & 0 & 0 & 0 \\
-\dfrac{12E_I I_I}{L_I^{3}} & \dfrac{2A_{II}E_{II}}{L_{II}} + \dfrac{12E_I I_I}{L_I^{3}} & 0 & -\dfrac{6E_I I_I}{L_I^{2}} & 0 \\
0 & 0 & \dfrac{96E_{II}I_{II}}{L_{II}^{3}} + \dfrac{A_I E_I}{L_I} & -\dfrac{24E_{II}I_{II}}{L_{II}^{2}} & -\dfrac{96E_{II}I_{II}}{L_{II}^{3}} \\
\dfrac{6E_I I_I}{L_I^{2}} & -\dfrac{6E_I I_I}{L_I^{2}} & -\dfrac{24E_{II}I_{II}}{L_{II}^{2}} - \dfrac{96E_{II}I_{textII}}{L_{II}^{3}} & \dfrac{8E_{II}I_{II}}{L_{II}} + \dfrac{4E_I I_I}{L_I} & \dfrac{24E_{II}I_{II}}{L_{II}^{2}} \\
0 & 0 & -\dfrac{96E_{II}I_{II}}{L_{II}^{3}} & \dfrac{24E_{II}I_{II}}{L_{II}^{2}} & \dfrac{96E_{II}I_{II}}{L_{II}^{3}}
\end{bmatrix}
\begin{bmatrix}
u_{1X} \\ u_{2X} \\ u_{1Z} \\ \varphi_{2Y} \\ u_{3Z}
\end{bmatrix}
$$

$$
=
\begin{bmatrix}
-u_0 \\ 0 \\ 0 \\ 0 \\ 0
\end{bmatrix},
\tag{5.77}
$$

from which the solution matrix can be obtained as:

$$
\begin{bmatrix}
u_{1X} \\ u_{2X} \\ u_{2Z} \\ \varphi_{2Y} \\ u_{3Z}
\end{bmatrix}
=
\begin{bmatrix}
-u_0 \\[4pt]
-\dfrac{\left(3E_I^{2}I_I^{2}L_{II}^{2} + 6E_I E_{II}I_I I_{II}L_I L_{II}\right)u_0}{3E_I^{2}I_I^{2}L_{II}^{2} + \left(2A_{II}E_I E_{II}I_I L_I^{3} + 6E_I E_{II}I_I I_{II}L_I\right)L_{II} + A_{II}E_{II}^{2}I_{II}L_I^{4}} \\[8pt]
0 \\[4pt]
\dfrac{3A_{II}E_I E_{II}I_I L_I^{2}L_{II}u_0}{3E_I^{2}I_I^{2}L_{II}^{2} + \left(2A_{II}E_I E_{II}I_I L_I^{3} + 6E_I E_{II}I_I I_{II}L_I\right)L_{II} + A_{II}E_{II}^{2}I_{II}L_I^{4}} \\[8pt]
-\dfrac{3A_{II}E_I E_{II}I_I L_I^{2}L_{II}^{2}u_0}{12E_I^{2}I_I^{2}L_{II}^{2} + \left(8A_{II}E_I E_{II}I_I L_I^{3} + 24E_I E_{II}I_I I_{II}L_I\right)L_{II} + 4A_{II}E_{II}^{2}I_{II}L_I^{4}}
\end{bmatrix}.
$$

$$\tag{5.78}$$

Post-processing based on the non-reduced system allows to calculate the reactions, i.e.,

$$
F_{1X}^{R} = \frac{\left(6A_{II}E_I^{2}E_{II}I_I^{2}L_{II} + 12A_{II}E_I E_{II}^{2}I_I I_{II}L_I\right)u_0}{3E_I^{2}I_I^{2}L_{II}^{2} + \left(2A_{II}E_I E_{II}I_I L_I^{3} + 6E_I E_{II}I_I I_{II}L_I\right)L_{II} + A_{II}E_{II}^{2}I_{II}L_I^{4}}
\tag{5.79}
$$

$$
M_{1Y}^{R} = \frac{\left(6A_{II}E_I^{2}E_{II}I_I^{2}L_I L_{II} + 6A_{II}E_I E_{II}^{2}I_I I_{II}L_I^{2}\right)u_0}{3E_I^{2}I_I^{2}L_{II}^{2} + \left(2A_{II}E_I E_{II}I_I L_I^{3} + 6E_I E_{II}I_I I_{II}L_I\right)L_{II} + A_{II}E_{II}^{2}I_{II}L_I^{4}},
\tag{5.80}
$$

which gives after some transformations the same representations as in Eqs. (2.310) and (2.311). Thus, the calibration curve can be calculated as given in Eq. (2.317).

3.27 Finite element post-processing of strains

From Tables 3.2 and 3.8 we can derive the following relation for the total strain in the second (horizontal) element ($L_{II}^{*} = \frac{L_{II}}{2}$):

$$\varepsilon_{\text{II}}\big|_{z=\pm\frac{h_{\text{II}}}{2}} = \frac{1}{L_{\text{II}}^{*}}\underbrace{(u_{3X}-u_{2X})}_{=0} + \left(\left[+\frac{6}{L_{\text{II}}^{*2}}-\frac{12X}{L_{\text{II}}^{*3}}\right]\underbrace{u_{2Z}}_{=0} + \left[-\frac{4}{L_{\text{II}}^{*}}+\frac{6X}{L_{\text{II}}^{*2}}\right]\varphi_{2Y}\right.$$

$$\left.+\left[-\frac{6}{L_{\text{II}}^{2*}}+\frac{12X}{L_{\text{II}}^{*3}}\right]u_{3Z}+\left[-\frac{2}{L_{\text{II}}^{*}}+\frac{6X}{L_{\text{II}}^{*2}}\right]\underbrace{\varphi_{3Y}}_{=0}\right)z$$

$$\tag{5.81}$$

$$=-\frac{2}{L_{\text{II}}}u_{2X}+\left(\frac{4}{L_{\text{II}}}\varphi_{2Y}+\frac{24}{L_{\text{II}}^{2}}u_{3Z}\right)z \tag{5.82}$$

$$=\frac{1}{\frac{2}{3}\frac{E_{\text{II}}I_{\text{II}}L_{\text{I}}}{E_{\text{I}}I_{\text{I}}L_{\text{II}}}+1+\frac{1}{12}\frac{h_{\text{II}}^{2}}{L_{\text{I}}^{2}}}\times\frac{h_{\text{II}}}{L_{\text{I}}}\times\left(\mp\frac{1}{2}+\frac{1}{12}\left(\frac{h_{\text{II}}}{L_{\text{I}}}\right)\right)\varepsilon_{\text{sp}}. \tag{5.83}$$

References

1. Fisher DM, Bubsey RT, Srawley JE (1966) Design and use of a displacement gage for crack extension measurements. NASA TN-D-3724. https://ntrs.nasa.gov/archive/nasa/casi.ntrs.nasa.gov/19670001426.pdf. Cited 8 Mai 2020
2. Javanbakht Z, Öchsner A (2018) Computational statics revision course. Springer, Cham
3. Öchsner A (2014) Elasto-plasticity of frame structure elements: modeling and simulation of rods and beams. Springer, Berlin

Index

© The Editor(s) (if applicable) and The Author(s), under exclusive license
to Springer Nature Switzerland AG 2021
A. Öchsner, *A Project-Based Introduction to Computational Statics*,
https://doi.org/10.1007/978-3-030-58771-0

Printed in the United States
by Baker & Taylor Publisher Services